BALANCED SCIENCE 1

Mary Jones

Geoff Jones

Phillip Marchington

David Acaster

CAMBRIDGE
UNIVERSITY PRESS

PUBLISHED BY THE PRESS SYNDICATE OF THE UNIVERSITY OF CAMBRIDGE
The Pitt Building, Trumpington Street, Cambridge CB2 1RP, United Kingdom

CAMBRIDGE UNIVERSITY PRESS
The Edinburgh Building, Cambridge CB2 2RU, United Kingdom
40 West 20th Street, New York, NY 10011-4211, USA
10 Stamford Road, Oakleigh, Melbourne 3166, Australia

©Cambridge University Press 1998
Artwork ©Geoff Jones (unless credited otherwise)

First published 1990
Second edition 1998

Printed in the United Kingdom at the University Press, Cambridge

A catalogue record for this book is available from the British Library

ISBN 0 521 59979 2

Prepared for the publishers by Stenton Associates

Cover photograph: Gymnast/Telegraph Colour Library

CONTENTS

How to use this book

HOW TO USE THIS BOOK

The two volumes of *Balanced Science* cover the material required by most of the double and single certificate Science GCSE courses and the National Curriculum. However, some specialist topics required by some modules of modular GCSE courses are not included. The material is organised into main subject areas. Each has its own colour key which is shown on the contents list on page 3 and on the contents lists appearing at the beginning of each subject area.

Each subject area is divided into **topics**, most of which are arranged on two facing pages. The topics are numbered consecutively throughout the two books, so that they form one possible sequence for teaching Science in the fourth and fifth years of secondary school. However, many other combinations of subject material are possible, and it is easy to use the topics in almost any order. For example, the two topics of *Osmosis* and *Osmosis in Living Cells* are included in the subject area *Air and Water*, but they could equally well be taught within the subject area *Cells and Transport*.

Core text is material suitable for students of a wide ability range. However, not all of the core text will be suitable for all syllabuses and teachers should be aware of this.

Investigations which the authors have found to be particularly useful or enjoyable have been chosen. The instructions are detailed enough to be followed by most students without further help. However, instructions for drawing up results tables and graphs have been deliberately omitted in several cases, so that the student's ability to do this can be assessed.
In some places measurements are shown as fractions rather than decimals. This has been done deliberately to make it easier for students to understand.

> The material is organised into main **subject areas**. Each has its own colour key which is shown on the contents list on page 3 and on the contents lists appearing at the beginning of each subject area.

92 RESPIRATION

Respiration is a process which occurs in all living cells. Respiration releases energy from food.

Every living cell respires

Each living cell in every living organism needs energy. Energy is needed to drive chemical reactions in the cell. It is needed for movement. It is needed for building up large molecules from small ones. If a cell cannot get enough energy it dies. Cells get their energy from organic molecules such as glucose. The chemical process by which energy is released from glucose and other organic molecules is called **respiration**. Every cell needs energy. Each cell must release its own energy from glucose. So each cell must respire. Every cell in your body respires. Every living cell in the world respires.

Fig. 92.1 Florence Griffith-Joyner ('Flo-Jo') winning the 100 m at the Seoul Olympics. The energy which she is using comes from respiration in her muscle cells.

INVESTIGATION 92.1

Getting energy out of a peanut

1 Spear a peanut on the end of a mounted needle. Be careful – it is easy to break the peanut.
2 Put some cold water into a boiling tube. Support the boiling tube in a clamp on a retort stand, with the base of the tube about 30 cm above the bench top. Take the temperature of the water and record it.
3 Set light to the peanut by holding it in a Bunsen burner flame. (Keep the Bunsen burner well away from the boiling tube.) When the peanut is burning hold it under the boiling tube. Keep it there until it stops burning.
4 Immediately take the temperature of the water and record it.

Fig. 92.2 Burning a peanut

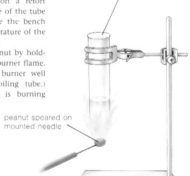

boiling tube containing water

peanut speared on mounted needle

Questions

1 Why should you keep the Bunsen burner away from the tube of water?
2 By how much did the temperature of the water rise?
3 What type of chemical reaction was occurring as the peanut burnt?
4 Put these words in the right order, and join them with arrows, to show where the heat energy in the water came from:

energy in peanut molecules

energy in glucose in peanut leaf

sunlight energy

heat energy in water

5 Put the following words over two of the arrows in your answer to question 4:

photosynthesis   oxidation
6 Do you think all the energy from your peanut went into the water? If not, explain what else might have happened to it.
7 How could you improve the design of this experiment to make sure that more of the peanut energy went into the water?

— EXTENSION —
8 People often want to know exactly how much energy there is in a particular kind of food. The amount of energy is measured in kilojoules.
4.18 J of energy will raise the temperature of 1 g of water by 1 °C.
Design a method for finding out the amount of energy per gram in a particular type of food.

Respiration is an oxidation reaction

You can do an experiment to release energy from food if you do Investigation 92.1. When a peanut burns, the energy in the peanut is released as heat energy. But how do your cells release energy from food such as peanuts?

Obviously, you do not burn peanuts inside your cells! But you do something very similar. The chemical reactions of burning and respiration are very like each other.

First, think about what happens when you burn the peanut. The peanut contains organic molecules, such as fats and sugars, which contain energy. When you set light to the peanut you start off a chemical reaction between these molecules and oxygen in the air. The peanut molecules undergo an **oxidation** reaction. They combine with oxygen. As they do so, the energy in them is released as **heat energy**.

When you eat peanuts, your digestive system breaks the peanut into its individual molecules. Your blood system then takes these molecules to your cells. Inside your cells the molecules combine with oxygen. They undergo an oxidation reaction. This is respiration. The energy in the molecules is released. But, unlike the burning peanut, much of the energy is *not* released as heat energy. It is released much more gently and gradually, and stored in the cell.

So, the burning of a peanut, and the respiration of 'peanut molecules' in your cells are both oxidation reactions. In both of them the peanut molecules combine with oxygen. In both of them the energy in the peanut molecules is released. But in your cells the reaction is much more gentle and controlled.

EXTENSION
ATP is the energy currency in cells

Respiration releases energy from food. Each cell must do this for itself. Every living cell respires to release the energy it needs. The energy released in respiration is not used directly for movement or any of the other activities of the cell. It is used to make a chemical called **ATP**. ATP is short for adenosine triphosphate. ATP, like glucose, contains chemical energy.

chemical energy in glucose → chemical energy in ATP

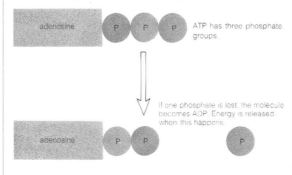

ATP has three phosphate groups.

If one phosphate is lost, the molecule becomes ADP. Energy is released when this happens.

Fig. 92.3 ATP and ADP

ATP is the ideal energy currency in a cell. The energy in an ATP molecule can be released from it very quickly – much more quickly than from a glucose molecule. The energy is released by breaking ATP down to **ADP**. ADP is short for adenosine diphosphate. Another good reason for using ATP as an energy supply is that one ATP molecule contains a much smaller amount of energy than one glucose molecule. If a cell needs just a small amount of energy, then it can break down just the right number of ATP molecules. The amount in a glucose molecule might be too much, and energy would be wasted. Each cell produces its own ATP by the process of respiration. ATP is not transported from cell to cell.

The respiration equation

Respiration is a chemical reaction. The word equation for the reaction with glucose is:

glucose + oxygen $\longrightarrow$ carbon dioxide + water + energy

The balanced molecular equation for the reaction is:

$$C_6H_{12}O_6 + 6O_2 \longrightarrow 6CO_2 + 6H_2O + energy$$

Questions

1 Respiration is a chemical reaction.
a Where does it take place?
b What type of chemical reaction is it?
c Why is it so important to living cells?

2 List two similarities, and one difference, between the burning of a peanut and the respiration of 'peanut molecules' in your cells.

3 a Write down the word equation for respiration.
b Write down the balanced molecular equation for respiration.

4 Respiration releases energy from food. Explain how the energy came to be in the food.

EXTENSION

5 a What is ATP?
b Why do cells use ATP as an energy store?
c A muscle cell uses glucose to provide energy for movement.
i List all the energy changes involved in this process, beginning with energy in sunlight.
ii Energy is 'lost' at each transfer. What do you think happens to this 'lost' energy at each stage?

223

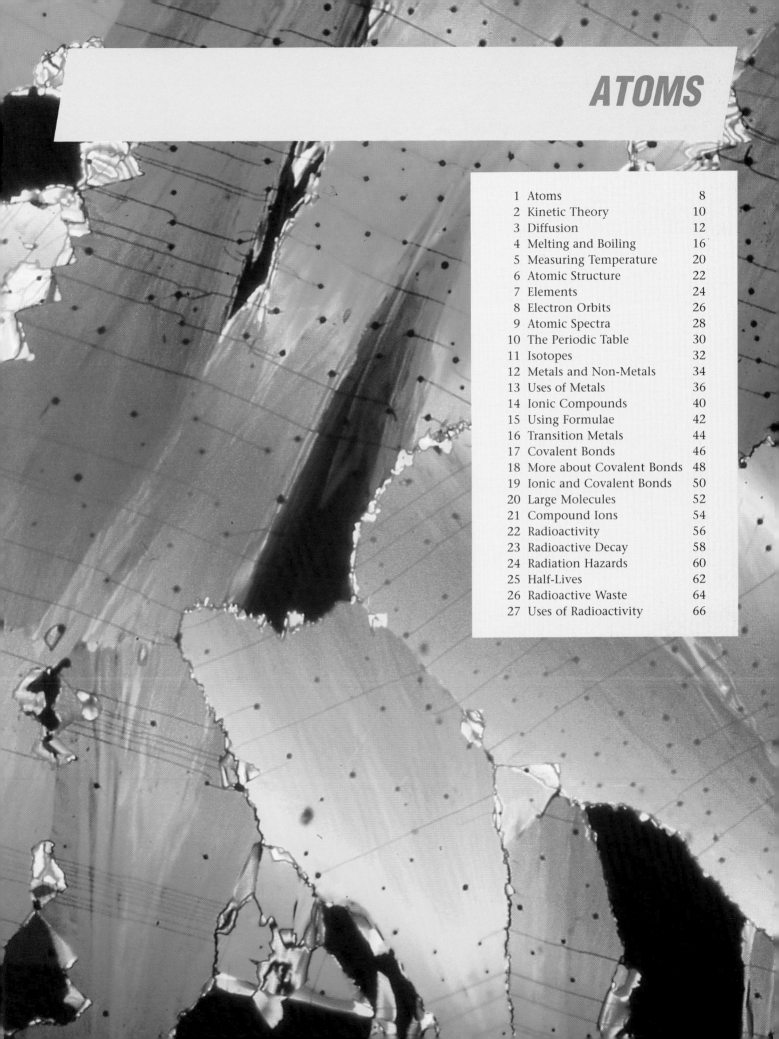

ATOMS

1 ATOMS All material is made from tiny particles, called atoms.

All substances are made of atoms

Everything that you see around you is made out of tiny particles called **atoms**. This page, the ink on it, you and your chair are all made of atoms. Atoms are remarkably small. The head of a pin contains about 60 000 000 000 000 000 000 atoms. It is difficult to imagine anything quite so small as an atom. The ancient Greeks believed that nothing smaller than an atom could exist, so they gave them the name 'atomos', meaning 'indivisible'. Atoms sometimes exist singly, and sometimes in groups. These groups of atoms are also known as **molecules**.

The way substances behave suggests that they are made of tiny particles

Atoms are far too small to be seen. Yet we know they must be there, because of the way that substances behave.

Crystals of many materials have regular shapes. The crystals of a particular substance always have the same shape. One explanation for this is that crystals are built up of tiny particles, put together in a regular way. These tiny particles are atoms.

Fig. 1.1 Sodium chloride (salt) crystals. The actual size of the biggest crystal in this enlarged photograph is about 0.3 mm across.

A crystal such as the one in the photograph contains millions upon millions of atoms. If a small piece is chipped away from a salt crystal, for example, the piece is still salt. Even the very smallest, microscopic piece you can chip away is still salt.

Fig. 1.2 A salt crystal. Sodium chloride crystals are cube-shaped. This is because they are made of particles which are arranged in a regular, cubic pattern. The diagram shows the particles in a very, very tiny piece of sodium chloride.

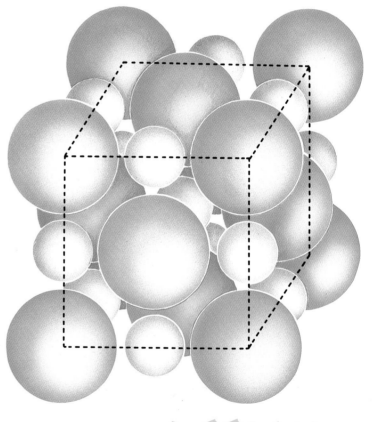

DID YOU KNOW?

If you divided a single drop of water so that everyone in the world had an equal share, everyone would still get about one million million molecules.

Brownian motion is evidence for the existence of particles

In 1827, an Oxford scientist, Robert Brown, was studying pollen grains through a microscope. The pollen grains were in liquid, and Brown was surprised to see that the movement was jerky. Other scientists became interested in Brown's observations. Many suggestions were made to try to explain them. Some people thought that the pollen might be moving of its own accord. Others suggested that it was being knocked around by something invisible in the liquid in which it was floating.

You can see this effect with the apparatus in Figure 1.4, using smoke grains instead of pollen grains.

No-one could possibly imagine that smoke grains are alive. The jerky movement can best be explained by imagining tiny air molecules around the smoke grains, bumping into them. This knocks the smoke grains around in jerky movements. The air molecules are much too small to be seen. They must be moving around very quickly.

This jerky movement of the pollen grains or smoke grains is called **Brownian motion**, after Robert Brown. It is strong evidence that all substances are made up of molecules, in constant motion, and much too small to be seen.

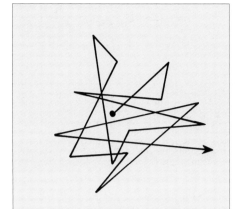

Fig. 1.3 Brownian motion. Small particles like smoke particles seem to move randomly, constantly changing direction.

INVESTIGATION 1.1

Using a smoke cell to see Brownian motion

A smoke cell is a small glass container in which you can trap smoke grains, and then watch them through a microscope.

1 Make sure that you understand the smoke cell apparatus. The smoke cell itself (see Figure 1.4) should be clean, because light must be able to shine through the sides of it. The light comes from a small bulb. It is focused onto the smoke cell through a cylindrical glass lens.

2 Set up a microscope.

3 Now trap some smoke in the smoke cell, in the following way. Set light to one end of a waxed paper straw. If you hold the burning straw at an angle, you can make the smoke pour out of the lower end and into the smoke cell. Quickly place a cover slip over the smoke in the smoke cell.

4 Put the smoke cell apparatus on the stage of the microscope. Switch on the light. Focus on the contents of the smoke cell.

You should be able to see small specks of light dancing around. These are the smoke grains. They look bright because the light reflects off them.

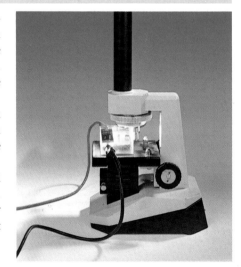

Fig. 1.4 The smoke cell apparatus

Fig. 1.5 Using a smoke cell

look down through microscope

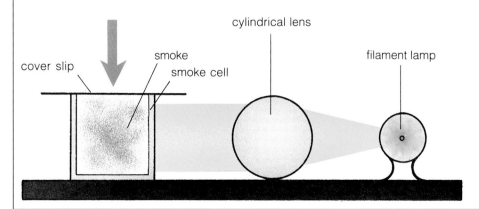

cover slip • smoke • smoke cell • cylindrical lens • filament lamp

Questions

1 a What can you see in the smoke cell?

  b What is in the cell which cannot be seen?

2 Why are the smoke grains dancing around?

3 What is the name for this movement?

2 KINETIC THEORY

All material is made from small moving particles called atoms. Materials can be solid, liquid or gaseous, depending on the arrangement and freedom of movement of these particles.

There are three states of matter

All material is made from tiny particles. These particles are constantly moving. The kinetic theory uses this idea of tiny moving particles to explain the different forms that material can take. 'Kinetic' means 'to do with movement'.

The three states of matter are solid, liquid and gas.

The first state - solid

In this state, matter tends to keep its shape. If it is squashed or stretched enough, it will change shape slightly. Usually, any change in volume is too small to be noticed. The particles are not moving around, although they are vibrating very slightly. Normally they vibrate about fixed positions. If a solid is heated the particles start to vibrate more.

Fig. 2.1a

The second state - liquid

In this state matter will flow. It will take up the shape of any container it is put in. The liquid normally fills a container from the bottom up. It has a fixed volume. If the liquid is squeezed it will change shape, but the volume hardly changes at all. The particles, like those in a solid, are vibrating. However, in a liquid the particles are free to move around each other. If a liquid is heated, the particles move faster.

Fig. 2.1b

The third state - gas

In this state matter will take up the shape of a container and fill it. The volume of the gas depends on the size of its container. If the gas is squashed it will change both volume and shape. The particles are free to move around, and do not often meet each other. The particles whizz around very quickly. If heated they move even faster.

Fig. 2.1c

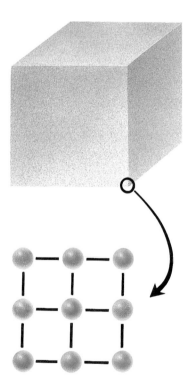

The particles are fixed in position. The forces between particles are strong. The particles cannot move past each other. They are close together.

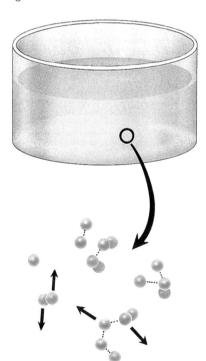

The particles can move past each other. They are joined together in small groups. They are not as close together as in a solid. The forces between the particles are not so strong.

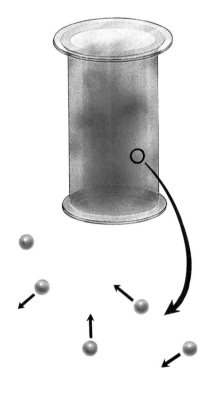

There are hardly any forces between the particles. They are a long way apart. The particles are moving quickly and so they spread out. If squashed, they move closer together.

A model for the kinetic theory

A platform is attached to an electric motor. The motor vibrates the platform up and down. The speed of the vibrations can be controlled.

Fig. 2.2

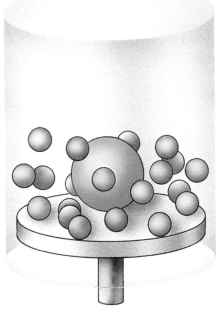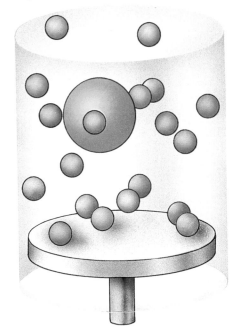

When the platform is still the small balls are like the particles in a solid. They are in fixed positions. If a larger ball is placed on the small ones, it sits on top of them as it would on a solid.

If the platform is made to vibrate gently, the small balls vibrate. But they still stay in the same place. This is what happens when a solid warms up. The particles vibrate, but stay in the same place.

If the platform vibrates faster, the balls start to move faster. It is as though the solid is being heated. When the balls begin to bounce above the level of the platform they are behaving like the particles in a liquid. Now the larger ball is surrounded by the smaller ones. It is as though it has 'sunk' into the liquid.

DID YOU KNOW?

If a gas is heated to a very high temperature, the molecules and then the atoms break apart. At tens of thousands of degrees Celsius, a plasma is formed. The Sun is a plasma. A fluorescent tube contains a plasma.

If the platform vibrates very fast, the balls fly around the whole container. They bounce off the walls. Now they are behaving like the particles in a gas. The large ball is knocked about by the small balls. It is behaving like the smoke grains in the smoke cell experiment.

In this model the average speed of the balls depends on the speed of vibration of the motor. In a substance, the average speed of the particles depends on the temperature. The temperature of a material indicates how quickly the particles are moving.

Fig. 2.3 The Sun is a ball of plasma. The red and blue patches are sunspots.

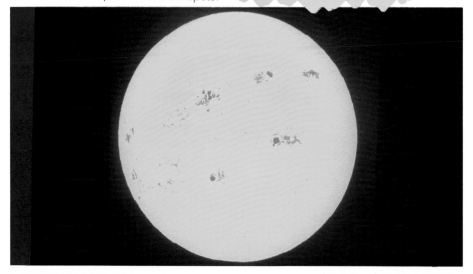

Questions

1 a Name the three states of matter.
  b Describe the arrangement and behaviour of particles in each of the three states.
  c For each of these three states, give one example of a substance which is normally in this state at room temperature.

3 DIFFUSION

Diffusion is the spreading out of particles. It provides further evidence for the kinetic theory.

Moving particles spread around

A smell will slowly spread across a room. We can explain this by imagining that the smell is made of moving particles of a smelly substance. The molecules move around, filling the room. This movement is called **diffusion**. Diffusion can be defined as *the movement of particles from a place where there is a high concentration of them, to a place where there is a lower concentration of them.* Diffusion tends to spread the particles out evenly.

You can watch diffusion happening if you use a substance, such as potassium permanganate, which has coloured particles. If a crystal of potassium permanganate is carefully dropped into a beaker of water, it dissolves. The particles in the crystal separate from each other and slowly move through the water. The colour only spreads slowly. The particles keep bumping into the water molecules. They do not travel in straight lines. Eventually the colour fills the whole beaker, but this takes a very long time.

Diffusion happens faster in gases than in liquids

In a gas the molecules are not so close together as in a liquid. If a coloured gas is mixed with a clear one, the colour spreads. This happens faster than in a liquid, because fewer particles get in the way. You can watch this happening with bromine, as it diffuses through air. Bromine is a brown gas. It covers about 2 cm in 100 s. If it diffuses into a vacuum, it goes even faster, because there are no particles to get in the way. It then travels 20 km in 100 s!

Fig. 3.1 Potassium permanganate diffusing in water. A crystal has just been dropped into the gas jar on the left. The potassium permanganate has been diffusing for about 30 minutes in the centre jar, and for 24 hours in the jar on the right.

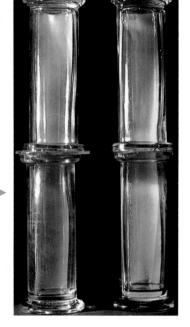

Fig. 3.2 An experiment showing the diffusion of bromine gas. On the left, the two gas jars are separated by a glass lid. The lower one contains bromine, and the upper one contains air. On the right, the lid has been removed. The bromine and air diffuse into one another.

Questions

1 Using examples, and diagrams if they help, explain how each of the following supports the idea that everything is made up of particles:
   a Brownian motion
   b crystal structure
   c diffusion

INVESTIGATION 3.1

How quickly do scent particles move?

If a bottle of perfume is opened, the smell spreads across a room. Design an experiment to find out how quickly the smell spreads from a particular type of perfume.

You will need to consider how you will decide when the scent has reached a particular part of the room. Would the same group of experimenters always get the same results? Would you get the same results in a different room? Or in the same room on a different day? Try to take these problems into account when you design your experiment.

Get your experiment checked, and then carry it out. Record your results in the way you think best. Discuss what your results suggest to you about the speed at which scent particles (molecules) move.

Diffusion of two gases

This experiment will be demonstrated for you, as the liquids used should not be touched.

A piece of cotton wool is soaked in hydrochloric acid. A second piece of cotton wool is soaked in ammonia solution. The two pieces of cotton wool are pushed into the ends of a long glass tube. Rubber bungs are then pushed in, to seal the ends of the tube.

The hydrochloric acid gives off hydrogen chloride gas. The ammonia solution gives off ammonia gas. (Both of these gases smell very unpleasant, and should not be breathed in in large quantities.)

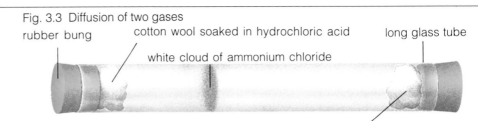

Fig. 3.3 Diffusion of two gases
rubber bung · cotton wool soaked in hydrochloric acid · long glass tube · white cloud of ammonium chloride · cotton wool soaked in ammonia solution

Questions

1 Hydrogen chloride and ammonia react together to form a white substance called ammonium chloride. Nearest which end of the tube does the ammonium chloride form?

2 How had the two gases travelled along the tube?

3 Which gas travelled faster?

4 The molecules of ammonia are smaller and lighter than the molecules of hydrogen chloride. What does this experiment suggest about how the size and mass of its molecules might affect the speed of diffusion of a gas?

5 If this experiment could be repeated at a higher temperature, would you expect it to take a longer or shorter time for the ammonium chloride to form? Explain your answer.

How small are potassium permanganate particles?

1 Measure 10 cm³ of water into a test tube. Add a few crystals of potassium permanganate and stir to dissolve.

2 Using a syringe, take exactly 1 cm³ of this solution and put it into a second tube. Add 9 cm³ of water to this tube, to make the total up to 10 cm³. You have diluted the original solution by 10 times. Put your two solutions side by side in a test tube rack.

3 Now dilute the second solution by 10 times, by taking exactly 1 cm³ of it, and adding it to 9 cm³ of water in a third tube. Add this tube to the row in the test tube rack.

4 Continue diluting the solution by 10 times, until you have a tube in which you can only just see the colour.

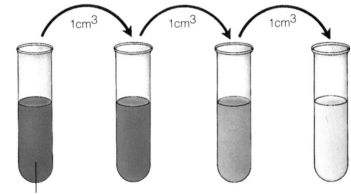

1cm³   1cm³   1cm³

potassium permanganate solution made by dissolving a few crystals in 10cm³ of water

Fig. 3.4 Making serial dilutions of a potassium permanganate solution

Questions

Think about the number of potassium permanganate particles in each of your tubes. You began with a lot of particles in your first tube – all the ones that were in the crystals that you added to the water. You mixed them thoroughly into the water, and then took **one tenth** of them out to put into the second tube. So the second tube contained only one tenth as many potassium permanganate particles as the first one.

1 How many times fewer particles are there in the third tube than in the first tube?

2 How many times fewer particles are there in the last tube than in the first tube?

3 In your last, very faintly coloured tube the colour is still evenly spread through the water. So there must still be at least a few thousand potassium permanganate particles there. If you imagine that there are a thousand potassium permanganate particles in this tube, can you work out how many there must have been in the first tube? (You need to do some multiplications by 10 – lots of them.)

4 What does this experiment tell you about the size of potassium permanganate particles?

Questions

1. A petri dish was filled with agar jelly, in which some starch solution was dissolved. Two holes were cut in the agar jelly. Water was put into one hole. A solution of amylase was put into the other hole. Amylase is an enzyme which digests starch. It changes starch into sugar.

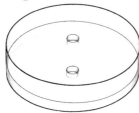

After a day, iodine solution was poured over the agar jelly in the dish. Iodine solution turns blue-black when in contact with starch. It does not change colour when in contact with sugar. The results are shown below.

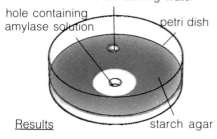

hole containing water
hole containing amylase solution
petri dish
Results
starch agar

a Why did most of the agar jelly turn blue-black when iodine solution was poured over it?

b Why did the part of the jelly around the hole which contained amylase solution not turn black?

c The area which did not turn blue-black was roughly circular in shape. Explain why. Use the word 'diffusion' in your answer.

d Why do you think that water was put into one of the holes in this experiment?

2. Divide part of a page in your book into three columns, headed solid, liquid and gas. Copy each of the following words and statements into the correct column. Some of them may belong in more than one column.

a water at room temperature
b salt at room temperature
c oxygen at room temperature
d made up of particles
e particles vibrate
f particles move around freely, a long way apart
g particles vibrate slightly, and are held together tightly

h particles held together, but move around each other freely
i fills a container from the bottom
j completely fills a container
k does not spread out in a container
l particles move faster when heated
m particles move much closer together when squashed

3. Explain the following:

a If you grow a salt crystal it will be cube-shaped.

b A solution of potassium permanganate can be diluted many times, and still look purple.

c If ammonia solution is spilt in a corner of a laboratory, the people nearest it will smell it several seconds before the people in the opposite corner.

d Pollen grains floating on water appear to be jigging around if seen through a microscope.

e A copper sulphate crystal dissolves in water to form a blue solution. This happens faster in warm water than in cold water.

EXTENSION

Cells can use energy to move substances by active transport

Cells may need to take in substances that are present in a low concentration. For example, a plant requires nitrate ions. Plant roots have fine, single-celled **root hairs**, which absorb water and minerals such as nitrate ions from the soil. The concentration of nitrate ions is often much lower in the soil than inside the root hair. So you would expect the nitrate ions to diffuse out of the root hair into the soil, down their concentration gradient.

But, despite this, the root hairs are still able to take in nitrate ions. They have special protein molecules, called **transporter proteins**, in their cell surface membranes, into

which nitrate ions fit perfectly. When a nitrate ion bumps into a transporter protein on the outer surface of the cell surface membrane, the cell uses energy to change the shape of the transporter and so push the nitrate ion into the cell (Figure 3.5). The energy comes from a molecule called **ATP**, which the cell makes by respiration (see Topic 92).

All living cells use active transport to move ions and other substances into or out of the cell, against their concentration gradient. They have many different transporter proteins in the cell surface membranes, each one shaped so that a particular ion or molecule fits perfectly.

1. The nitrate ion enters the transporter protein.

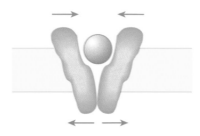

2. The transporter protein changes shape. The energy needed for this comes from ATP, produced in respiration in the cell.

3. The change of shape of the transporter protein pushes the nitrate ion into the cell.

Fig. 3.5   Active transport

Question

The manufacture of a transistor

Diffusion is a process of great importance in the manufacture and operation of electronic components, such as transistors. A transistor is an electronic switch. It is used to control the flow of electrons in a circuit.

A transistor is formed from a silicon crystal, in which there is a small percentage of atoms of other substances. These other atoms alter the electrical properties of the crystal. The other atoms are added to particular regions of the silicon, so that these regions have different properties. These regions are known as n-type and p-type semiconductors.

Transistors need to be very small so that they can respond to changes quickly. There are several ways of making them.

This is one method.

First, a wafer of n-type silicon is heated in oxygen. The oxygen forms a layer of silicon oxide on the surface of the silicon. Chemicals are then used to etch away the oxide in a small area, of diameter a. In this area, the pure silicon is exposed. This is called an n-type semi-conductor.

Next, the silicon wafer is placed in a hot boron atomosphere. The boron atoms cannot penetrate the silicon oxide layer, but they can diffuse into the exposed silicon. They enter the surface and spread a little way under the silicon oxide layer. Wherever the boron enters the silicon, a p-type semiconductor is formed. The p-type semiconductor has a diameter a little larger than a.

Now the silicon wafer is heated in oxygen again to cover it completely with an oxide layer. A new area, of diameter b, is etched in the oxide. The wafer is then exposed to hot phosphorus gas. Phosphorus atoms diffuse into the exposed surface, but not through the silicon oxide. Where the phosphorus atoms enter the silicon, an n-type semiconductor is formed. The phosphorus atoms spread a little way under the oxide layer.

The junctions between the n-type and p-type semiconductors are very important if the transistor is to work well. By making them like this, the junctions are protected underneath a layer of silicon oxide.

Finally, electrical contacts are made by depositing metallic vapour on certain parts of the transistor and heating it. So now the whole transistor is covered and protected by either an oxide layer or the electrical contacts. This prevents impurity atoms in the air from diffusing into the junctions and spoiling the transistor.

Fig. 3.6

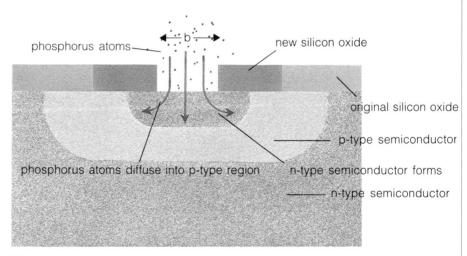

Questions

1 Why is it a good idea to make transistors small?

2 a How is a layer of silicon oxide made to form on the surface of the silicon wafer?

b Explain how this layer helps in the next stage of the process.

3 Why are the new semiconductor regions bigger than the exposed areas on the silicon wafer?

4 The higher the temperature while the boron is entering the silicon wafer, the bigger the diameter of the p-type semiconductor region. Why?

5 What use is the oxide layer in the completed transistor?

MELTING AND BOILING

As a substance is heated it melts and then boils. The melting and boiling points of a pure substance are always the same, under the same conditions.

Heating substances makes the particles move faster

When a solid is heated, the particles (atoms or molecules) vibrate faster and faster. The energy of motion, or **kinetic energy**, of the particles increases.

Eventually, they have so much energy that they can begin to break away from each other. They begin to separate and move more freely. The solid **melts**, turning into a liquid.

If you go on heating the liquid, the heat energy makes the particles move even faster. As their kinetic energy increases, the clusters of particles begin to break apart. The individual particles move further away from each other. The liquid boils, turning into a gas. Boiling is when heat causes bubbles of vapour to form within a liquid.

Fig. 4.1 Atoms or molecules move faster when a substance is heated.

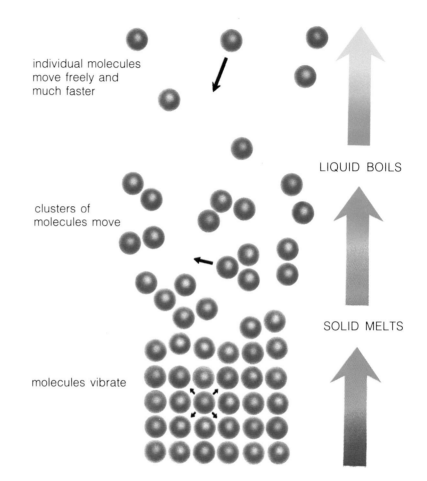

individual molecules move freely and much faster

clusters of molecules move

molecules vibrate

LIQUID BOILS

SOLID MELTS

melting → ← freezing

evaporating → ← condensing

ice

water

steam

Fig. 4.2 Water changes from solid, through liquid, to gas when heated.

A pure substance has a constant melting and boiling point

When a solid melts, or a liquid boils, the bonds which hold the particles together are broken. The stronger the bonds are, the more energy is needed to break them. The stronger the bonds are, the higher the temperature must be to break them.

The strength of the bond between particles is not the same for different substances. The bonds will usually be broken at different temperatures. So different substances melt and boil at different temperatures.

A pure substance will always melt or boil at the same temperature under the same conditions. Pure water, for example, always melts at 0 °C and boils at 100 °C, at normal atmospheric pressure.

Some materials change directly from a solid to a gas

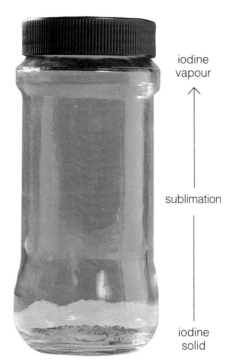

iodine
vapour

sublimation

iodine
solid

Fig. 4.3 Iodine changes directly from a solid to a gas when heated.

Some materials may change directly from a solid to a gas. This is called **sublimation**. Solid carbon dioxide changes to carbon dioxide gas as it warms up. It is called 'dry ice', and is often used to make 'mist' in theatre productions. Another substance which sublimes is iodine. Iodine crystals change to a purple gas when heated.

Melting and boiling points can be used for checking purity or temperature

If you have a sample of a substance and you want to find out if it is pure, you can measure its melting or boiling point. If it is not what it should be, then you know that your substance is not pure.

If you have a sample of a substance which you know is pure, you can use it to check a temperature. How could you do this?

Temperature does not rise while a substance is melting or boiling

When a solid is heated, its particles move faster and faster. The temperature of the solid steadily rises.

When the temperature reaches the melting point of the solid, the bonds between the particles start to break. While this is happening the temperature of the substance does not rise. You have to carry on heating until the bonds are all broken. Although you are putting in extra heat energy, the temperature does not rise. The extra energy is used to break the bonds.

When the bonds are broken, the solid melts and becomes a liquid. Now the temperature begins to rise again. The same thing happens while the liquid is boiling. When the temperature of the liquid has reached boiling point, it stays the same until the liquid has boiled. Once it is a gas the temperature can begin to rise again.

Question

A bowl of crushed ice was taken out of a freezer, and left on a laboratory bench to thaw. Its temperature was taken every 5 min for 1 h. The temperature of the laboratory was 20 °C.

The results are shown in the table below.

Time (min)	0	5	10	15	20	25	30	35	40	45	50	55	60
Temp. (°C)	−10	−5	−1	0	0	0	2	9	14	18	19	20	20

  a Draw a line graph to show these results.
  b What was the temperature inside the freezer?
  c What was happening to the ice between 15 and 25 min after being taken out of the freezer?
  d Why did the temperature not change between these times?
  e Was the ice pure or did it contain impurities? Explain your answer.
  f Would you expect the temperature of the water to continue rising above 20 °C? Explain your answer.
  g A sample of benzene that had been cooled was put into a tube and heated in a water bath. Its temperature was taken every 5 min for 1 h. The results are shown below.

Time (min)	0	5	10	15	20	25	30	35	40	45	50	55	60
Temp. (°C)	5	5	5	13	28	45	60	73	78	80	80	80	80

  h Plot these results as a line graph.
  i What are the melting and boiling points of benzene?

When salt is put on roads, it stops ice from forming

Pure water freezes, and pure ice melts, at 0 °C. If there is an impurity in the water, it freezes at a lower temperature. An impurity lowers the freezing point of a substance.

If salt is spread on roads, the salt and water mixture will have a lower freezing point than pure water. Its freezing point is below 0 °C. So ice will not form until the temperature drops well below 0 °C. The more salt is added, the lower the freezing point becomes. You could put so much salt on the roads that no ice would form until the temperature became –18 °C! But this would not be a good idea, because salt on cars speeds up rusting. It also kills plants growing on the roadside verges. Salt is usually spread on roads when ice or snow is expected, to prevent the formation of a slippery surface. It is usually rock salt which has grit mixed with it. The grit helps to increase friction on the road. This gives tyres a better grip, even if some ice does form.

INVESTIGATION 4.1

Cooling wax

This experiment tests your powers or observation. You have probably seen melted wax cooling before. This time, really watch it carefully!

1 Put some wax into a test tube, and stand the tube in a beaker of boiling water. Leave it until the wax has melted.

2 Now let the wax cool down slowly. Watch it carefully. Record anything that appears to be interesting.

3 Try to explain everything that you see. You will need to think about particles (molecules) of wax.

Fig. 4.4 A gritting lorry. Salt and grit spread on roads prevent ice forming, and improve grip.

INVESTIGATION 4.2

Measuring the melting point of a solid

1 Set up the apparatus shown in Figure 4.5.

2 Draw a results chart, so that you can fill in your readings from instruction 4.

3 Heat the water until the solid melts.

4 *Work quickly.* Take the tube, containing the melted substance, out of the beaker of hot water. Put it into a clamp on a retort stand. Record its temperature every 30 s. Do this until the substance has completely solidified.

5 Plot a cooling curve for the substance. Put time on the horizontal axis, and temperature on the vertical axis.

Questions

1 Why do you think that a water bath was used to heat the substance?

2 This substance has a melting point below 100 °C. Would this method work for a solid with a melting point higher than this? How could you adapt the apparatus to make it suitable for a solid with a higher melting point?

3 Why was the tube supported in a clamp as it cooled, and not left in the beaker of water?

4 What is the melting point of this substance?

5 Find out what the melting point of a pure sample of this substance should be. Was your sample pure?

6 Why is this experiment not accurate? Suggest some ways in which it could be made more accurate.

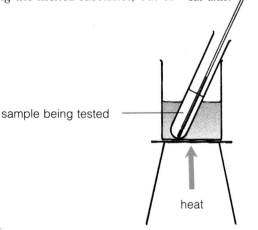

sample being tested

heat

solid heated in water bath

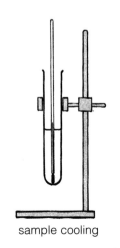

sample cooling

Fig. 4.5

Questions

1 The graph shows the melting and boiling points of five substances.
  a Which substance has the lowest boiling point?
  b What is the melting point of:
    i ethanol **ii** mercury **iii** oxygen?

Room temperature is around 20 °C.
  c Which of these substances is/are solid at room temperature?
  d Which is/are liquid at room temperature?
  e Which is/are gaseous at room temperature?

2 A sample of water is tested to see if it is pure. It is found to boil at 104 °C.
  a Is the water pure?
  b Will it freeze at 0 °C, above this, or below this temperature? Explain.

3 Cockroaches can be frozen in ice and revived without any apparent harm. They have a type of antifreeze in their blood and cells.
  a How might ice crystals damage a cockroach's cells?
  b Why might it be an advantage for a cockroach to have antifreeze in its body?

4 a If a pond is covered with a layer of ice, where in the pond would the warmest water be found?
  b What would the temperature of this water be?
  c If there were fish in the pond, why would it be important to keep a hole in the ice?
  d A concrete-lined pond may crack in a cold winter. Why?

5 Ethylene glycol is used as an antifreeze in car engines. When it is mixed with water, it lowers the freezing point of the water. The mixture is used in the car engine's cooling system. The chart shows the freezing points of mixtures of ethylene glycol and water.
  a Plot this information as a line graph:

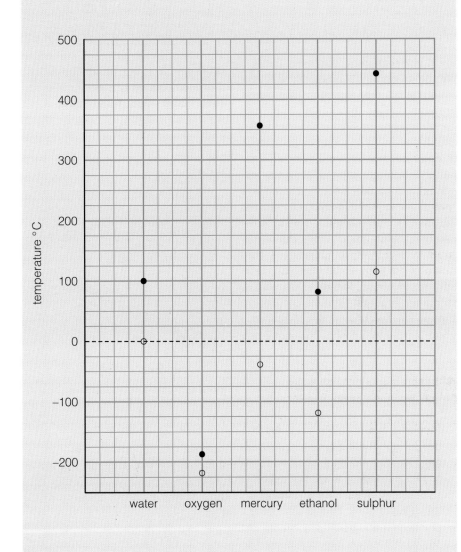

- Boiling Point
- Melting Point

% water	100	90	80	70	60	50	40
% ethylene glycol	0	10	20	30	40	50	60
freezing point (°C)	0	–4	–9	–16	–24	–34	–47

b When water changes to ice, it expands. How could this damage a car's cooling system?
c What is the freezing point of pure water?
d What is the freezing point of a 45: 55 water:ethylene glycol mixture?
e What mixture would you use in your car radiator if the lowest winter temperature was expected to be –26 °C?
f Your car's cooling system has a capacity of 7000 cm³. Use your answer to question **e** to calculate how much ethylene glycol you need to add to the cooling water.

MEASURING TEMPERATURE

Temperature is a measure of the average speed of movement of the particles in a substance. The faster they are moving, the higher the temperature.

The human sense of temperature is not very reliable

We have nerve endings in our skin which sense temperature. These send messages to the brain. The brain uses these messages to decide whether whatever is touching our skin is hot, warm or cold. But this is not a very accurate sense. It tends to *compare* temperatures rather than measure them. If you have been outside on a very cold winter's day, you may feel really warm when you go into a room. But that same room, at the same temperature, would feel cool to you if you had just been outside in hot sunshine.

Temperatures can be measured by comparing them with fixed points

Temperatures are measured by comparing them with fixed points. The two fixed points most often used are the melting and boiling points of pure water. On the **Celsius** scale of temperature, the lower fixed point is the melting point of pure ice at normal atmospheric pressure. This is called **0 °C**. The upper fixed point is the boiling point of pure water. This is **100 °C**. The gap in between these two fixed points is divided into 100 equal intervals, or **degrees**. If the temperature of something is 50 °Celsius, then it is halfway between the temperature of melting ice and boiling water. This temperature scale is called Celsius after the Swedish scientist who invented it. But it is sometimes also called centigrade, because it uses 100 intervals between the melting and boiling points of water.

Liquid-in-glass thermometers use the expansion of liquid to measure temperature

The diagram shows the sort of thermometer you use in a laboratory. As the liquid inside gets hotter it takes up more room. The liquid moves up the narrow capillary tube. Because the tube is extremely narrow, a small change of temperature makes the liquid move a long way up the tube. The walls of the bulb are very thin, so that heat goes through quickly. The walls of the stem are thicker, so that they are strong.

Fig. 5.1a   A liquid-in-glass laboratory thermometer

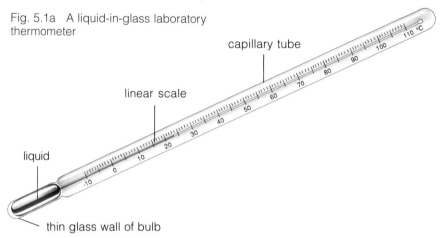

Fig. 5.1b   A clinical thermometer

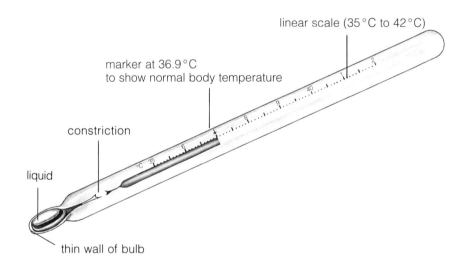

Clinical thermometers are used for measuring human temperature

The clinical thermometer is also a liquid-in-glass thermometer. It has a constriction in the capillary tube. As the temperature rises, the liquid is forced past the constriction. When the thermometer cools, the liquid does not go back into the bulb. This means that a temperature can be measured in the mouth, and the thermometer taken out to be read. The reading stays the same until the liquid is shaken down past the constriction. This must be done before it is used again.

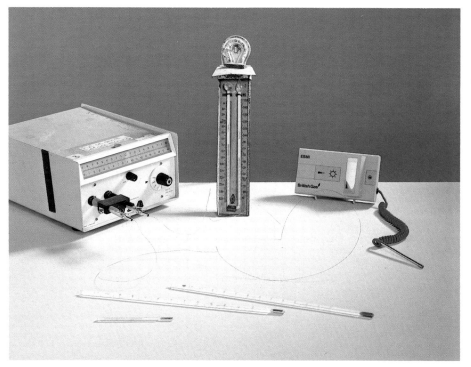

Fig. 5.2 A variety of thermometers. At the back left is a galvanometer and thermocouple. The wires connected to the galvanometer are joined to a different type of wire – you can see two junctions, where they are twisted together. These junctions are **thermocouples**. They produce a voltage when heated, which gives a reading on the galvanometer. One of the junctions is put in melting ice to provide a reference reading for 0 °C, while the other is used to measure the temperature you want to know. Thermocouples give accurate readings, work at higher temperatures than many other types of thermometer, and can be used to take temperatures of very small things. The other thermometers, moving clockwise from the galvanometer, are: a maximum/minimum thermometer; an electronic digital thermometer; a thermometer containing alcohol; one containing mercury; and a clinical thermometer.

Questions

1 Think carefully about this one! You may come up with some surprising answers. You might like to check them by experiment afterwards – but you will need be very observant.

a When the bulb of a thermometer is suddenly placed in a hot liquid, which part of the thermometer gets hot first?

b Which part of the thermometer will expand first?

c What happens if the bulb increases in volume, but the liquid in the thermometer does not?

d What happens to the 'reading' on the thermometer at the moment it is placed in the hot liquid?

2 Why is a thermometer a better measure of temperature than your skin?

3 Two liquids often used in thermometers are mercury and alcohol.

Some of their properties are listed in the table below.

a You have been asked to choose a liquid to put in a thermometer. Which of the properties in the table are relevant to your choice?

b For each of the following uses, say which of the two liquids you would prefer to have in your thermometer. In each case, give reasons for your choice.

i measuring the temperature of a stew.

ii measuring body temperature.

iii measuring the temperature at the South Pole.

iv for general laboratory use.

v to check the temperature inside a fridge.

vi to check the temperature inside a freezer.

4 The kelvin scale of temperature is used for many scientific calculations and measurements. Each kelvin is the same size as 1 °C. But the 'starting points' are different. The kelvin scale begins at a much lower temperature than the Celsius scale. 0 K is –273 °C. (Notice that there is no ° when writing temperatures in the kelvin scale.)

What is the melting point of ice in kelvin?

— *EXTENSION* —

	mercury	alcohol
boiling point	365 °C	78.5 °C
freezing point	–39 °C	–117 °C
colour	silver	clear, but can be dyed
cost	expensive	cheap
conducts heat	well	not so well
toxicity	poisonous	not poisonous in small amounts
metal or not	metal	non-metal
conducts electricity	well	an insulator
flammability	nonflammable	flammable
density	high	low
surface tension	high	not so high
degree of expansion when heated	average	large

6 ATOMIC STRUCTURE

Atoms themselves are made up of even smaller particles. The most important of these are protons, neutrons and electrons.

Atoms contain protons, neutrons and electrons

nucleus, containing neutrons and protons

region where electrons are found

Fig. 6.1 An atom

Atoms are made up of even smaller particles. There is a **nucleus** in the centre of each atom. The nucleus can contain **protons** and **neutrons**. Around the nucleus is the rest of the atom, where **electrons** are most likely to be found. Protons and neutrons have about the same mass. Electrons are about 2000 times lighter.

Protons have a small positive electrical charge. Electrons have an equal but opposite (negative) charge. The number of protons and electrons in an atom are exactly equal, so the two equal and opposite charges cancel out. Atoms have no overall charge.

Questions

1   If an atom has 10 protons, how many electrons does it have?
2   Which particle in an atom carries a positive charge?
3   Which particle in an atom carries a negative charge?

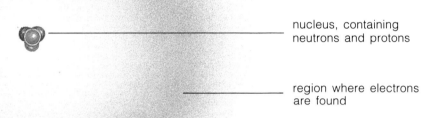

Protons, neutrons and electrons

Particle	Relative mass	Relative charge
proton	1	+1
neutron	1	0
electron	1/2000	-1

DID YOU KNOW?

Although atoms are made from electrons, protons and neutrons, these are not the fundamental building blocks of the atom. There are at least 37 of these fundamental particles. There are many kinds of leptons, quarks, antiquarks, photons, gravitons, bosons and gluons. Some of these particles have not actually been detected yet, but physicists think they must exist because of the way that atoms behave.

The history of atomic structure

The Ancient Greeks were the first people to think of matter as being made of tiny particles. They imagined these particles to be like solid balls and this idea of the atom remained until the early 1900s. Around this time, evidence emerged that atoms contained at least two kinds of matter, some of it with a positive charge and some with a negative charge. A new model of the atom was suggested – a ball of positively charged 'dough' in which negatively charged electrons were dotted around like currants.

Fig. 6.2a   The ancient Greeks imagined that atoms were like solid balls.

Fig. 6.2b   Around 90 years ago it was suggested that an atom was rather like a plum pudding.

The Rutherford experiment

Ernest Rutherford was born in New Zealand in 1871. He won a Nobel Prize in 1908 for work on radioactivity. While continuing this work in Manchester, England, he made a discovery which he, and all other scientists at the time, found quite amazing.

He and two colleagues, Geiger and Marsden, were carrying out an experiment in which they shot alpha particles at a very thin piece of gold foil, in a vacuum. The apparatus is shown in Figure 6.3a.

Most of the alpha particles went straight through (A), some went through but changed direction slightly (B) and an even smaller number actually bounced back (C).

This suggested to Rutherford that the atom must be mainly space, and that the positive charge was not spread around, but in the centre. There was a positively charged central core, made of particles called protons, with the negatively charged electrons around the outside. He put forward this new model of the atom in 1911, and the modern view of the atom is very similar.

Fig. 6.3b   Particles can pass through the gold because most of an atom is space.

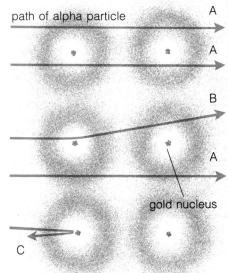

path of alpha particle

gold nucleus

Rutherford was astonished at these results. He wrote: 'It was quite the most incredible event that ever happened to me in my life. It was as incredible as if you fired a 15-inch shell at a piece of tissue paper and it came back and hit you. On consideration, I realised that this scattering backwards must be the results of a single collision, and when I made calculations I saw that it was impossible to get anything of that order of magnitude unless you took a system in which the greater part of the mass of the atom was concentrated in a minute nucleus.'

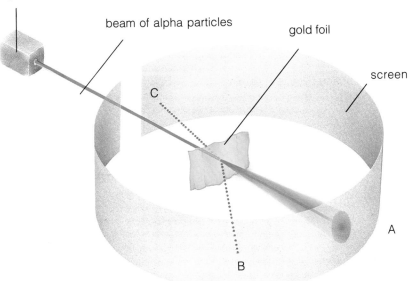

alpha particle source

beam of alpha particles

gold foil

screen

Fig. 6.3a   Rutherford's experiment. The entire apparatus was enclosed in a vacuum chamber. Why do you think this was necessary?

 **7 ELEMENTS**

An element is a substance made of atoms which all contain the same number of protons.

There are over 100 different kinds of atom

Over 100 different kinds of atoms exist. They are different from each other because they do not have the same numbers of protons, neutrons and electrons.

Atoms with the same number of protons as each other behave in the same way chemically. Atoms with different numbers of protons behave differently. There are about 90 naturally occurring types of atoms. The remainder are made by humans.

Fig. 7.1   Some examples of elements. Beginning at the back right, and working clockwise: chlorine (in the gas jar), chromium, sulphur, iodine, zinc, carbon, and mercury.

An element is a substance whose atoms all have the same number of protons

A substance made from atoms which all have the same number of protons is called an **element**. As there are about 90 naturally occurring kinds of atoms, there are about 90 naturally occurring elements. You can see them all listed in the Periodic Table, on page 31. Each element has its own symbol. Hydrogen, for example, has the symbol **H**. Helium has the symbol **He**.

The atomic number of an element is the number of protons in each atom

The number of protons an atom contains is called its **atomic number**. All the atoms of an element contain the same number of protons so they all have the same atomic number. The element with the smallest number of protons is hydrogen. It has just one proton, so its atomic number is one. Helium has two protons, so its atomic number is two.

You sometimes need to write the atomic number of an element when you write its symbol. You show it like this: $_1$**H**. This shows that the atomic number of hydrogen is one. Helium has the atomic number two so it is written $_2$**He**.

Fig. 7.2   A hydrogen atom.

Atomic number = number of protons = 1

Mass number = number of protons + number of neutrons = 1

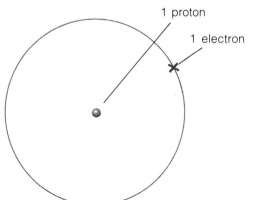

Fig. 7.3   A helium atom.

Atomic number = number of protons = 2

Mass number = number of protons + number of neutrons = 4

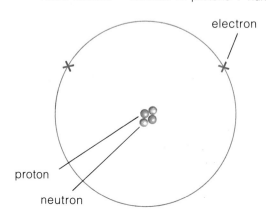

The mass number of an atom is the total number of protons and neutrons it contains

The **mass number** of an atom is the total number of 'heavy' particles that it contains. If you look at the table on page 22, you will see that the 'heavy' particles are neutrons and protons. A hydrogen atom contains just one proton, one electron and no neutrons, so its mass number is one. It is the smallest atom.

The next largest atom is the **helium** atom. A helium atom contains two protons, two neutrons and two electrons. As it has two protons, its atomic number is two. It has four 'heavy' particles – two protons and two neutrons – so its mass number is four.

If you want to show the mass number of an element, you write it like this: **¹H** or **⁴He**.

Fig. 7.5 Using symbols to show atomic number and mass number.

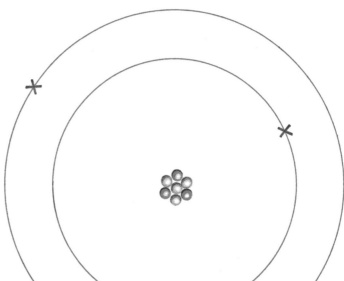

Fig. 7.4 A lithium atom. Lithium is the next largest atom after helium. What is its atomic number? What is its mass number?

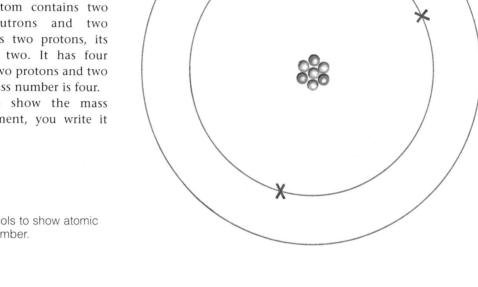

$^{235}_{92}U$

This is the mass number. It tells you the total number of protons and neutrons in one atom of the element.

This is the atomic number. It tells you how many protons there are in one atom of the element.

Questions

1 What is an element? Give three examples of elements.

2 Use the table on page 27 to write down the symbols of these elements:
   a carbon b chlorine c oxygen
   d sodium e potassium f helium
   g nitrogen h neon i argon j boron
   k sulphur l magnesium

3 a What is meant by the term atomic number?

b What is the atomic number of:
   i hydrogen
   ii lithium
   iii uranium?

c What is meant by the term mass number?

d For each of the following examples, give (a) the atomic number and (b) the mass number:
   i $^{35}_{17}Cl$ ii $^{12}_{6}C$ iii $^{14}_{6}C$ iv $^{16}_{8}O$

4 Copy and complete the following table.

Mass number	20	40			28
Atomic number	10			13	
Number of neutrons		22			
Symbol	Ne	Ar	$^{24}_{12}Mg$	$^{27}_{13}Al$	$^{28}_{14}Si$

8 ELECTRON ORBITS

The electrons of an atom are arranged in orbits around the nucleus. Each orbit can only hold a certain number of electrons.

Electrons are arranged in orbits

You have seen how different kinds of atoms have different numbers of protons. A hydrogen atom has one proton. A lithium atom has three protons. A sodium atom has 11 protons. The number of protons in an atom is called its atomic number. The number of protons in an atom determines what sort of atom it is.

The number of electrons in an atom is the same as the number of protons. So a hydrogen atom has one electron. A lithium atom has three electrons. A sodium atom has 11 electrons. The electrons are arranged in **orbits** around the nucleus of the atom. The orbits are sometimes called shells.

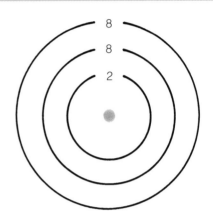

Fig. 8.1 Electron orbits. The inner orbit can hold up to two electrons. The next two can each hold eight.

Each orbit can only hold a certain number of electrons

The electron orbits are filled up from the inside outwards. The first orbit can hold up to two electrons. The second orbit can hold up to eight electrons. The third orbit can also hold up to eight electrons.

A lithium atom has three protons and three electrons. Two electrons are in the first orbit, so this orbit is full. The third electron is in the second orbit. A sodium atom has 11 protons and 11 electrons. Two electrons fill up the first orbit. Eight electrons fill up the second orbit. The last electron goes into the third orbit.

The arrangement of electrons is called the electron configuration

The way in which the electrons are arranged in an atom is called its **electron configuration**. You can show the electron configuration by drawing the atom. Or you can show it by numbers. For example, the electron configuration of a lithium atom is **Li (2,1)**. This means that lithium has two electrons in its inner orbit, and one electron in the second orbit. The electron configuration of a sodium atom is **Na (2,8,1)**.

The electron configuration of an atom determines how that atom will react with other atoms. You will find much more about this later in the book.

Fig. 8.2 The electron configurations of three atoms.

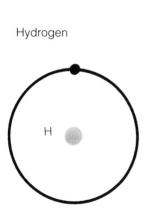

Hydrogen

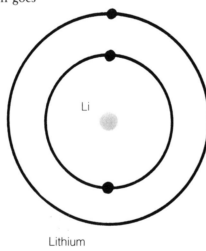

Lithium

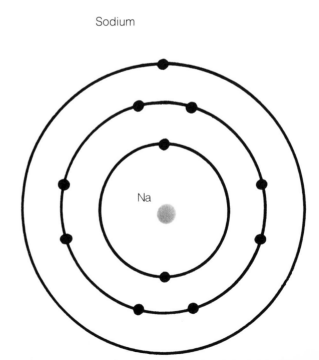

Sodium

Element	Symbol	Atomic number (number of protons)	Mass number (number of protons + neutrons)	Electron configuration
hydrogen	H	1	1	1
helium	He	2	4	2
lithium	Li	3	7	2,1
beryllium	Be	4	9	2,2
boron	B	5	11	2,3
carbon	C	6	12	2,4
nitrogen	N	7	14	2,5
oxygen	O	8	16	2,6
fluorine	F	9	19	2,7
neon	Ne	10	20	2,8
sodium	Na	11	23	2,8,1
magnesium	Mg	12	24	2,8,2
aluminium	Al	13	27	2,8,3
silicon	Si	14	28	2,8,4
phosphorus	P	15	31	2,8,5
sulphur	S	16	32	2,8,6
chlorine	Cl	17	35	2,8,7
argon	Ar	18	40	2,8,8
potassium	K	19	39	2,8,8,1
calcium	Ca	20	40	2,8,8,2

Table 8.1   The first twenty elements.

Questions

1   Draw the following atoms to show the number of electrons in each of their electron orbits:
    a carbon b oxygen c neon
    d chlorine e argon f potassium
2   Name three elements whose atoms have full outer electron orbits.
3   Name three elements whose atoms have only one electron in their outer orbit.
4   Name three elements whose atoms have outer orbits which are one electron short of being full.

9 ATOMIC SPECTRA

The electrons in an atom are arranged in orbits. Movement of electrons between these orbits may result in the absorption or emission of light.

Heating atoms can make them emit light

Figure 9.1 shows how the structure of a sodium atom can change. The electrons are arranged in shells around the nucleus. This diagram shows the atom changing from its **lowest energy state**. This is the state in which the arrangement of electrons is most stable.

If the atom is heated, the heat energy may allow an electron to move out into a higher orbit than normal. We can say that the electron has been 'excited'. This is an unstable situation. The electron very quickly returns to its normal orbit. But as it does so it releases the extra energy it has been given. This energy is given out as light.

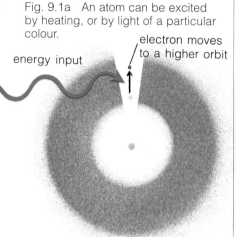

Fig. 9.1a An atom can be excited by heating, or by light of a particular colour.

energy input

electron moves to a higher orbit

A particular energy change produces a particular colour of light

When an electron falls between two particular orbits in a particular kind of atom, the energy given out is always the same. The colour of the light given out, or **emitted**, depends on these energy changes. So whenever an electron falls between the same two orbits in a particular kind of atom, the same colour of light is emitted.

Figure 9.2 shows the colours of light emitted by a sodium atom when it is heated. This is called the **line emission spectrum** of sodium. Notice that there are several coloured lines. Each line represents an energy change in the atom. The two yellow lines are the strongest. This is the yellow light that comes from sodium street lamps. You can also see this colour if you hold a small amount of a sodium compound in a blue Bunsen burner flame.

Fig. 9.2 Sodium emission spectrum. The colours produced by excited sodium atoms can be separated into lines. The emission spectrum is produced when the atoms release energy.

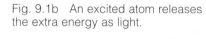

Fig. 9.1b An excited atom releases the extra energy as light.

electron moves to a lower orbit

energy released

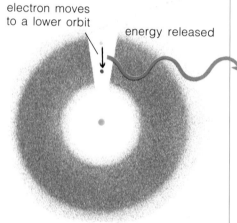

Fig. 9.3 Sodium chloride in a Bunsen burner flame emits a yellow light which is the same as that seen in sodium street lamps.

EXTENSION

The light from stars gives us information about them

Stars are a long way off. Light from our own Sun, 150 million km away, takes 8½ minutes to reach us. Light from the next nearest star, Proxima Centauri, takes over four years! When you look at the North Star, the light entering your eye has taken 700 years to arrive. You are seeing the star as it was 700 years ago.

It is very unlikely that anyone will ever be able to go to a star. But we can learn a lot from their light.

You have seen how atoms can **emit** light if they are given particular sorts of energy. This process also happens in reverse. Atoms can **absorb** light, if it is of the particular colour that makes these energy changes happen. The energy of the light is used to excite the electrons in the atom. Each kind of atom can absorb particular colours of light.

If we look at the light from a particular star, we can find which colours are missing from its spectrum. Figure 9.4 shows an example from the Sun. This is called an **absorption spectrum**. The colours are missing because particular atoms in the Sun have absorbed them. So we can work out which kinds of atom are present in the Sun.

Space is not as empty as many people think. If we look at the light from a more distant star, the light may have passed through clouds of gas deep in space. These gas clouds, known as **nebulas**, might only have ten hydrogen atoms in every 1 cm^3. But the clouds are enormous, so the overall number of atoms can be large. As well as single atoms, the gas clouds contain more complex groups of atoms. These can absorb colours from the starlight. In 1937, the atom pair CN was discovered out in space, from an absorption spectrum.

Hydrogen clouds around a nebula can absorb all the ultraviolet light from a star. The gas cannot absorb all this energy for ever, and much of it is released again as particular colours.

The study of spectra is called spectroscopy. Spectroscopy can tell us about the atoms in space, which we will never be able to reach. It is the energy changes within the atom that enable us to identify them.

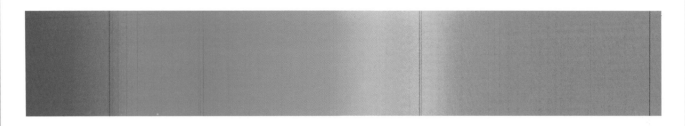

Fig. 9.4 The Sun's absorption spectrum. The dark lines represent 'missing' colours. These tell us which atoms are present in the Sun.

DID YOU KNOW?

Helium was identified in the Sun, using spectroscopy, *before* it was discovered on Earth.

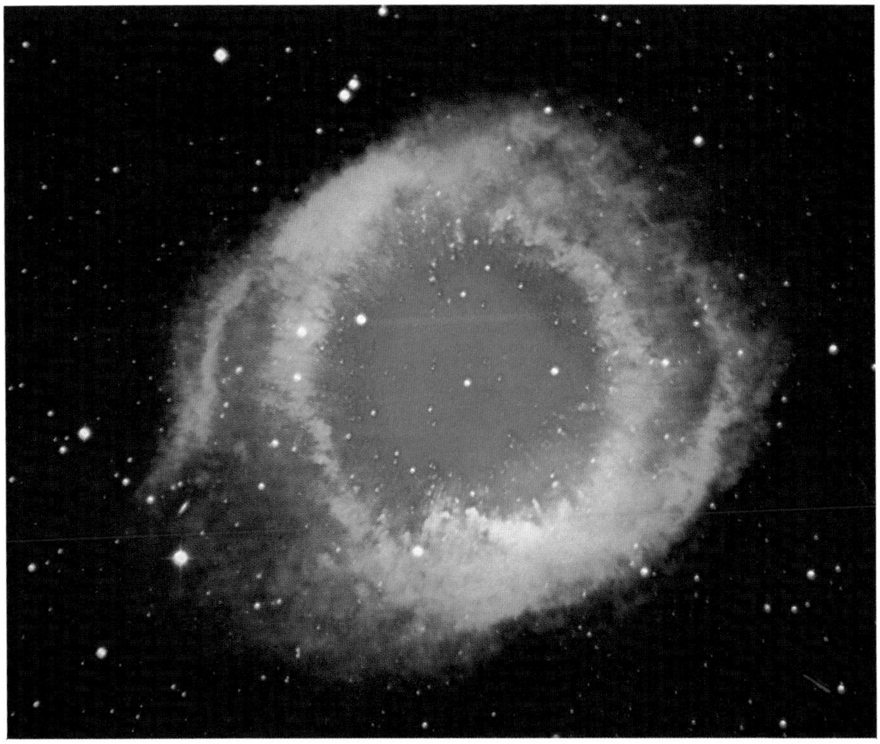

Fig. 9.5 A gas cloud, or nebula. This is the Helix planetary nebula photographed from Australia. It is 38 million million km across. In the centre is a white dwarf star. The blue-green parts contain mostly oxygen and nitrogen. The pink parts contain hydrogen.

Questions

1 Explain why some kinds of atoms give out light when they are heated.

2 a What is an absorption spectrum?

b How can absorption spectra give us information about distant stars and galaxies?

EXTENSION

10 THE PERIODIC TABLE

The different kinds of elements can be arranged in a pattern, according to the structure of their atoms and the way in which they behave.

Elements can be arranged into groups

Each element behaves in a different way from every other element. We say that their **properties** are different. But there are also similarities in the way that some elements behave.

For example, you will see in Topic 195 that **lithium**, **sodium**, and **potassium** have very similar properties. They are all reactive metals. They are often called the **alkali metals**.

Fluorine, **chlorine**, **bromine** and **iodine** also show great similarities in the way they behave. They are often known as the **halogens**.

Elements with a similar arrangement of electrons in their outer orbit behave in a similar way

When chemists first arranged elements into groups, they did not know why some elements showed similarities in the way they behaved. Now we know that it is to do with the number of electrons they have in their outer orbit. Elements with the same number of electrons in their outer orbit behave in a similar way.

For example, lithium, sodium and potassium all have **one** electron in their outer orbit. Fluorine, chlorine, bromine and iodine all have **seven** electrons in their outer orbit. Lithium, sodium and potassium belong to **Group 1**, because their outer orbit contains one electron. Fluorine, chlorine, bromine and iodine belong to **Group 7**, because their outer orbit contains seven electrons.

The Periodic Table shows all the elements arranged in Groups

You can see these Groups in the Periodic Table. The elements in each Group are arranged vertically. The element with the smallest atomic number is at the top of the Group, and the one with the largest atomic number is at the bottom.

The horizontal rows in the Periodic Table are called **Periods**. Look at Period 2. It begins with lithium, then beryllium, then boron. These elements are arranged in increasing atomic number. (Remember – the atomic number is the number of protons in an atom.)

The Periodic Table can suggest how reactive an element will be

Atoms are most stable when their outer electron orbit is full. The Group 0 elements have eight electrons in their outer orbit. The orbit is full. So Group 0 elements are very **unreactive**. They are sometimes called the **noble gases**.

Group 7 elements have seven electrons in their outer orbit. They only need one more electron to fill this orbit. So they readily take electrons from other atoms, to fill up their outer orbit and become stable. This makes them very **reactive** elements. You will find more about this in Topic 196.

Group 6 elements have six electrons in their outer orbit. They need two more

electrons to fill this orbit. Like Group 7 elements, they will take electrons from other atoms. But they do this less readily, so they are less reactive than Group 7 elements.

At the other end of the Periodic Table, Group 1 elements have only one electron in their outer orbit. They can have a full outer orbit by losing this electron. They do this very easily, giving up their electron to other atoms. So Group 1 elements are very reactive.

Apart from the noble gases, the most reactive elements are near the left- and right-hand sides of the Periodic Table.

Metals are on the left-hand side of the Periodic Table

The zig-zag line separates metals from non-metals. All the elements on the left-hand side of the line are metals. All the elements on the right-hand side of the line are non-metals.

Metals are elements which tend to lose electrons. The metals have atoms which can most easily end up with full electron orbits by losing electrons.

The most reactive metals are the ones nearest the left-hand side of the Periodic Table, such as sodium and potassium. Why do you think that these are the most reactive metals?

Transition elements have more complex electron arrangements

The middle of the Periodic Table is taken up with the **transition elements**. You will find out more about some of these elements on page 44. These elements do not fit into one of the eight groups. They have more complex electron arrangements.

GROUP

1	2													3	4	5	6	7	0
						H 1													**He 2**
Li 3	Be 4													B 5	C 6	N 7	O 8	F 9	Ne 10
Na 11	Mg 12													Al 13	Si 14	P 15	S 16	Cl 17	Ar 18
K 19	Ca 20	Sc 21	Ti 22	V 23	Cr 24	Mn 25	Fe 26	Co 27	Ni 28	Cu 29	Zn 30			Ga 31	Ge 32	As 33	Se 34	Br 35	Kr 36
Rb 37	Sr 38	Y 39	Zr 40	Nb 41	Mo 42	Tc 43	Ru 44	Rh 45	Pd 46	Ag 47	Cd 48			In 49	Sn 50	Sb 51	Te 52	I 53	Xe 54
Cs 55	Ba 56	La 57	Hf 72	Ta 73	W 74	Re 75	Os 76	Ir 77	Pt 78	Au 79	Hg 80			Tl 81	Pb 82	Bi 83	Po 84	At 85	Rn 86
Fr 87	Ra 88	Ac 89																	

Ce 58	Pr 59	Nd 60	Pm 61	Sm 62	Eu 63	Gd 64	Tb 65	Dy 66	Ho 67	Er 68	Tm 69	Yb 70	Lu 71
Th 90	Pa 91	U 92	Np 93	Pu 94	Am 95	Cm 96	Bk 97	Cf 98	Es 99	Fm 100	Md 101	No 102	Lr 103

PERIOD 1–7 (rows above)

KEY

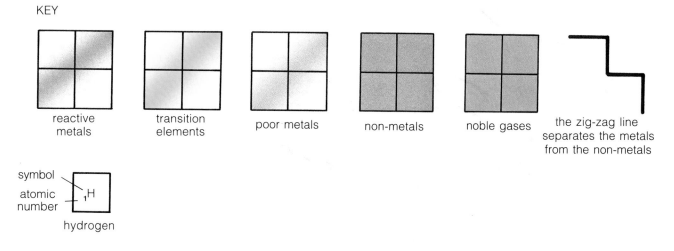

reactive metals | transition elements | poor metals | non-metals | noble gases | the zig-zag line separates the metals from the non-metals

symbol
atomic number — $_1$H
hydrogen

Fig. 10.1 The Periodic Table. The Periodic Table lists all the elements in order of their atomic number. The elements are arranged in Groups, according to the number of electrons in their outer orbit. For example, Group 1 elements – Li, Na, K, Rb, Cs and Fr – all have one electron in their outer orbit. All the elements beyond uranium (atomic number 92) have been made by people. They do not occur naturally on Earth.

Questions

1 What is the atomic number of :
   a sodium
   b oxygen
   c copper?

2 Which element has the atomic number of:
   a 17
   b 6
   c 11?

3 How many electrons are there in the outer orbit of:
   a a magnesium atom
   b a neon atom
   c a nitrogen atom?

4 a Why are the elements in Group 1 very reactive elements?
   b Why are the elements in Group 7 very reactive elements?
   c Why are the elements in Group 0 very unreactive elements?

— EXTENSION —

DID YOU KNOW?

On Earth the rarest naturally occurring element is astatine. There is thought to be only 0.16 g in the whole of the Earth's crust.

31

11 ISOTOPES

Isotopes are atoms of the same element, but with different numbers of neutrons.

Isotopes of an element have the same atomic number, but different mass numbers

All atoms with the same number of protons belong to the same element. They behave in exactly the same way in chemical reactions. As well as having the same number of protons, they also have the same number of electrons. For example, all hydrogen atoms have one proton and one electron. The atomic number is one.

Most hydrogen atoms have no neutrons so the mass number is one. But about one hydrogen atom in 10 000 is heavier than this. It has a neutron in its nucleus. This form of hydrogen still has one proton. It still has an atomic number of one. It is chemically identical to the most common form. It occupies the same place in the Periodic Table. For this reason it is called an **isotope** of hydrogen. ('Isotope' means

'same place'.) This isotope is sometimes called **deuterium**. Hydrogen and deuterium are both isotopes of hydrogen. They have the same atomic number, but different mass numbers. A

deuterium atom is heavier than a hydrogen atom, because of its extra neutron.

Fig. 11.1a   A hydrogen atom

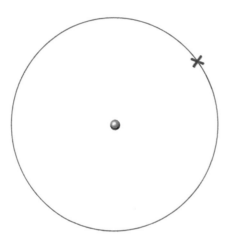

Fig. 11.1b   A deuterium atom

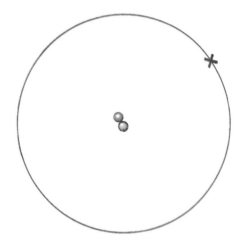

Carbon also has isotopes

Hydrogen is not the only element to have isotopes. Many elements have isotopes. Carbon has several isotopes. The 'normal' carbon atom has six protons and six neutrons. It has a mass

number of 12, and its symbol is ^{12}C. It is called carbon twelve. Another isotope has eight neutrons. It still has six protons, or it would not be carbon. Its mass number is 14 and its symbol is ^{14}C. It is called carbon fourteen.

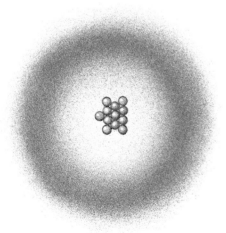

Fig. 11.2a   A carbon 12 atom has 6 protons and 6 neutrons.

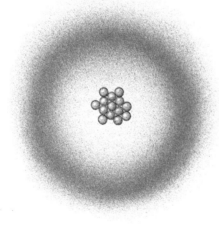

Fig. 11.2b   A carbon 14 atom has 6 protons and 8 neutrons.

Mass number is the atomic mass of a particular isotope of an element

When we talk about the atomic mass of an element, we can mean two things. We might mean the atomic mass of a particular isotope of that element. Or we might mean the average atomic mass of all the isotopes present in a particular sample of that element.

Take the element carbon, for example. The commonest isotope of carbon has six protons and six neutrons. We can say that its mass number is 12. Mass number refers to the atomic mass of a particular isotope of an element. If you say 'the mass number of carbon is 12', you really mean 'the mass number of the commonest isotope of carbon is 12'.

Relative atomic mass is an average for the different isotopes

In a sample of carbon, there will be atoms with different mass numbers. If you took 100 carbon atoms, you would probably find that 99 of them were carbon 12 atoms. One of them would probably be a carbon 13 atom. If you sorted through millions of carbon atoms, you might find a carbon 14 atom.

So if you found the *average* mass of a random sample of 100 carbon atoms, it would be a little bigger than 12. The few heavier atoms would make the average mass about 12.01. This average mass number of a random sample of an element, taking into account the different isotopes in it, is called the **relative atomic mass**. Because the relative atomic mass of an element is an average, it is often not a whole number.

Chlorine, for example, has two common isotopes. They are chlorine 35 and chlorine 37. In a sample of chlorine, there are nearly three times as many chlorine 35 atoms as chlorine 37 atoms. This makes the average mass of the atoms 35.5. The relative atomic mass, or A_r, of chlorine is 35.5.

Table 11.1 Some relative atomic masses

element	relative atomic mass A_r	most common isotope	percentage occurrence
chlorine	35.5	^{35}Cl	75
barium	137.34	^{138}Ba	72
germanium	72.69	^{74}Ge	36.5
mercury	200.5	^{202}Hg	30
strontium	87.62	^{88}Sr	82.6
thallium	204.3	^{205}Tl	70.5

Questions

1 a What do all isotopes of a particular element have in common?

b How do isotopes of a particular element differ from each other?

2 Three isotopes of magnesium are ^{24}Mg, ^{25}Mg and ^{26}Mg.

a Use the Periodic Table on page 31 to find the atomic number of magnesium.

b Every magnesium atom has the same number of protons. What is this number?

c What is the total number of protons and neutrons in a ^{24}Mg atom?

d How many neutrons are there in a ^{26}Mg atom?

> ### DID YOU KNOW?
> Two elements share the record for the highest number of known isotopes. Both xenon and caesium have 36.

e The relative atomic mass of magnesium is 24.3. Why is this not a whole number?

f Which is the most common isotope of magnesium?

3 In 1000 thallium atoms, there are 705 atoms of ^{205}Tl.

a What is the total number of neutrons and protons in these 705 thallium atoms?

b The remaining 295 atoms are ^{203}Tl atoms. What is the total number of neutrons and protons in these 295 thallium atoms?

c What is the average number of protons and neutrons in the 1000 thallium atoms?

d What is the A_r (relative atomic mass) of thallium?

4 Natural copper contains two isotopes, ^{63}Cu and ^{65}Cu. The atomic number of copper is 29.

a How many protons are there in a copper atom?

b How many neutrons are there in each of the two isotopes of copper?

c In naturally occurring copper, 69% of the atoms are ^{63}Cu, and the remainder are ^{65}Cu. What is the relative atomic mass of copper?

12 METALS AND NON-METALS

All elements can be put into two groups – metals or non-metals. Four-fifths of all elements are metals.

Most elements are metals

You will remember that there are about 90 naturally occurring elements. The Periodic Table on page 31 shows these elements arranged in Groups. All the elements to the left of the zig-zag line are **metals**, and all those to the right are **non-metals**. There are far more metallic elements than non-metallic ones. Around four-fifths of all the different elements are metals.

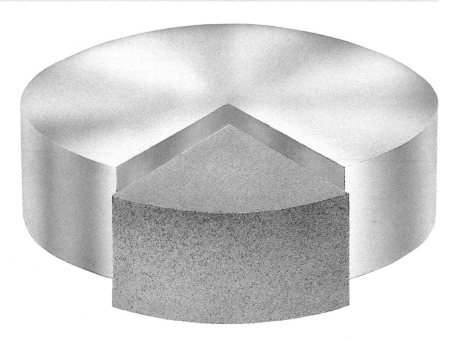

Fig. 12.1 Around four-fifths of all the known elements are metals. But metals only make up less than one quarter of the materials found on Earth, because many metals are very rare.

Metals and non-metals have different properties

Metals have certain properties in common. They are usually **shiny**. However, some metals, like iron, look dull because they have a covering of metal oxide on their surface. If you rub this off with sandpaper you can see the shiny metal beneath.

All metals will **conduct electricity**. Most non-metals will not. One exception is carbon, which is a non-metal but *can* conduct electricity.

Most metals **melt** when heated. Some of the more reactive ones, such as magnesium, also burn with a flame when you heat them. It is hard to generalise about non-metals, but some, like sulphur, phosphorus and carbon, **burn** when heated in air.

When substances burn in air, they form compounds called oxides. When a metal burns it forms an oxide that is **basic**. When a non-metal burns it forms an oxide that is **acidic**.

Most metals are **malleable**. This means that they can be beaten or rolled into different shapes. Non-metals tend to be brittle, which means that they break easily.

Fig. 12.2 Sulphur and chromium. Sulphur is a non-metal and chromium is a metal.

Question

You have probably seen examples of metals and non-metals in previous lessons. If so, try this question from memory. Copy this chart, and fill in the spaces. In each case, there are two statements to choose from.

Substance	Metal or non-metal?	Dull or shiny?	Good or poor conductor of electricity and heat?	Melts or burns when heated?	Brittle or malleable?	Is the oxide acidic or basic?
Silver						
Carbon						
Magnesium						
Copper						
Sulphur						

Metals have their special properties because of the atoms of which they are made

Metallic elements have atoms which lose some of their electrons quite easily. The reason they tend to do this is explained on page 40. In a piece of metal, the atoms are packed together in a regular pattern. They are packed so tightly that some of the electrons become detached from their atoms. These 'free' electrons move around in the spaces between the atoms. They are shared between the atoms in the piece of metal.

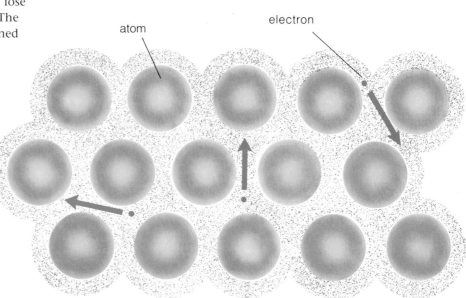

Fig. 12.3 How metals conduct heat and electricity. The atoms in a metal are packed in a regular pattern called a **lattice**. Some of the electrons can move freely in the spaces between the atoms. The freely-moving electrons can transfer heat and electricity through a piece of metal.

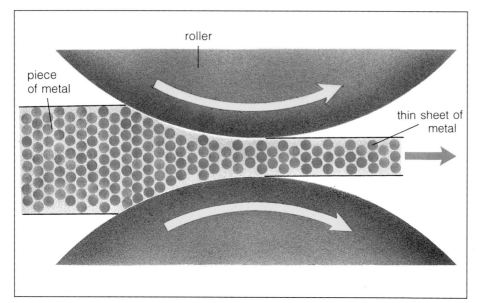

Fig. 12.4 Rolling a sheet of metal. Metals such as aluminium can be rolled into thin sheets, because the regular layers of atoms can slide over one another.

The electrons carry a negative charge. The metal atoms carry a positive charge, because they now have more protons than electrons. The positive and negative charges attract each other, so the whole arrangement is held closely together.

The free electrons explain why metals are so good at conducting electricity. The electrons in metals can readily be made to move in a particular direction. This movement is what we call an electric current.

Heat is conducted through a material by the movement of its particles. The free electrons in a metal can easily vibrate when they get hotter, and bump into neighbouring electrons, making them vibrate as well. In this way heat is quickly passed from one end of a strip of metal to the other.

When a piece of metal is hammered it can easily change shape as the layers of atoms slide over each other. This is why metals are malleable. For the same reason, many metals can be drawn out into thin wires. They are said to be **ductile**.

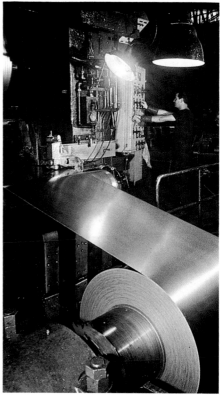

Fig. 12.5 Rolling sheet metal

13 USES OF METALS

The properties of metals make them very useful to us for making all sorts of things. Mixtures of two or more metals, called alloys, are also widely used in constructing all kinds of objects.

Each metal has its own properties

Although all metals share certain properties, no two metals are identical. These individual properties make different metals suitable for different uses. Sometimes there is a need for a substance which has a combination of properties not found in any one metal. By mixing two or more metals together a new substance called an **alloy** can be made. Sometimes an alloy consists of one or more metals mixed with a tiny amount of non-metal, e.g. steel. This new substance will have a different set of physical properties from either of its 'parent' metals.

Only some of the most commonly used metals and alloys are mentioned on these two pages. Engineers are constantly developing new alloys for new purposes.

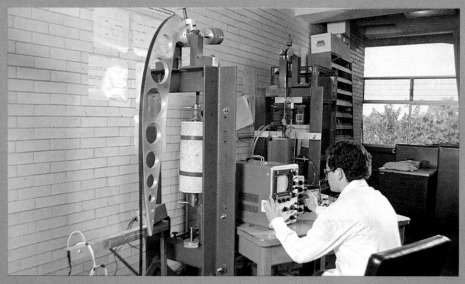

Fig. 13.1 A technician in a metallurgy laboratory, testing the strength of metal structures.

Uses of some pure metals

One metal which you have almost certainly made use of is **aluminium**. Aluminium is a light metal which can easily be rolled out into very thin, yet strong sheets. It is also very resistant to corrosion. This makes it suitable for milk bottle tops. It does not melt easily and can withstand high temperatures, so it is used for making containers for prepacked foods, and for making foil for covering food as it cooks. Many

drink cans are made of aluminium as it is non-corrosive, light and yet strong.

Copper is especially good at conducting electricity and is ductile. This means that it can be pulled out into a thin wire. So copper is used for making electrical wires.

Gold, **silver** and **platinum** are used for making jewellery. They are all very unreactive metals, so they do not tarnish easily and stay shiny and attractive. They may also be used for coating parts in electrical switches.

Mercury is an unusual metal because it is liquid at room temperature. It is used in thermometers.

Lead is a heavy metal which is extremely malleable. This makes it suitable for making waterproof edgings to roofs such as around chimneys. Before it was known to be poisonous, lead was used for making water pipes and toy soldiers. Lead is also used for shielding from radiation.

Fig. 13.2 Objects made from pure metals. They include lead, aluminium, iron, tungsten, silver, gold and copper. Can you identify which is which? What properties of each metal make them suitable for these purposes?

Uses of some alloys

The commonest alloy in use today is **steel**. Steel is an alloy of **iron** and **carbon**. The carbon makes the iron much stronger. There are different kinds of steel. They can have varying amounts of carbon in them, and many kinds have other elements as well. Steel which contains only iron and carbon is called **mild steel**. This is the sort of steel which is normally used for building large structures such as bridges and cars. For making cutlery, **nickel** and **chromium** are added as well. This makes the steel even harder and stops it rusting. It is called **stainless steel**.

The first alloy discovered by humans was **bronze**. Bronze is an alloy of **copper** and **tin**, and was used in the Bronze Age for making weapons. Later, when iron was discovered, bronze weapons rapidly became outdated because they were not very strong and did not stay sharp for long. But we still use bronze for statues because it is a very attractive alloy, and does not rust.

Solder is an alloy of two metals with low melting points – **tin** and **lead**. The alloy has an even lower melting point than either of its two 'parent' metals, and is used for joining metals together, such as wires.

Fig. 13.3 Objects made from alloys. They include brass, solder, stainless steel, aluminium alloy, and high-tensile steel. Can you identify which is which? Which metals are present in each alloy? What properties of each alloy make it suitable for the purpose shown?

INVESTIGATION 13.1

Comparing the strength of different metals

You are going to test wires made from different metals, to find out how strong they are. You may also be able to test different thicknesses of wire of the same metal.

1 Read through the experiment and draw up a suitable results chart. You will need to record the type of metal used, its thickness, and the weight that it supported.

2 Put one piece of wire aside – you are not going to test this until the end of your experiment. It is best to choose a medium thick wire of a metal of which you have several other thicknesses.

3 Choose your first wire. Record the metal it is made from, and its thickness, in your table. Make it into a hook, by bending a short length of the wire round a pencil.

4 Mount your hook as shown in the diagram. Now hang masses on it until it gives way. (It is important to decide when you think the hook has given way, and to do the same each time.) Record the maximum weight it supported.

5 Now do the same with other metals, or with other thicknesses of the same wire. It is important that you make your hook in exactly the same way each time.

Questions

1 For each metal that you tested, plot a line graph of weight supported (vertical axis) against thickness of wire (horizontal axis).

2 Why was it important to bend each of your hooks by the same amount when you made them?

3 Which was the strongest hook?

4 Which was the strongest metal?

5 Use your graphs to predict the weight that your untested piece of wire could support. Write this down and then test it. How close was your prediction?

6 How accurate do you think your results are? Can you suggest any ways in which this experiment might be improved?

hook made from wire to be tested

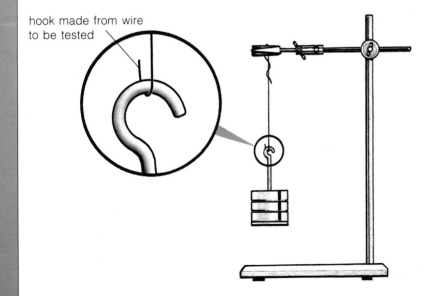

Fig. 13.4

Breaking a wire

This experiment will probably be demonstrated, because when a wire suddenly breaks the ends can fly outwards with considerable force!

Watch carefully as a wire is loaded with weights. The weights will be added until the wire snaps. This experiment can be repeated with different types of wire.

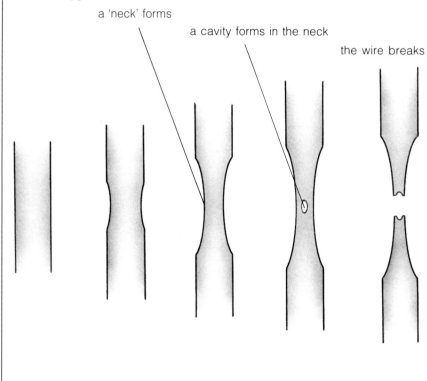

a 'neck' forms

a cavity forms in the neck

the wire breaks

Fig. 13.5 Breaking a wire

Questions

1 Does the wire snap suddenly, or can you predict when it will snap? If so, how can you predict it?

2 Are the two halves of the broken wire, added together, longer than the original?

3 Does the type of metal used make much difference to the result?

4 Is the strongest wire the same as that found in the hook experiment (Investigation 13.1)?

5 Arrange the wires in order of strength. Do the same for your results from the hook experiment. Are the orders the same?

6 Were your doing the same thing to the wires in this experiment, and in the hook experiment? Try to think what is happening to the metal atoms, and describe it.

7 Look very carefully at the broken ends of the wire. Use a hand lens. You may be able to see a small hole in each broken end. The diagram shows how this forms.

What must be happening to the layer of atoms in the wire as this happens?

Questions

1 Which of these elements are metals, and which non-metals?
 a iron b nickel
 c carbon d oxygen
 e gold f helium
 g magnesium h sulphur
 i mercury j tin

2 a Make a list of four properties of metals.
 b Name two metals which are lacking one or more of these properties, and explain why they are unusual.

3 Why is copper a good conductor of electricity?

4 a What is an alloy?
 b Name two alloys. For each list its main uses, and the properties which make it suitable for these uses.

5 Lead and tin can be mixed together to form solder. The chart shows the melting points of alloys of different proportions of tin and lead.

 a Plot a line graph to show this data. Put percentage of tin on the horizontal axis, and melting point on the vertical axis.

 b Which has the lowest melting point – tin or lead?

 c What percentages of lead and tin would you mix to get an alloy with the lowest possible melting point? What would this melting point be?

 d Why is an alloy of tin and lead used for solder, rather than pure tin or pure lead?

 e Below what temperature would an alloy containing equal quantities of tin and lead be solid?

% lead	100	90	80	70	60	50	40	30	20	10	0
% tin	0	10	20	30	40	50	60	70	80	90	100
melting point (°C)	327	304	280	257	234	211	188	194	209	223	237

Dental Fillings

Teeth have evolved over millions of years to become perfect biting and chewing machines. The outer layer is made of enamel, a dense, brittle material composed of the mineral calcium phosphate. Beneath the enamel is a layer of dentine. This is a bone-like substance, softer than enamel, and containing protein as well as calcium phosphate. Some of the properties of enamel and dentine are shown in the table below.

	compressive strength (MPa)	elastic modulus (GPa)	hardness (Knoop number)
enamel	900	47	343
dentine	300	9	68

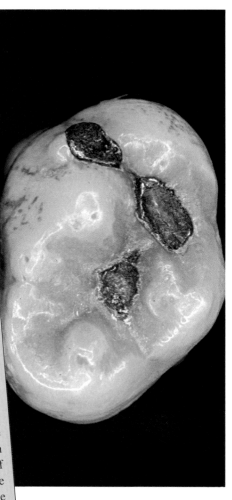

The compressive strength shows the maximum force that the material can take pressing down on it. The elastic modulus shows the amount by which the material can 'give'. The lower it is, the more the material can 'give'. When biting, the very hard enamel on the surface of the tooth is able to withstand very high compressive forces. Although it is brittle, the elasticity of the dentine beneath it cushions it, and makes it much less likely to break.

If a tooth decays the decayed part is removed and a filling is put in. The material used for the filling must have similar properties to the tooth. It must also be resistant to the warm, moist, salty, and sometimes acidic conditions inside the human mouth. Also, the material must be able to be moulded to fit the cavity exactly, bind tightly to the tooth and then harden quickly. Ideally, it should also look as 'tooth-like' as possible.

Several different filling materials have been developed. Perhaps the most familiar is **mercury amalgam**. This is an alloy of 12 % tin, 3 % copper, 0.2 % zinc, 35 % silver and 50 % mercury. It is mixed just before the filling is put in, and stays soft and pliable for long enough for the dentist to pack and mould it into the cavity. As it hardens, it expands slightly, which helps to hold it firmly in the tooth. The first material to be used for fillings was **gold**. This can be worked to make it soft enough to put into the tooth cavity, and it then hardens inside the mouth. However, pure gold always remains rather soft, and a gold-silver-copper alloy makes a better filling. Gold is a very unreactive metal so has high corrosion resistance. However, it is also very expensive. For fillings at the front of the mouth, appearance may be very important. Acrylic polymers have been developed which can match the colour of teeth perfectly, but these do not yet have such good mechanical properties as the metal alloys.

1   From what materials are enamel and dentine composed?
2   Describe the different properties of dentine and enamel. Explain how they enable a tooth to perform its function efficiently.
3   Why is it advantageous to have a layer of enamel on the outside of the dentine?
4   What properties are needed in a dental filling material?
5   What advantages does mercury amalgam have over gold, when used as a dental filling?
6   Acrylic polymers are still not widely used for fillings. Why is this?

14 IONIC COMPOUNDS

Atoms may gain or lose electrons, to become ions. Positive and negative ions are strongly attracted to each other.

Some atoms tend to lose electrons

You have seen in the last few pages that the atoms of metallic elements tend to lose electrons. Why is this?

Figures 14.1 and 14.2 show the structure of the atoms of sodium and neon. Neon has 10 protons and 10 electrons. The electrons are arranged in two complete shells. Neon is a very **stable** atom. It has a complete outer shell and no tendency to gain or lose electrons.

Sodium, however, has 11 protons and 11 electrons. It has two complete electron shells, and a single electron in its outer shell. This is a very **unstable** arrangement. The sodium atom has a strong tendency to lose this single electron, so that it ends up with a complete outer shell. When a sodium atom loses its outer electron, it becomes an **ion** (Figure 14.3).

An ion is an atom or group of atoms with an overall electrical charge. The sodium ion has a positive charge, because it still has its original 11 protons, but only 10 electrons. An ion with a positive charge is called a **cation**.

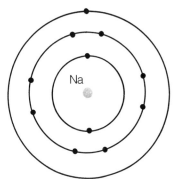

Fig. 14.2 A sodium atom. There is a single electron in the outer orbit. If the atom loses this electron, it will end up with a full outer orbit, like neon.

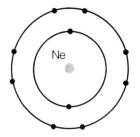

Fig. 14.1 A neon atom. The outer electron orbit is complete. This makes neon a very stable, unreactive element.

	charge
11 protons	= +11
11 electrons	= −11
total	= 0

	charge
11 protons	= +11
10 electrons	= −10
total	= +1

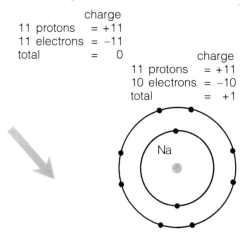

Fig. 14.3 A sodium ion. When a sodium atom loses its lone electron, it becomes a positive ion. It has a positive charge, because it now has one more proton than it has electrons.

Some atoms tend to gain electrons

Figures 14.4 and 14.5 show an argon atom and a chlorine atom. Argon, like neon, has a complete outer shell of electrons. It is a very stable atom, with no tendency to lose or gain electrons. Chlorine, however, has an outer shell which needs one more electron to complete it. So chlorine atoms have a strong tendency to gain electrons. When a chlorine atom gains an electron, it ends up with 17 protons and 18 electrons. This gives it an overall negative charge. It becomes a negatively charged ion, or an **anion**.

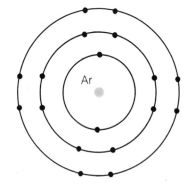

Fig. 14.4 An argon atom. Like neon, argon has a full outer electron orbit. This makes argon a very unreactive, stable element.

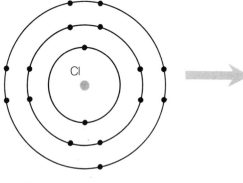

	charge
17 protons	= +17
17 electrons	= −17
total	= 0

Fig. 14.5 A chlorine atom. This has seven electrons in its outer orbit. It needs one more electron to fill up this orbit.

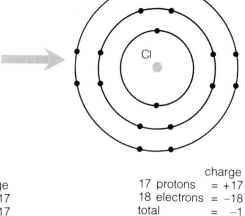

	charge
17 protons	= +17
18 electrons	= −18
total	= −1

Fig. 14.6 A chloride ion. This is a chlorine atom which has gained an electron to complete its outer orbit. It has a negative charge, because it has one more electron than it has protons.

Sodium and chlorine combine to make sodium chloride

What is likely to happen when sodium and chlorine atoms come together? A sodium atom can readily lose an electron, while a chlorine atom can readily gain one. It looks as though both of them can become stable if each sodium atom gives up an electron to a chlorine atom.

You can see what really happens if you watch sodium metal burning in chlorine. The sodium burns brightly and a white solid forms on the side of the jar. This solid is common salt. Its correct name is **sodium chloride**.

As sodium burns in chlorine, the sodium atoms lose electrons and the chlorine atoms gain them. The sodium atoms become positive ions, while the chlorine atoms become negative ions. The positive and negative ions are very strongly attracted to each other so they arrange themselves into a pattern like that shown in Figure 14.8. This is called a **lattice**. A grain of salt contains millions upon millions of sodium and chloride ions arranged like this. The force of attraction between the positive and negative ions is very strong. It is called an **ionic bond**.

Sodium chloride is a compound

You will remember that a substance which is made from just one kind of atom is called an **element**. Sodium and chlorine are both elements. Sodium chloride contains two elements, sodium and chlorine. A substance which is made of two or more kinds of atoms or ions that have joined together is called a **compound**. Sodium chloride is a compound. Because it is made of ions, it is called an **ionic compound**.

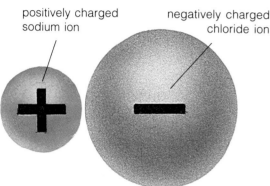

positively charged sodium ion

negatively charged chloride ion

Fig. 14.7 Sodium chloride. The positive and negative charges on the two ions hold them tightly together.

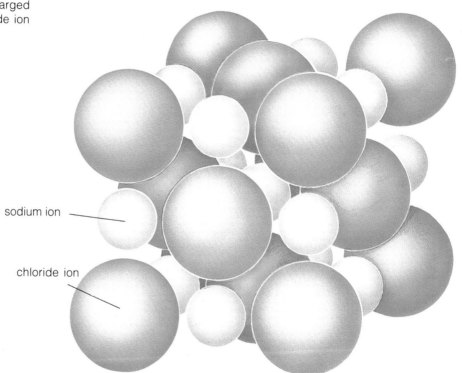

sodium ion

chloride ion

A pair of ions is the smallest part of sodium chloride which can exist

Sodium ions and chloride ions combine to form the compound sodium chloride. The smallest part of a sodium chloride crystal which would still *be* sodium chloride would be one sodium ion and one chloride ion. It is unlikely that a single sodium ion and a single chloride ion would ever find themselves alone together! But it is possible, and this would be the smallest part of the compound which still has the properties of sodium chloride.

Fig. 14.8 A sodium chloride lattice. The sodium and chloride ions arrange themselves in a regular pattern, or lattice. Can you pick out the cubic shape? Salt crystals are cube-shaped because of this arrangement of ions.

Questions

1 For each of the following, write a definition, and give an example:
   a atom **b** ion **c** cation **d** anion
2 Use the information about electron configuration in Table 8.1 to predict whether each of these elements will form positive or negative ions:
   a fluorine **b** potassium **c** lithium

15 USING FORMULAE

The name of a compound tells you which elements it contains. Formulae also tell you the relative numbers of the different kinds of atoms or ions in the compound.

The name of a compound often tells you which elements it contains

The name 'sodium chloride' tells you that this compound contains sodium and chlorine. When naming an ionic compound, the name of the positive ion always comes first. Its name is just the same as the name of the uncharged atom. We talk about 'sodium atoms' and 'sodium ions'.

The name of the negative ion comes last. You have probably realised that its name is not quite the same as the name of the uncharged atom. We talk about 'chlorine atoms' but 'chloride ions'. So the compound of sodium and chlorine is called sodium chloride. This rule almost always works. Table 15.1 gives you some more examples.

Some ions have a double charge

Figure 15.1 shows a magnesium atom. It has two electrons in its outer ring. If it loses these two electrons, it ends up with a complete outer ring. When it does this, it ends up with *two* more protons than electrons. It has a double positive charge. The formula for a magnesium ion shows this double charge. It is written **Mg²⁺**. Another common positive ion with a double charge is calcium, **Ca²⁺**.

Negative ions can also have double charges. Figure 15.3 shows an oxygen atom. It has six electrons in its outer shell and needs two more to complete it. When it gains them it becomes an oxide ion, **O²⁻**.

Question

If you had a tiny bit of NaCl containing only 1 million Na⁺ ions, how many Cl⁻ ions would it contain?

Elements	Positive ion	Negative ion	Name of compound
sodium, chlorine	sodium	chloride	sodium chloride
potassium, iodine	potassium	iodide	potassium iodide
magnesium, oxygen	magnesium	oxide	magnesium oxide
iron, sulphur	iron	sulphide	iron sulphide

Table 15.1   Some ionic compounds

The formula for sodium chloride is NaCl

The symbol for an atom of sodium is Na. When it loses an electron and becomes an ion, it has a positive charge. So the symbol for a sodium ion is **Na⁺**. The symbol for a chlorine atom is Cl. Chloride ions have a negative charge, so the symbol for a chloride ion is **Cl⁻**.

In sodium chloride there is one chloride ion for every sodium ion. The overall charge is nil, because the positive charges on the sodium ions balance the negative charges on the chloride ions. So the formula for sodium chloride is **NaCl**. The formula does not show the charges on the sodium or chloride ions, because when they are together the charges cancel each other out.

Every compound has a formula. The formula tells you a lot. It tells you what *kind* of atoms there are in the compound, and it tells you *how many* of each sort there are combined in that compound.

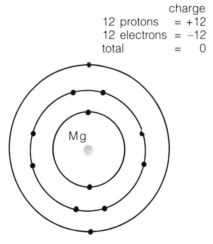

```
                    charge
12 protons       = +12
12 electrons     = −12
total            =   0
```

Fig. 15.1   A magnesium atom. It has two electrons in its outer orbit.

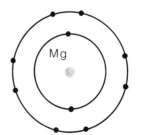

```
                    charge
12 protons       = +12
10 electrons     = −10
total            =  +2
```

Fig. 15.2   A magnesium ion, Mg²⁺. The two outer electrons are lost, leaving the ion with a complete outer orbit.

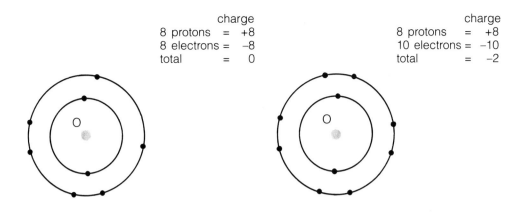

charge
8 protons = +8
8 electrons = −8
total = 0

charge
8 protons = +8
10 electrons = −10
total = −2

Fig. 15.3 An oxygen atom

Fig. 15.4 An oxide ion, O^{2-}. An oxygen atom can achieve a full outer electron orbit by gaining two electrons.

The numbers of positive and negative ions in an ionic compound are not always equal

What happens if magnesium and chlorine react together? Magnesium atoms tend to lose electrons, becoming positive ions. Each magnesium atom loses two electrons, becoming Mg^{2+}. Chlorine atoms tend to gain electrons, becoming negative ions. Each chlorine atom gains one electron, becoming Cl^-.

So *one* magnesium atom can supply *two* chlorine atoms with the electrons they need. So when magnesium burns in chlorine, the atoms form a structure in which there is *one* magnesium ion for every *two* chloride ions.

The formula for magnesium chloride shows this. It is written **MgCl₂**. The small $_2$ after the Cl tells you that there are two chloride ions in the compound magnesium chloride. There is no number after the Mg. This means that there is one Mg ion in the compound. Whenever no number is written after a symbol in a formula, this means that there is only one atom or ion of that element in a molecule. So MgCl₂ tells you that, in magnesium chloride, there are two chloride ions for every one magnesium ion.

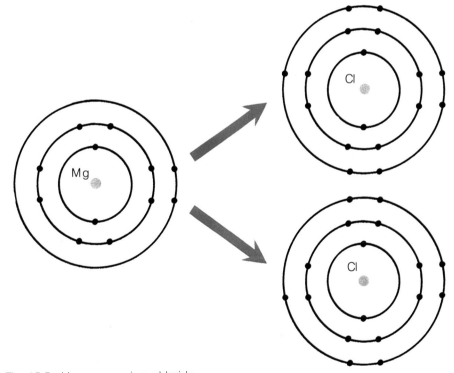

Fig. 15.5 How magnesium chloride forms. If one magnesium atom gives one electron to each of two chlorine atoms, then all three end up with complete outer electron orbits.

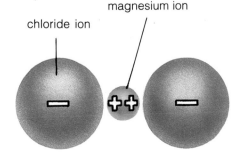

chloride ion

magnesium ion

Fig. 15.6 Magnesium chloride, MgCl₂. There is one magnesium ion for every two chloride ions.

Question

If you had a tiny bit of magnesium chloride, containing only 1 million Mg^{2+} ions, how many Cl^- ions would it contain?

16 TRANSITION METALS

The transition metals are the elements which fill the centre block in the Periodic Table. Most of them can form more than one kind of ion.

Most transition metals can form more than one kind of ion

The transition metals are the ones in the centre block of the Periodic Table (page 31). They include many common metals, such as iron and copper.

The atoms of transition metals, like all metals, can become more stable by losing electrons and forming positive ions. But most transition metals can become stable by losing **different numbers** of electrons. This means that they can form *more than one kind of ion*.

Copper is a typical example. The copper atom can become stable by losing either one or two electrons. If it loses one electron, it forms an ion with a single positive charge, **Cu⁺**. This ion is called **copper (I)**. If it loses two electrons, it forms an ion with a double positive charge, **Cu²⁺**. This ion is called **copper (II)**.

Figure 16.2 shows how these ions may combine with chloride ions. Each chloride ion has a single negative charge, Cl^-. Copper (I) ions have a single positive charge, so these two types of ion will combine together in equal numbers to form the ionic compound **CuCl**. This compound is called **copper (I) chloride**. Copper (II) ions, however, have a double positive charge, Cu^{2+}. So *two* chloride ions are needed to combine with each copper (II) ion. The ionic compound formed has the formula **CuCl₂**. It is called **copper (II) chloride**.

Fig. 16.1 The salts of transition elements are often coloured. The green powder is nickel (II) chloride; the bluer one is copper (II) chloride.

A copper (I) ion has a single positive charge

A chloride ion has a single negative charge

A copper (II) ion has a double positive charge

CuCl, copper (I) chloride. One copper (I) ion combined with one chloride ion balances their charges.

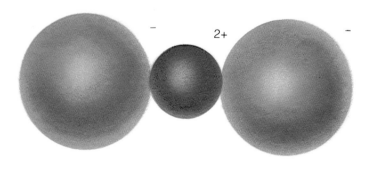

CuCl₂, copper (II) chloride. One copper (II) ion combined with two chloride ions balances their charges.

Fig. 16.2 Copper (I) chloride and copper (II) chloride

44

Transition metals form coloured compounds

Most of the compounds of transition elements are coloured. The colours are often different for the different kinds of ions formed by one metal. Copper (II) compounds, for example, are often blue, while copper (I) compounds are often white or green.

Iron is another important transition metal. It forms two sorts of ions, **Fe^{2+}** and **Fe^{3+}**. These are called **iron (II)** and **iron (III)**. Iron (II) compounds are usually pale green, while iron (III) compounds are usually yellow or brown.

Table 16.1 Some elements which form ions.

Element	Symbol	Ion	Name of ion
Potassium	K	K$^+$	Potassium
Sodium	Na	Na$^+$	Sodium
Lithium	Li	Li$^+$	Lithium
Magnesium	Mg	Mg^{2+}	Magnesium
Calcium	Ca	Ca^{2+}	Calcium
Iron	Fe	Fe^{2+}	Iron (II)
		Fe^{3+}	Iron (III)
Copper	Cu	Cu$^+$	Copper (I)
		Cu^{2+}	Copper (II)
Zinc	Zn	Zn^{2+}	Zinc
Aluminium	Al	Al^{3+}	Aluminium
Lead	Pb	Pb^{2+}	Lead
Fluorine	F	F$^-$	Fluoride
Chlorine	Cl	Cl$^-$	Chloride
Bromine	Br	Br$^-$	Bromide
Iodine	I	I$^-$	Iodide
Oxygen	O	O^{2-}	Oxide
Sulphur	S	S^{2-}	Sulphide

Questions

1 a What is the symbol for a sodium atom?

b What is the symbol for a sodium ion?

c Does a sodium atom lose or gain electrons when it becomes an ion?

d How many electrons does it lose or gain?

2 a What is the symbol for a magnesium ion?

b How many electrons does a magnesium atom lose when it becomes an ion?

c What is the symbol for an oxygen ion?

d How many electrons does an oxygen atom gain when it becomes an ion?

e How many magnesium atoms will be needed to supply the electrons for one oxygen atom when they become ions?

f The compound formed when magnesium ions and oxygen ions combine together is called magnesium oxide. Write down the formula for magnesium oxide.

3 a How many electrons does a chlorine atom gain when it becomes a chloride ion?

b How many electrons does a calcium atom lose when it becomes an ion?

c How many chlorine atoms can be supplied with electrons by one calcium atom?

d Write down the formula for calcium chloride.

4 a Potassium atoms lose one electron when they become ions. Write down the symbol for a potassium ion.

b Sulphur atoms gain two electrons when they become ions. Write down the symbol for a sulphur ion.

c How many potassium atoms will be needed to supply the electrons for one sulphur ion?

d Write down the formula for potassium sulphide.

5 Explain why:

a Sodium atoms form positive ions, but chlorine atoms form negative ions.

b A potassium ion has a charge of $^+$, but a magnesium ion has a charge of $2+$.

c The ions in sodium chloride are held closely together.

d A sodium ion is written as Na$^+$, and a chlorine ion is written as Cl$^-$, but the ionic compound sodium chloride is written as NaCl.

6 Name the following compounds (refer to the Periodic Table if you need to):
a KCl **b** MgO **c** Li$_2$S **d** CaF$_2$ **e** Na$_2$S **f** PbCl$_2$ **g** SrO **h** RbF

---

7 The formula for aluminium fluoride is AlF$_3$.

a How many fluoride ions are there for every aluminium ion?

b How many electrons does one fluorine atom gain when it becomes an ion?

c From your answers to parts **a** and **b**, work out how many electrons one aluminium atom loses when it becomes an ion.

d Write down the symbol for an aluminium ion.

e What is the symbol for an oxide ion?

f If powdered aluminium is heated strongly in oxygen, the aluminium and oxygen atoms can react together to form aluminium oxide. The formula for aluminium oxide is Al$_2$O$_3$. Explain clearly why the aluminium and oxide ions combine together in these proportions. You may like to include a diagram in your answer.

8 Work out and write down formulae for these ionic compounds:
a calcium chloride
b calcium oxide
c potassium sulphide
d lithium oxide
e lithium bromide
f magnesium fluoride

9 Name these compounds of transition metals:
a Cu$_2$O **b** CuO **c** FeO **d** Fe$_2$O$_3$ **e** CuCl$_2$ **f** FeCl$_3$

10 Write formulae for:
a iron (II) chloride
b copper (I) sulphide
c copper (II) chloride
d copper (I) iodide
e iron (II) sulphide

11 Use the Periodic Table on page 31 to explain the following:
a All elements in Group 1 form an ion with one positive charge.
b All elements in Group 7 form an ion with one negative charge.

— EXTENSION —

17 COVALENT BONDS

Atoms can often share electrons to achieve a stable electron structure. This sharing holds the atoms closely together, and is called a covalent bond.

Atoms can fill their outer shells by sharing electrons with other atoms

You have seen that atoms tend to lose or gain electrons in order to end up with full outer electron orbits. If they completely lose or gain an electron, they become ions. Positive and negative ions are strongly attracted together, forming ionic bonds between themselves. Compounds formed in this way are called ionic compounds.

But how do compounds form when neither type of atom in the compound can lose electrons and form a positive ion? For example, bromine and chlorine form a compound with the formula BrCl. Yet both bromine atoms and chlorine atoms form ions by gaining electrons. They both form ions

with a negative charge. If they are both gaining electrons, where can they be getting them from? Anyway, as both bromide and chloride ions are negatively charged, they would push each other apart – because if two particles are *both* negatively charged (or both positively charged) they repel each other. Clearly, BrCl cannot be an ionic compound. There must be another way that atoms can be held together.

In ionic compounds atoms fill their outer orbits by losing or gaining electrons. But there is another way in which they can do this. Atoms can fill their outer orbits by **sharing** electrons with other atoms.

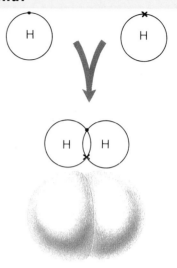

Fig. 17.1 A hydrogen molecule, H_2. Two hydrogen atoms can achieve complete outer electron shells by sharing electrons.

Hydrogen atoms pair up and share electrons

Figure 17.1 shows how hydrogen atoms share electrons with each other. Hydrogen atoms only have one electron and one proton. To fill their electron orbit, they need one more electron. If two hydrogen atoms share their single electrons with each other then each has, in effect, a full electron orbit. The two hydrogen atoms are held closely together when they share electrons like this. The bond between them is called a **covalent bond**.

The pair of bonded hydrogen atoms is a hydrogen **molecule**. It is the smallest part of the element hydrogen which can exist on its own. The formula for a hydrogen molecule is H_2, showing that the hydrogen atoms are linked in pairs.

Fig. 17.2 A chlorine molecule, Cl_2. One pair of electrons is shared.

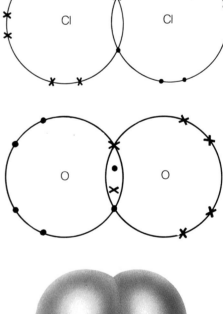

Fig. 17.3 A nitrogen molecule, N_2. Three pairs of electrons are shared between the two nitrogen atoms, so the bond between them is a triple bond.

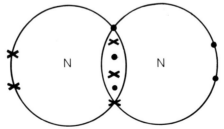

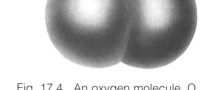

Fig. 17.4 An oxygen molecule, O_2. Two pairs of electrons are shared, so the bond between the oxygen atoms is a double bond.

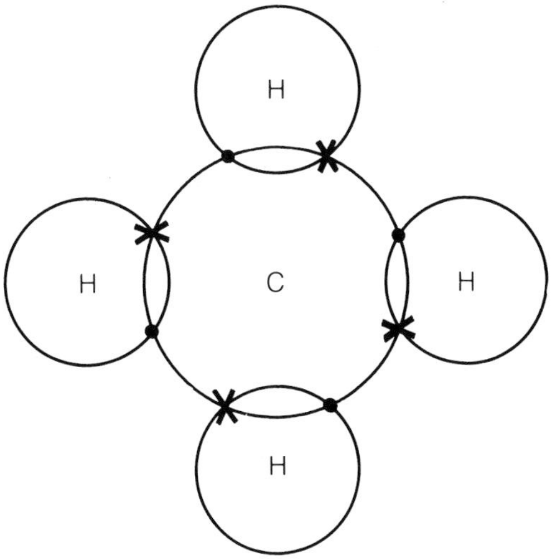

Fig. 17.5   A methane molecule, CH_4

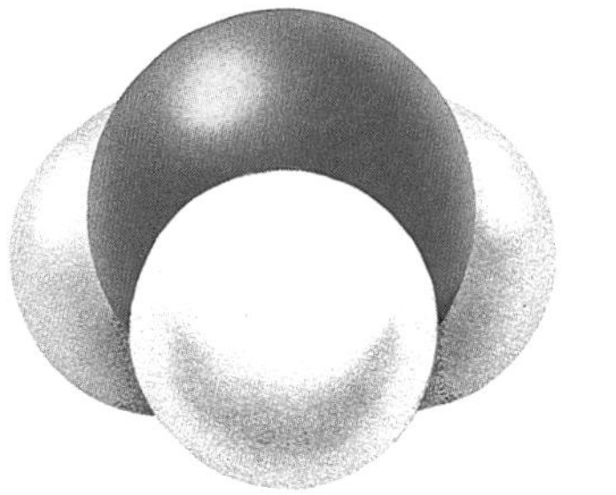

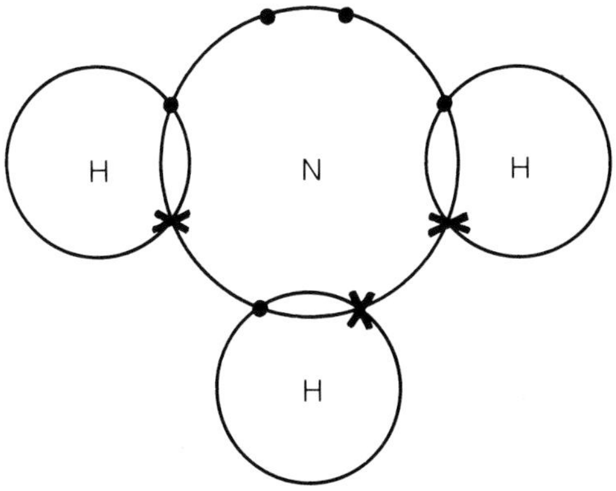

Fig. 17.6   An ammonia molecule, NH_3

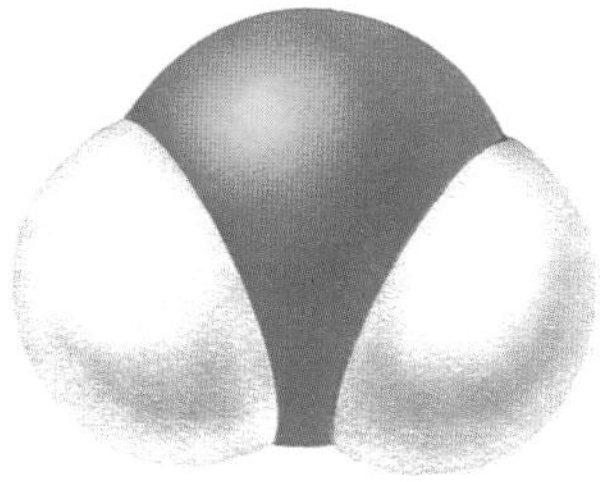

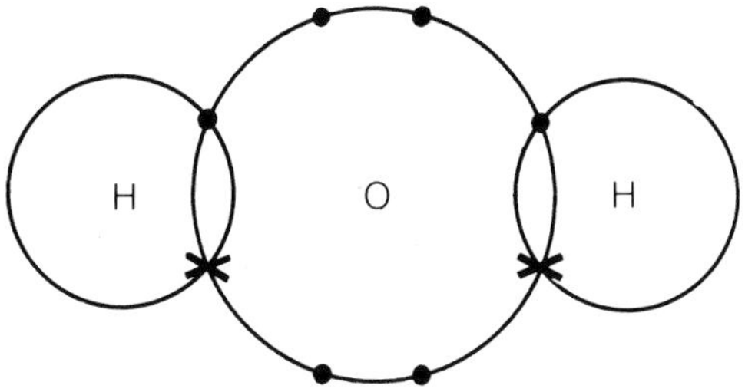

Fig. 17.7   A water molecule, H_2O

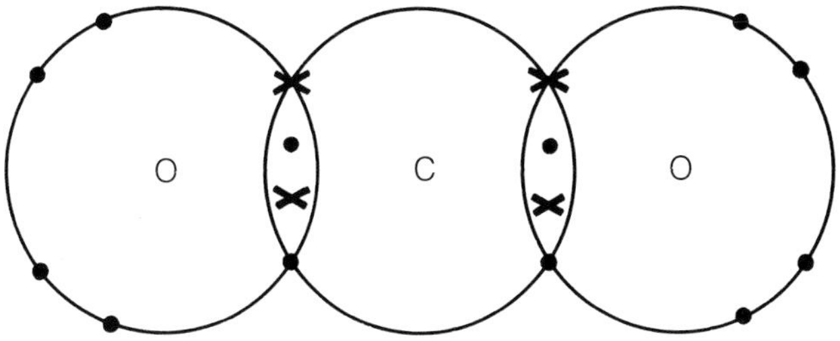

Fig. 17.8   A carbon dioxide molecule, CO_2

Many non-metal atoms form covalent bonds

Figures 17.1 to 17.4 show some other examples of atoms which form covalent bonds. Sometimes they do this with other atoms like themselves. Oxygen gas, for example, is made up of pairs of oxygen atoms linked with covalent bonds. Its formula is O_2. Chlorine gas is also made of pairs of chlorine atoms, forming chlorine molecules, Cl_2.

Atoms can also form covalent bonds with different sorts of atoms. When they do this, a **covalent compound** is formed. Figures 17.5 to 17.8 show some examples of covalent compounds. **Water, ammonia, carbon dioxide** and **methane** are four very important covalent compounds.

One pair of shared electrons is one covalent bond

You can see from the diagrams of the covalent compounds that they share their electrons in pairs. Each *pair* of electrons shared is *one* covalent bond. Hydrogen molecules, for example, have one bond, because they share one pair of electrons.

But oxygen atoms share *two* pairs of electrons when they form oxygen molecules. So they have *two* bonds between them. This arrangement is often called a **double bond**.

Nitrogen atoms share even more electrons. Each nitrogen atom has five electrons in its outer shell, and needs three more to make a stable outer orbit of eight electrons. So two nitrogen atoms share *three* pairs of electrons to make nitrogen molecules, N_2. This is sometimes called a **triple bond**.

18 MORE ABOUT COVALENT BONDS

There is an enormous variety of covalent compounds.

What are you made of?

Did you know that living things are made almost entirely of six elements? Apart from small traces of other elements, you consist almost entirely of compounds made from carbon, hydrogen, nitrogen, oxygen, phosphorus, and sulphur. How can something as complex as a living animal be made from such a simple set of substances? The answer is that these six elements can produce an almost endless variety of covalent compounds. The diagrams on this page show three more covalently bonded substances made from the same elements as those shown in Topic 17.

Two elements can join in different ways to make different compounds

Figure 18.1 is a diagram of the bonding in a molecule of **hydrogen peroxide**. The molecule is made of hydrogen atoms and oxygen atoms, just like the water molecule on the previous page. But while a water molecule consists of two hydrogen atoms and one oxygen atom and so has the formula H_2O, a hydrogen peroxide molecule consists of two hydrogen atoms and two oxygen atoms and so has the formula H_2O_2.

Although they are made of the same elements, hydrogen peroxide and water are different compounds. They behave differently, or we could say they have **different properties**. Note that in the diagram of hydrogen peroxide every atom shares electrons and achieves a full outer orbit. Note also that each atom started with its correct number of outer electrons – one in the case of hydrogen and six in the case of oxygen.

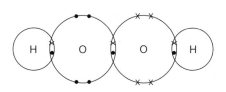

Figure 18.1 Hydrogen peroxide molecule H_2O_2

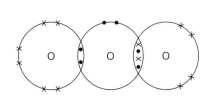

Figure 18.2 Ozone molecule O_3

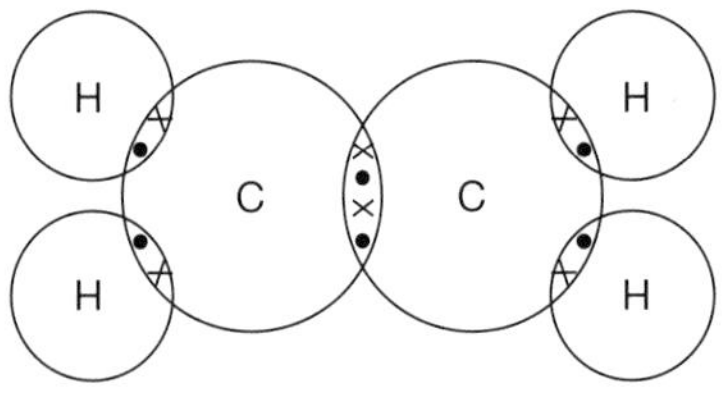

Figure 18.3 Ethene molecule C_2H_4

Atoms of a single element can also bond in different ways

Having the atoms bonded in a different way can also give us different forms of elements. Figure 18.2 shows three oxygen atoms bonded together. This substance is called **ozone** and has the formula O_3. Note that it is an element – each molecule is made of one type of atom only. This form of oxygen is not very common in the air we breathe at ground level but there is a layer of it high in the atmosphere. The bonding diagram looks a bit strange when examined closely. The bond on the left is made by two electrons both from the same oxygen atom, but the rules are still not broken – each oxygen atom begins with six outer electrons and achieves a share in eight.

Question

1   Figure 18.3 shows a molecule of ethene. Look at the diagram and answer the following questions.

a How many atoms are there in an ethene molecule?

b Which elements are present in ethene?

c Which compound on the previous page was made from these elements?

d Will these two compounds made from the same elements have exactly the same properties?

e What is the formula of ethene?

f How many outer electrons did each atom in the ethene molecule have before it bonded?

g How many electrons does bonding give each atom a share in?

h Where in this molecule are there single covalent bonds?

i Where in this molecule is there a double covalent bond?

Some advice on drawing covalent bonding diagrams

Sometimes you may be asked to draw diagrams to show the structure of a molecule of a covalent compound. If you follow these rules, you are more likely to get it right!

1 Check that every atom has a filled outer orbit. In this diagram of 'ammonia' the hydrogen atoms do not – so it must be wrong.

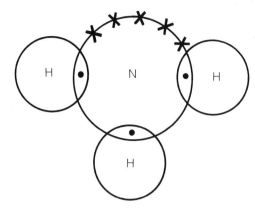

2 Check that each atom has started off with the correct number of electrons. Dot and cross diagrams help you to check this. In this diagram of 'ammonia' the nitrogen has begun with six outer electrons, but it should have five.

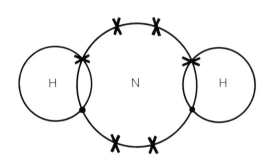

3 Even if you cannot see the solution immediately, follow rules 1 and 2 and try – it is surprising how much easier it can become!

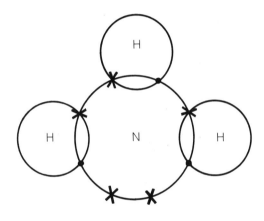

Questions

1 From the molecules shown in Topic 17, name:
   a two elements
   b two compounds
   c one element which exists as a gas at room temperature
   d one compound which exists as a liquid at room temperature
   e one covalent element whose atoms are held together by double bonds
   f one compound where the molecules contain two different sorts of atoms
   g the compound with the formula NH_3
   h one compound which has four bonds

2 Below are the atomic numbers of six elements:

Element	Atomic number
H	1
Cl	17
C	6
N	7
O	8
S	16

   a Draw electronic structures for the atoms of these six elements.

   b Draw diagrams to show how electrons are shared to make molecules of the following covalent compounds (show outer electrons only):

   i HCl
   ii CCl_4
   iii NCl_3
   iv OCl_2

   c Now try the same for these molecules – they are quite a bit harder!

   i CO (hint – unequal sharing)
   ii S_8 (hint – ring-shaped)
   iii C_2H_2 (hint – all four atoms are in a straight line).

EXTENSION

19 IONIC AND COVALENT PROPERTIES

Ionic and covalent compounds tend to behave differently from each other.

INVESTIGATION 19.1

Comparing ionic and covalent compounds

You are going to do a set of experiments on six compounds. A, B and C are ionic compounds. D, E and F are covalent compounds. When you have completed the investigation, you should be able to make generalisations about the properties of ionic and covalent compounds.

1 Read through the investigation. Then design a results chart which you can use as you perform the experiments. Ask for help if you are not sure how to do this.

2 WEAR SAFETY SPECTACLES. Place a little of substance A on a piece of ceramic paper. Heat in a blue Bunsen burner flame using tongs. Does it melt or burn easily? Observe what happens, and fill in the results in

your chart. Now do the same for the other five compounds.

3 Put some water in a test tube. Add a little of substance A and shake. Does A dissolve in water? Fill in your result in the chart.

4 If A does dissolve, use the circuit provided to see if the solution conducts electricity. Fill in your result in the chart.

5 Repeat steps **3** and **4** for the other five substances.

6 Repeat steps **3** and **4**, but use 1,1,1-trichloroethane instead of water. This is a liquid which does not contain water. (Your teacher may demonstrate this part of the investigation for you.)

Questions

It is not always possible to draw conclusions about behaviour which are true for all ionic and covalent compounds. Some of your results may not fit neatly into a pattern. But you should be able to see enough of a pattern to answer the following questions.

1 Which are easier to melt – ionic or covalent substances?

2 Are ionic substances more likely to dissolve in water, or · in a non-aqueous solvent (one which does not contain water)?

3 Are covalent substances more likely to dissolve in water, or in a non-aqueous solvent?

4 Do solutions of ionic substances conduct electricity?

5 Do solutions of covalent substances conduct electricity?

Question

Fig. 19.1 Three of these aqueous solutions are of covalent compounds, and three of ionic compounds. Which are which? Why? What is slightly unusual about the three covalent compounds?

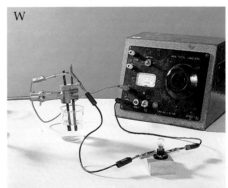

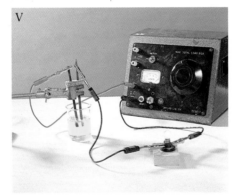

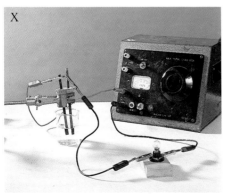

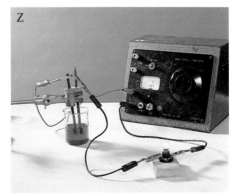

An explanation of the properties of ionic and covalent compounds

MELTING POINTS

In an ionic compound all the ions in the lattice are held firmly together. The bonds are strong. If the compound is heated the heat energy makes the particles vibrate. But it takes a lot of heat energy to break the particles completely apart. This means that ionic compounds need a lot of heat to change them from solids to liquids. In other words they have **high melting points**. All ionic compounds are solids at room temperature.

In a covalent substance the atoms in each molecule are held firmly together by covalent bonds. But the individual molecules are held together by only weak forces between them. This means that not very much heat is needed to separate the molecules. Quite low temperatures are enough to change the substance from a solid to a liquid or gas. In other words covalent substances have **low melting** and **boiling points**. Many covalent substances are gases at room temperature.

SOLUBILITY IN WATER

In an ionic compound, the charges on the ions attract water molecules to them. This means that they will **dissolve in water**. You will find more about how substances dissolve in water on page 169.

In a covalent substance the molecules often do not attract water molecules as much. This means that they often **do not dissolve in water**, but they can **dissolve in non-aqueous solvents**.

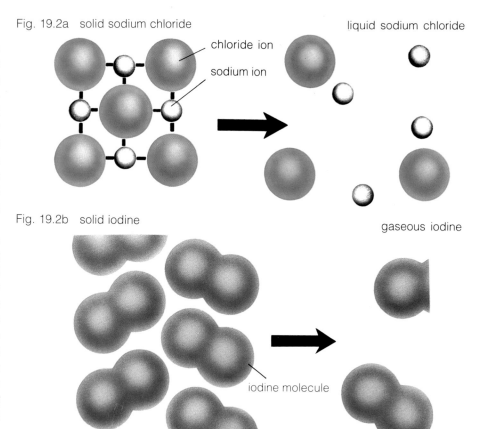

Fig. 19.2a   solid sodium chloride   liquid sodium chloride

chloride ion

sodium ion

Fig. 19.2b   solid iodine   gaseous iodine

iodine molecule

Fig. 19.3a   Sodium chloride readily dissolves in water.

Fig. 19.3b   Iodine hardly dissolves at all in water.

CONDUCTION OF ELECTRICITY

An electric current is a flow of charged particles. A solution of an ionic compound has freely-moving charged particles. This means it **can easily conduct electricity**.

Covalent substances **cannot conduct electricity**.

Fig. 19.4a   A solution of sodium chloride in water conducts electricity.

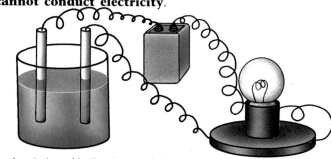

Fig. 19.4b   A solution of iodine in petrol does not conduct electricity.

20 LARGE MOLECULES

Some covalent compounds have molecules that are extremely large.

The atoms of some substances form giant molecules

Some substances don't seem to have any of the three types of structure you have met so far. They are made of non-metals, so they are not metallic or ionic. They have extremely high melting and boiling points, so they can't be covalent compounds with small molecules. The atoms in these substances are held together by covalent bonds, but they are not held in small groups. Instead, the atoms are held together in vast arrays to form giant molecules.

Diamond is one example of a substance with a giant molecular structure. Every atom is the same. They are all carbon atoms, so diamond is one form of the element carbon. Figure 20.1 shows a small part of a diamond crystal. Each atom is held to the four atoms around it by strong covalent bonds. There are no weak points in this structure. That is

why diamonds are so hard. Since a lot of energy is required to break each diamond atom away from all those around it, diamond has a very high melting point.

Graphite also has a giant molecular structure (Figure 20.2). Once again every atom is a carbon atom, so graphite is another form of the element carbon. Graphite is used for pencil lead, so pencil 'lead' and diamonds are made from different forms of the same substance. In graphite each carbon atom is bonded to three other carbon atoms around it by strong covalent bonds to form layers of carbon atoms. Like diamond this means graphite has a very high melting point. But the layers are not strongly bonded to each other, so they can slide over each other. That is what is happening when you use a

pencil – layers of graphite atoms slide off the end of your pencil onto the paper, leaving a black line on the paper. The structure of graphite allows electrons to move along the layers, so graphite conducts electricity, which is very unusual for a non-metal.

Figure 20.3 shows a third example of a giant covalent structure. This is a compound, silicon dioxide, familiar to us as sand and quartz. As with the two previous examples the diagram can show only a few atoms to give some impression of the structure, but even a single grain of sand contains billions of billions of atoms bonded in this way! Since covalent bonds are strong, silicon dioxide is hard and has a high melting point.

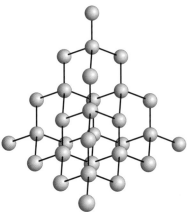

Fig. 20.1 The structure of diamond

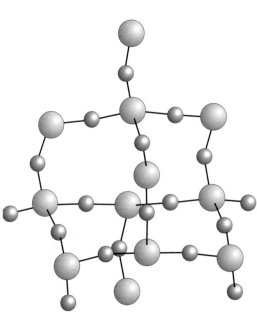

Fig. 20.3 The structure of silicon dioxide

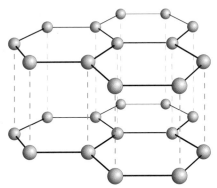

Fig. 20.2 The structure of graphite

All substances containing atoms that are held together by covalent bonds are known as covalent substances, but we can now sub-divide this category further. In some of these substances the bonds hold the atoms in vast arrays or lattices. These substances are said to have a **giant covalent** or **giant molecular** structure. Diamond and the other substances shown on page 52 are examples of giant covalent or giant molecular structures. In other covalent substances the bonds hold the atoms in small (or comparatively small) groups. These substances are said to have **simple molecular** structures. Water, ammonia, and the other substances shown in Topics 17 and 18 are examples of simple molecular structures. Some substances with simple molecular structures can have quite big molecules, for example the DNA molecule in Figure 20.5, but even a molecule consisting of thousands of atoms is tiny compared to a true giant molecule like silicon dioxide or graphite.

Question

1 Find out:
  a more about why graphite conducts electricity
  b about another form of carbon called buckminsterfullerene
  c why diamond is denser than graphite
  d a more accurate estimate of the number of atoms in a grain of sand.

Many biological molecules are very big

You will have found when you did Investigation 19.1, that not all substances fit neatly into one of these two groups. In particular, there are many covalent substances which have some properties more like those of ionic compounds.

Very big covalent molecules tend to be solid at room temperature, and have high melting and boiling points. Sugar is a good example. The large mass of the sugar molecules means that a lot of heat energy is needed to get them moving freely.

Some covalent compounds are soluble in water. This is because they do actually have small charges on them. These charges are not as large as those on ions. They are called **dipoles**, and you will find more about them on page 163. Sugar molecules, for example, have small positive and negative charges on them, so they will dissolve in water.

All the big biological molecules – proteins, carbohydrates, and fats – are covalent. But because they are big molecules, they are often solid at room temperature. Also many of them have dipoles, so they will dissolve in water.

— EXTENSION —

Fig. 20.5 Many biological molecules contain thousands of atoms. This is a computer graphic representation of a tiny part of a DNA molecule. Most of the atoms in the molecule are bonded covalently. DNA is the genetic material found in all living cells.

Fig. 20.4 A dog's fur is made of keratin, which is a protein. Proteins are covalent compounds consisting of big molecules. Keratin is insoluble in water, but many proteins, such as haemoglobin, are soluble in water.

21 COMPOUND IONS

Ions can be formed from groups of atoms. These groups are called compound ions.

A compound ion is an ion formed from a group of atoms

You may remember that the definition of an ion is **an atom or group of atoms with an overall positive or negative charge**. Ions have charges because the number of protons and electrons is unequal. A positive ion has more protons than electrons. A negative ion has more electrons than protons. So far, all the ions you have met in this book have been formed from single atoms. But there are many important ions which are formed from **groups** of atoms. Some of them are shown in Figures 21.1 to 21.5. The atoms in the group are joined together by sharing electrons – in other words, they are held together by covalent bonds. But, even by sharing, every atom in the group doesn't quite get a complete outer electron shell. The group of atoms still needs to gain or lose an electron or two in order to end up with all the outer shells completely filled.

Look at the diagram of the **carbonate ion**. The three oxygen atoms fill their outer shells by sharing electrons with the carbon atom. But for *all* the atoms in the group to end up with complete outer shells, the group still needs to gain two more electrons. When it does this, the total number of electrons in the group is two more than the total number of protons. So the group ends up with a double negative charge. It becomes a **negative ion**. The carbonate ion is one example of a compound ion. A **compound ion** is a group of atoms with an overall positive or negative charge.

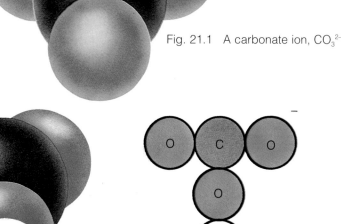

Fig. 21.1 A carbonate ion, CO_3^{2-}

Fig. 21.2 A hydrogencarbonate ion, HCO_3^-

Fig. 21.3 A sulphate ion, SO_4^{2-}

Fig. 21.4 A nitrate ion, NO_3^-

Fig. 21.5 An ammonium ion, NH_4^+

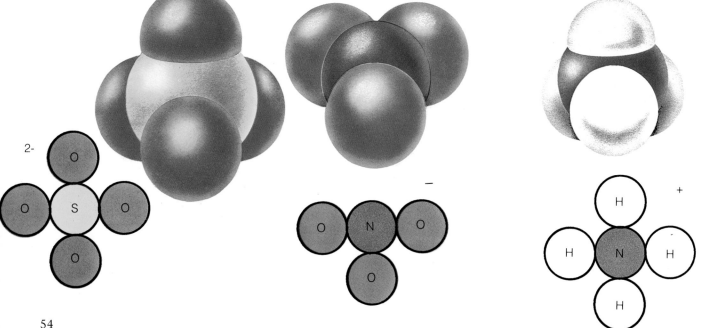

Compound ions form compounds in the same way as simple ions

Compound ions combine with other ions to form ionic compounds. You can work out their formulae in the same way as before. For example, the carbonate ion, CO_3^{2-}, has a double negative charge. A sodium ion, Na^+, has a single positive charge. So if a lattice is formed with two sodium ions for every one carbonate ion, the charges will cancel out. The formula for this ionic compound is **Na_2CO_3**. Its name is **sodium carbonate**.

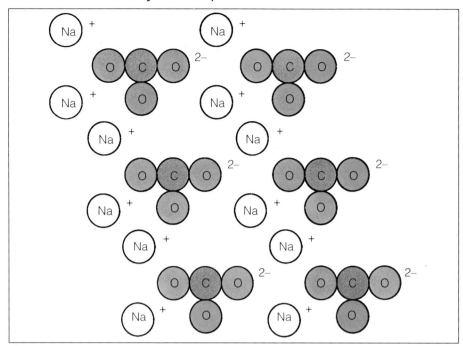

Fig. 21.6 Sodium carbonate, Na_2CO_3. The Na^+ and CO_3^{2-} ions form a lattice, in which there are two sodium ions for every carbonate ion. The ions are held strongly together by the attraction between the positive and negative charges.

Brackets are needed to show more than one compound ion in a formula

Most of the common compound ions have negative charges, but the **ammonium** ion, NH_4^+, has a single positive charge. In ammonium chloride, since the chloride ion has a single negative charge, one ammonium ion is needed to balance the charge on one chloride ion. The formula is written as **NH_4Cl**. But in ammonium sulphate, the sulphate ion has a double negative charge. This means that *two* ammonium ions are needed to balance the charge on one sulphate ion.

How can you write a formula showing that there are two NH_4^+ ions in a molecule? You could write NH_4NH_4, but that is rather untidy. You certainly could not write NH_{42} because that would mean that there were 42 hydrogen atoms in the molecule! The only way to show two ammonium ions is to write **$(NH_4^+)_2$**. The bracket shows that the $_2$ applies to everything inside the bracket. So the formula for ammonium sulphate is **$(NH_4^+)_2SO_4$**.

In **magnesium nitrate**, the magnesium ion has a double charge, Mg^{2+}. The nitrate ion has a single charge, NO_3^-. So two nitrate ions are needed to balance the charge on each magnesium ion. The formula for magnesium nitrate is written as **$Mg(NO_3)_2$**. The bracket is needed to show that the $_2$ applies to the whole of the nitrate ion, NO_3^-.

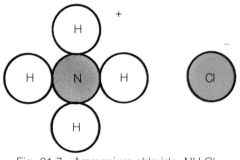

Fig. 21.7 Ammonium chloride, NH_4Cl

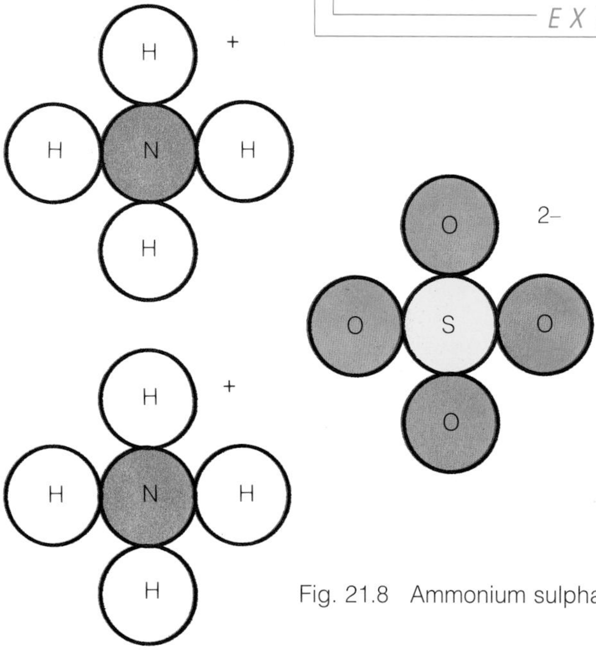

Fig. 21.8 Ammonium sulphate, $(NH_4)_2SO_4$

Questions

1 Name these compounds:
   a NH_4Cl **b** $Ca(NO_3)_2$ **c** $MgSO_4$
   d NaOH

2 Write formulae for these compounds:
   a sodium hydrogencarbonate
   b ammonium sulphate
   c calcium hydroxide
   d silver nitrate
   e calcium hydrogencarbonate
   f ammonium nitrate
   g copper (II) sulphate

E X T E N S I O N

Many isotopes are radioactive. The radiation they give off can cause ionisation.
The three types of ionising radiation are alpha, beta and gamma radiation.

Some isotopes produce ionising radiation

'Radiation' is something which is sent out, or 'radiated' from an object. Many isotopes produce radiation. The type of radiation they produce is **ionising radiation**.

Towards the end of the nineteenth century, a French scientist, Becquerel, discovered that uranium gave out radiation. The radiation blackened a photographic plate. Marie Curie, a Polish scientist married to a Frenchman, read about Becquerel's experiments. She did experiments of her own, and in 1898 discovered radioactive isotopes in pitchblende ore. By 1902, with the help of her husband, she had isolated radioactive isotopes of the elements radium and polonium. In 1903, Marie Curie, her husband Pierre, and Becquerel were awarded a Nobel Prize. Marie was awarded a second Nobel prize, for Chemistry, in 1911. But she paid highly for her fame. In 1934 she died as a result of her prolonged exposure to ionising radiation.

Unstable isotopes tend to be radioactive

Isotopes may be stable or unstable. Stable isotopes stay as they are. Carbon 12 is an example of a stable isotope. Unstable isotopes tend to change. When an atom of an unstable isotope changes it gives out ionising radiation. Carbon 14 is an example of an unstable isotope. An element or isotope which gives off radiation is said to be **radioactive**. Carbon 14 is a **radioactive isotope** of carbon, or **radioisotope**.

A radioisotope can also be called a **radionuclide**. A stable isotope can be made unstable by hitting it with neutrons.

stable	unstable (radioactive)
carbon 12	carbon 14
gold 197	gold 198
lead 208	lead 198

Table 22.1   Stable and unstable isotopes

Radiation can be detected because it causes ionisation

Humans have no sense which can detect ionising radiation. We must use instruments to detect it. Ionising radiation ionises atoms. As the radiation passes through material it removes electrons from atoms, producing ions. Ions, unlike atoms, have a charge. If we can detect this charge we can detect ionising radiation.

Figure 22.1 shows a **Geiger–Müller** tube. When ionising radiation enters the tube it ionises the argon gas. The electrons from the argon atoms go to the anode. The positive argon ions go to the cathode. This causes a tiny current to flow in the circuit. The current is amplified and detected on a counter.

Another way of detecting radiation is with **photographic film**. If radiation falls on a film, and the film is developed, it appears dark. Care must be taken not to allow any light to fall on to the film before it is developed.

Fig. 22.1   A Geiger–Müller tube

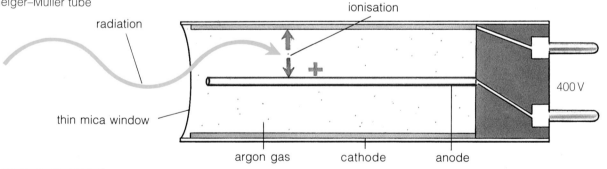

INVESTIGATION 22.1

Investigating the radiation levels from a gamma source

The photograph shows some apparatus which could be used to find out how the radiation from a gamma source varies with distance from the source.

Design an experiment to find out the answer to this problem, using some or all of the apparatus in the photograph. You will not be able to do the experiment yourself, but your teacher may demonstrate it to you.

Think about:
what you would measure
how you would measure it
how you would present your results.
Write down your ideas fully, so that someone could follow your instructions without having to ask you for any more help.

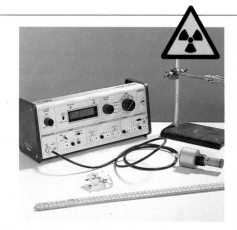

Alpha, beta and gamma radiation

There are three different types of ionising radiation which can be emitted by radioactive isotopes.

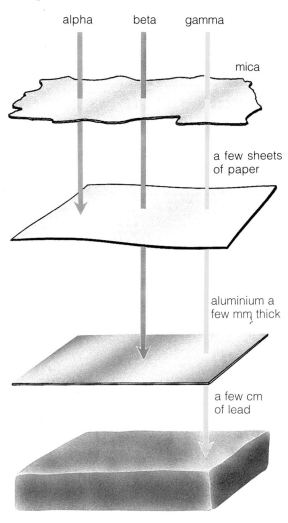

Fig. 22.2 The penetrating properties of ionising radiation.

Alpha radiation is made up of fast moving helium nuclei. The helium nuclei are called alpha particles. The particles have a positive charge. Alpha particles change direction if they pass through an electric or magnetic field. They are said to be **deflected** by the field. Alpha particles are quite easily stopped by thin materials. Even air will stop them. If you hold a Geiger–Müller tube more than a few centimetres from an alpha source, you will not be able to detect the radiation. It is stopped by the air.

Fig. 22.3 How alpha, beta and gamma radiation behave in a magnetic field. The magnetic field is coming up out of the page.

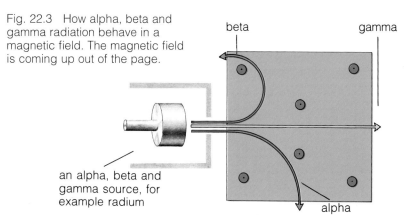

Fig. 22.4 How alpha, beta and gamma radiation behave in an electric field.

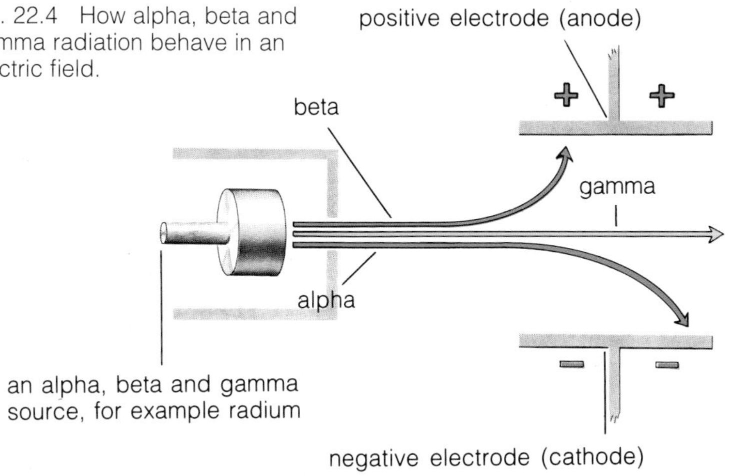

Beta radiation is made up of electrons moving at high speed. The electrons are called **beta particles**. They have a negative charge. Like alpha particles, they are deflected by electric and magnetic fields. But because they have a negative charge instead of a positive charge, they are deflected in the opposite direction. Beta particles are not stopped as easily as alpha particles. Beta particles can travel several metres in air.

Gamma radiation is not a stream of particles. It is a form of electromagnetic radiation. Gamma radiation does not carry a charge, so it is not deflected by electric or magnetic fields. Gamma radiation has very high energy. It can travel a very long way in air, and can even pass through several centimetres of lead or an even thicker piece of concrete.

INVESTIGATION 22.2

The effect of radiation on living organisms

You will be given some normal barley seeds, and some barley seeds which have been exposed to radiation. Barley seeds take about 7 to 10 days to germinate. Over the next few days, they will grow to a height of several centimetres.

Design an experiment, using these two types of barley seeds, to find out how radiation affects barley seeds. Write up your method in detail. (Don't forget to take timing into account, as the barley seeds may take longer to germinate and grow than you think.) Get your design checked by your teacher before you carry out your experiment.

Record your results in the way you think is best. Discuss what your results suggest to you about the way in which radiation affects barley seeds. How accurate do you think your results are? How would you improve your experiment if you could do it again?

23 RADIOACTIVE DECAY

No-one can tell exactly when an individual atom will decay. The decay of one atom to form another can be shown by nuclear equations.

Atoms produce radiation when their nuclei change

A radioactive atom is an unstable atom. Atoms must have the correct balance of neutrons and protons to be stable. Small atoms have similar numbers of neutrons and protons. The heaviest atoms have about 50% more neutrons than protons. No atom with more than 83 protons is stable. In an unstable atom, changes happen in the nucleus to make it more stable. These changes cause radiation and energy to be released.

For example, carbon 14 emits beta radiation. The nucleus of a carbon 14 atom contains six protons and eight neutrons. This is unstable. Sooner or later, one of the neutrons changes to a proton and an electron. The electron is emitted from the atom as a beta particle. But is the atom still a carbon atom? It now has seven protons in its nucleus. The element with seven protons in each of its atoms is **nitrogen**. So the carbon atom has become a nitrogen atom!

This process of changing from one element to another while emitting radiation is called **radioactive decay**. The nucleus you start with is called the **parent nuclide** and the product of the decay is the **daughter nuclide**. Nitrogen is stable because it has seven protons and seven neutrons, so the radioactive decay of carbon 14 stops at nitrogen. Figure 23.3 shows some more examples.

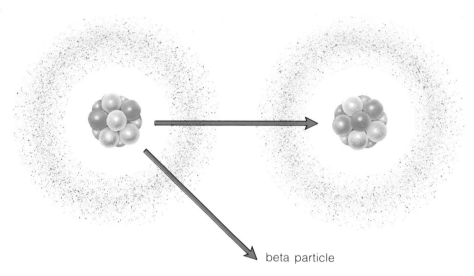

Fig. 23.1 Carbon 14 decay. One of the neutrons in the carbon atom changes to a proton and an electron. The electron is emitted as a beta particle.

beta particle

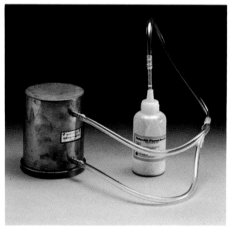

Fig. 23.2 Apparatus for measuring radioactive decay of radon gas. Squeezing the plastic container pushes the gas into the closed ionisation chamber, on the left. This has a central brass electrode, connected to a 16 volt supply. As the radon decays, it emits alpha particles, which produce a small current. A very sensitive ammeter measures this current, which depends on the number of radon atoms in the chamber. As the radon decays, the current drops, so a graph of current against time produces a decay curve.

Fig. 23.3 Radioactive decay of uranium. A uranium 238 atom decays to thorium 234. In the process it gives off two neutrons and two protons, which form an alpha particle.

Thorium 234 is not stable, either. It goes through 13 more decays, in which 7 alpha particles and 6 beta particles (electrons) are lost. Eventually, lead 206 is produced, which is stable and does not undergo radioactive decay.

All 14 steps make up the **radioactive decay series** for uranium 238. Material does not just vanish during radioactive decay. If you start with a sample of uranium 238 it will gradually change into other atoms.

Uranium 238 atom, containing 92 protons and 146 neutrons

Thorium 234 atom, with 90 protons and 144 neutrons

7 alpha particles are lost

a lead 206 atom, with 82 protons and 124 neutrons

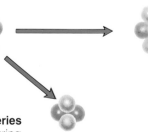

an alpha particle

6 beta particles are lost

Radioactive decay is a random process

The atoms of radioactive isotopes are unstable. Sooner or later they will decay, giving off radiation as they do so. This is a random process, and we cannot tell exactly when any particular atom will decay. But in a lump of radioactive material, there are millions upon millions of atoms. Some will decay quickly, and others will take longer. The **activity** of the material tells you how quickly the atoms are decaying.

Activity is measured in **becquerels**. If one atom decays every second then the activity is one becquerel (1 Bq). You can calculate the activity by finding the number of decays per second.

$$\text{activity} = \frac{\text{number of decays}}{\text{time taken in seconds}}$$

Remember that the activity tells you how often, on average, the atoms are turning into other atoms. Changing the physical conditions, for example temperature, will not change the activity. Changing the chemical conditions, for example pH, will not change the activity. This is particularly important when radioactive material is used in the body to diagnose different conditions (see Topic 27). Whatever the physical and chemical conditions inside the body, the activity of the sample will behave in a predictable way.

What *will* change the activity is the amount of radioactive material you have. The more atoms there are, the more likely it is that some will decay. In 1 kg of uranium 238, approximately 12 million atoms will decay to thorium 234 every second. The activity would be 12 MBq (Figure 23.4).

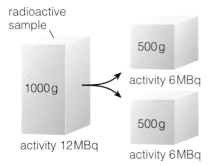

Fig. 23.4 The count from a radioactive sample depends on what isotope is in the sample, and how many atoms the sample contains.

Nuclear equations show what happens in a decay

The decay shown in Figure 23.3 can be represented by an equation:

nucleon number

$$^{238}_{92}\text{U} \longrightarrow ^{234}_{90}\text{Th} + ^{4}_{2}\text{He}$$

proton number

For each atom:
- The top number is the total number of **nucleons** (particles in the nucleus). As this is the number of protons and neutrons, it is the same as the **mass number**. In nuclear equations it is called the **nucleon** number.
- The bottom number is the number of protons. It is the same as the atomic number and is called the **proton** number.
- The number of particles in the nucleus less the number of protons must equal the number of neutrons so:

    number of neutrons = nucleon number – proton number

example: The thorium atom has 234 nucleons and 90 protons
    so it has 234 – 90 =144 neutrons.

In nuclear equations, the nucleon numbers and proton numbers on each side of the equation must be the same – the equation must be **balanced**. In this equation, the uranium atom loses 4 nucleons in the form of an alpha particle. This is a helium nucleus and has 2 neutrons and 2 protons. So the nucleon number goes down to 234 and the proton number goes down to 90 – a thorium atom. Checking the top and bottom numbers on each side of the equation: 238 = 234 + 4 and 92 = 90 + 2, so the equation is balanced.

In beta particle decay, an electron leaves the nucleus. A neutron changes into a proton (Figure 23.5), so the nucleon number stays the same but the proton number goes up. So, for beta decay, we just increase the proton number by one for each beta particle emitted. The decay of carbon to nitrogen is shown by this equation. We balance the top and bottom by saying that the proton number of the electron is –1.

$$^{14}_{6}\text{C} \longrightarrow ^{14}_{7}\text{N} + ^{0}_{-1}\text{e}$$

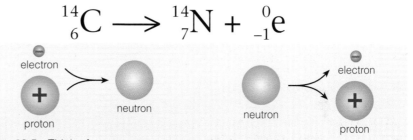

Fig. 23.5   Think of a neutron as a proton that has gained an electron to cancel its positive charge. Then you can see that if an electron is ejected from the nucleus of an atom, a neutron has changed into a proton.

- Gamma rays don't contain any particles, so they don't appear in a nuclear equation. All that happens is that the radioactive nucleus gives out some energy. For example, nickel 60 produces gamma rays.

$$^{60}_{28}\text{Ni} \longrightarrow ^{60}_{28}\text{Ni}$$

higher energy $\longrightarrow$ lower energy + gamma radiation

Question

1   For each atom listed below, calculate the number of neutrons. Plot a graph of number of neutrons against number of protons. What pattern do you notice for the proton/neutron balance for these stable atoms? How would the graph look if the numbers of protons and neutrons were always equal?

	C	Si	Ca	Fe	Y	Rh	I	Tm	Bi
mass number	12	28	40	56	89	103	127	169	209
atomic number	6	14	20	26	39	45	53	69	83

Radiation is all around us. Large doses of radiation can be very harmful to living things.

Everyone is constantly exposed to radiation

Radiation is all around you. Most is produced by natural substances, such as rocks. Some is made by humans. The normal level of radiation to which we are all exposed is called background radiation. Figure 24.2 shows the most important sources to which we are exposed.

Fig. 24.1 A film badge contains photographic film. If the wearer is exposed to radiation, it will affect the film. The badges are collected regularly, and the film is developed, to check if the wearer is being exposed to radiation.

Fig. 24.2 The main sources of radiation in the United Kingdom. About 87% of the radiation to which people are exposed is from natural sources.

The air we breathe in contains small amounts of radioactive isotopes. These produce radiation in our lungs. All types of food also contain radioactive isotopes. The main one is potassium 40.

Radioactive radon gas enters buildings from the ground. It gets trapped inside the building. So radiation levels from radon are much higher indoors than outside.

Many building materials contain radioactive isotopes which emit gamma radiation.

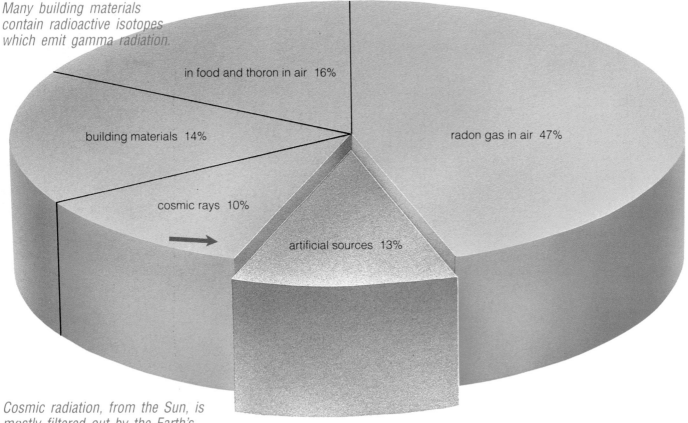

in food and thoron in air 16%
building materials 14%
cosmic rays 10%
artificial sources 13%
radon gas in air 47%

Cosmic radiation, from the Sun, is mostly filtered out by the Earth's atmosphere before it reaches the ground. Travellers in aeroplanes are exposed to much more cosmic radiation than people on the ground.

By far the largest human-made contribution is from X rays and other medical uses. Only 1 % is from sources such as nuclear waste, or accidents such as at Chernobyl.

Large doses of radiation can be harmful

Radiation damages living things because it damages the molecules in their cells. Alpha, beta and gamma radiation are all ionising radiations. They knock electrons away from atoms. This changes the way in which the atoms behave, and so changes the structure and behaviour of molecules in cells. If an organism receives a large dose of radiation, then a great many molecules may be damaged. This affects all sorts of processes going on in the body, and the person feels very ill. This is called **radiation sickness**.

The skin is also badly damaged with **radiation burns**.

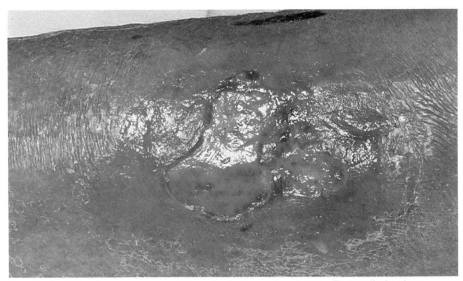

Fig. 24.3   This ulcer was caused by a high dose of radiation. The radiation has killed the skin cells. Eventually, the cells surrounding the ulcer will divide to form new ones, making a new layer of skin over the wound.

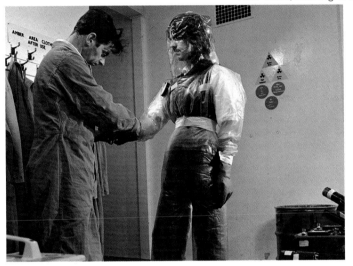

Fig. 24.4   A worker wearing protective clothing to protect his body from radiation.

Radiation can cause cancer

One very important molecule found in every cell is **DNA**. DNA is the chemical which carries inherited information from one generation to another. It is the DNA in your cells which gives instructions for everything which your cells do. If the DNA in a cell is damaged, then the cell may begin to behave very differently from normal. It is said to have **mutated**.

Radiation can cause mutations in cells by damaging the DNA. Sometimes this is not harmful. But sometimes the mutations cause the cell to divide uncontrollably. This may develop into **cancer**. Sometimes, the cancer does not develop until many years after exposure to the harmful radiation.

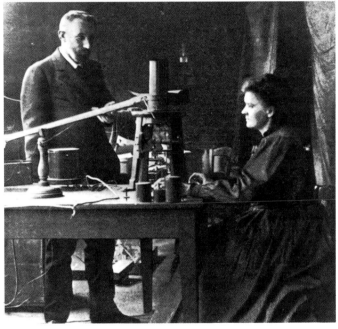

Fig. 24.5   Marie Curie was born in Poland in 1867. She worked on uranium, radium and radioactivity. The old unit of radioactivity, the curie, is named after her. 1 curie = 37 000 million becquerels (37 GBq).

Questions

1 a What is meant by the term 'background radiation'?
  b Make a list of the main causes of background radiation.
  c Where does most background radiation come from?
2   Give three ways in which someone's dosage of radiation might be made higher than average.
3   Alpha particles are heavy. They are quickly stopped by air, or by skin. Gamma rays, though, can pass easily through most materials.

a If an alpha source and a gamma source were held 20 cm from your hand, which would be the more dangerous to you?
  b If you swallowed an alpha source and a gamma source, which would be the more dangerous to you? Explain why.

25 HALF-LIVES

The large number of atoms involved in radioactive decay means that we can predict how fast decay will take place in a particular isotope.

The half-life of a radioactive isotope is the time taken for half of it to decay

The more unstable an isotope is, the more its atoms are likely to decay. If you have a sample of carbon 14, the atoms will stay around much longer than if you had a sample of the metal polonium 212. If you have 10g of carbon 14 and watch it for 5600 years, you will find that you have only 5g left.

The other half would have decayed to nitrogen. Polonium 212 takes 0.00000003 seconds for half of it to decay. The activity of a 10g polonium 212 sample would be much higher than that of a 10g carbon 14 sample.

The time taken for half of a particular radioactive isotope to decay is always the same. This is called its **half-life**.

The activity tells you how quickly the atoms are decaying, so a high activity means a short half-life. A sample of a more stable isotope has a lower activity and a longer half-life. Since the activity is not affected by physical and chemical changes, the half-life does not change either.

A half-life curve shows how the activity falls with time

A plot of activity against time falls quickly at first and then slows down. With fewer unstable atoms, the activity is less. This type of curve is called an **exponential curve**. After one half-life, half the atoms have not decayed. After two half-lives, one quarter of the original number of atoms is left.

Some half-lives

Thorium 232	13 900 000 000 years
Uranium 238	4510 000 000 years
Potassium 40	1260 000 000 years
Uranium 235	713 000 000 years
Radium 226	1622 years
Cobalt 60	5.26 years
Iodine 131	8.07 days
Radon 222	3.82 days

Working out half-lives

If the half-life of a radioactive substance is short, you can time how long it takes until half of it has decayed. For a less active isotope it isn't practical to wait for half its lifetime. It would take about 23 years for the activity of a sample of radium 226 to fall by just 1 per cent! You could start by plotting the beginning of a curve like Figure 25.1, and then predict when half of the isotope would have decayed. An even better way is to measure the activity and amount of isotope, from which it is possible to calculate the half-life. After allowing for the number of atoms, a greater activity means a shorter half-life.

INVESTIGATION 25.1

Using cubes to simulate radioactive decay

You will need at least 100 cubes, with one face different from the other five faces.

1 Draw a results chart.
2 Scatter the cubes on the bench top. Take away all the cubes which fall with the different face uppermost. These are the ones which have decayed. Record the number remaining.
3 Keep repeating this process, removing the 'decayed' cubes each time. Keep going until you run out of cubes.
4 Plot a graph of the number of cubes (y axis) against the number of throws (x axis). Join the points to make a smooth curve.

Questions

1 What can you say about the shape of the graph?
2 How many throws does it take before you remove half the cubes?
3 How many throws does it take to remove the next half?
4 What is the 'half-life' for this decay, measured in number of throws?

Fig. 25.1 A half-life curve. The half-life in this example is 25 seconds. Every 25 seconds, the amount of the radioactive material falls by one-half. This will also reduce the count rate by one-half every 25 seconds.

The radioactive material decays into another material.

In this example, the count rate has been adjusted to allow for the background count. Unless the background level is found and taken from the results, the curve would not be quite this shape. This shape of curve is called an exponential curve.

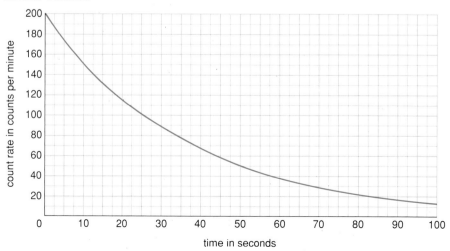

Radioactivity in rocks can be used to calculate the age of the Earth or samples of rock

Uranium 238, uranium 235 and thorium 232 are all found naturally on Earth. They each decay through a number of stages called a **radioactive decay series**. For all three isotopes the series ends in forms of lead that are stable. The half-lives of the daughter nuclei formed are all much shorter than the half-life of the isotope that started the series. The half-lives of uranium and

thorium control the time taken to decay to lead. By measuring the amounts of these isotopes and lead in samples of rock, you can calculate the time for which the isotopes have been decaying. If the alpha particles from the decay have been trapped in the rock as helium, then the relative amounts of uranium and helium can be used as the basis of this calculation.

Another useful isotope with a long half-life is potassium 40. The amount of this isotope is compared to the amount of stable argon 40 trapped as a gas. From these and other measurements, the age of the Earth has been estimated at 4 500 000 000 years. Measurements of meteorites and samples of rocks from the Moon give about the same age for the rest of the solar system.

Questions

Read the following passage and then answer the questions that follow.

Radon gas in the home

Radon gas is a naturally occurring radioactive gas. It is formed during the radioactive decay of natural material and has little to do with the nuclear industry. Most soil and rock contains some uranium. Uranium decays over a long period of time and one of the products of this decay is radium. Radium decays to radon gas, releasing alpha particles. As the half-life of uranium is very large, uranium acts as a 'reservoir', constantly topping up the radon levels. The half life of radon is 3.825 days.

The main source of radon in houses is from the ground. Brick and stone (especially granite) do contain uranium but don't generate much radon gas. Where there are no buildings, the radon from the ground disperses into the atmosphere, but in houses it can collect and concentrate. The levels of radon in the home are affected by many factors. The type of soil or rock in the ground affects the amount of radon produced. Radon levels are generally higher in south-west England but there are other 'hot-spots', such as parts of Northamptonshire and Scotland. This does not mean that all houses in these regions have high levels of radon, because it can vary from one house to the next. Houses also tend to draw more air (containing radon) in from the ground as warm air escapes from the top of the house, through chimneys and upstairs vents. So open upstairs windows can bring in more radon than they let out.

Radon is dangerous if we inhale the radioactive decay

products. Some of these products have very short half-lives and could increase the risk of lung cancer. Smoking is still the biggest cause of lung cancer, but in the USA it is thought that radon may cause 20 000 lung cancer deaths a year.

The activity of a radioactive source is measured in becquerels. Radon gas concentrations are measured in becquerels per cubic metre, or Bq/m^3. Most houses have levels of around 20 Bq/m^3. A lifetime exposure to levels of 200 Bq/m^3 increases the risk of cancer by a factor of ten, and this is the **action level** at which a householder should do something about radon. The National Radiological Protection Board will take measurements in homes at risk using plastic sensors. These are left in place for 3 months and then analysed to work out the average levels of radon gas. Peak readings are of no use since it is the long-term exposure that is important in assessing the risks.

If levels are high they can be reduced by preventing radon getting in. This can be done by sealing floors or by pressurising the whole house with a small air pump. Fitting small vents in downstairs windows encourages fresh air to come from outside, and not through the floor. The best method is to ventilate the ground under the floor. A hole or **sump** is created if the house has solid floors and the air (and radon with it) is extracted by a pump.

1 What is meant by the term 'half-life'? How long is the half-life of uranium?

2 Name the two nuclei that are formed when radium decays.

3 a Complete the nuclear decay equations.

$$^{226}_{?}\text{Ra} \longrightarrow \, ^{?}_{86}\text{Rn} + \, ^{4}_{2}\text{He}$$

$$^{?}_{86}\text{Rn} \longrightarrow \, ^{218}_{?}\text{Po} + \, ^{4}_{2}\text{He}$$

$$^{234}_{90}\text{Th} \longrightarrow \, ^{234}_{91}\text{Pa} + \, ^{?}_{?}?$$

b What is the other decay product from thorium?

c Why is this product less dangerous in the lungs than the product formed from radon?

4 A room measures 2 m by 3 m by 2.5 m.
  a What is the volume of the room?
  b Assuming average radon levels, what would be the total radon activity in the room?

5 In a sealed room with an activity of

800 Bq/m^3, how long would it take until radon activity fell to the action level?

6 The half-lives of some of the radon decay products are less than that of radon. What does that tell you about the activity of these products? Your friend says that if they don't last as long, they can't be as dangerous. Do you agree or disagree with this statement? Give your reasons.

26 RADIOACTIVE WASTE

Radioactive material can stay radioactive for a long time and must be disposed of carefully.

Nuclear power stations produce less waste than coal-fired power stations

A nuclear power station uses the energy released in nuclear reactions to generate electricity. (You can read more about this in Topic 239.) In one year a nuclear reactor produces around 20 tonnes of waste. About 50 000 tonnes of uranium have to be mined to make the fuel. To generate the same amount of electricity by burning coal would need 2 million tonnes of coal. A coal-fired power station produces up to 120 000 tonnes of ash, 200 000 tonnes of sulphur dioxide, 7 million tonnes of carbon dioxide and releases more radioactivity than the nuclear power station does.

Although there is not much waste from the nuclear power station, this waste is radioactive. Radioactive (nuclear) waste is also produced by nuclear reactors in other industries. Nuclear waste can be grouped into three classes.

Low-level waste comes from nuclear reactors in power stations, and also hospitals and industry. Most nuclear waste is low-level waste. It isn't very radioactive and the products have short half-lives, so it can just be buried in the ground. Only 1% of the radioactivity from nuclear waste is from low-level waste. A lot of it is paper or cloth so it is often burnt (in sealed containers!) to reduce its size. Liquids can be diluted and released into the sea.

Intermediate-level waste usually contains reactor parts and chemicals, and material contaminated by a reactor or from the reprocessing ('recycling') of nuclear fuel. The radioactivity must be contained so the waste is often covered in concrete. If the material has a long half-life it must be buried deep underground.

High-level waste is mainly liquid from fuel reprocessing. This is very radioactive and contains isotopes with long half-lives. As well as shielding, the waste needs to be cooled, because the radioactive decay releases energy. High-level waste can be **vitrified** by trapping it in special types of glass. It is then sealed in stainless steel tanks and can be stored deep underground. Most of the radioactivity from nuclear waste comes from high-level waste, although it only accounts for 3% of the volume.

Fig 26.1 High-level liquid waste is converted into solid glass blocks by a process called vitrification.

Fig 26.2 Spent fuel is transported in flasks which have walls one foot thick and are capable of withstanding the severest of accidents.

Spent fuel can be buried or reprocessed

When nuclear fuel is removed from a power station it is held in storage under water. Most of the radioactivity produced in the reactor stays in the fuel. Water helps trap the radiation and keep the fuel rods cool. The rods are stored in this way for up to 40 years. Some of the material will have to be stored for thousands of years. To be safe it will have to be buried very deep in areas where there is no danger of earthquakes or other natural disasters.

The fuel rods contain a lot of material that can be used again. In Europe the fuel is **reprocessed**. A large amount of uranium (96% or more) and a small amount of plutonium can be recovered. To do this, the fuel is dissolved in nitric acid and the uranium and plutonium are separated out from the waste material. This still leaves a small amount of high-level waste to be disposed of.

One possibility of reducing the time for which nuclear waste will have to be stored is to change the isotopes with long half-lives into shorter-lived ones. This is called **transmutation**. By bombarding the waste with neutrons from a reactor, isotopes can be converted to different ones. For example, iodine 129 has a half-life of 20 million years and can be converted to stable xenon.

Fig. 26.4   Chernobyl in 1991. A large concrete shelter or sarcophagus was constructed during 1986 in an attempt to contain radiation and seal the plant. This is the darker part of the structure in the centre of the picture.

Fig. 26.3   Chernobyl nuclear reactor after the explosion of 26 April 1986. More than 6 tonnes of nuclear material was released.

Nuclear accidents can happen

In 1986 an experiment at the nuclear power station at Chernobyl, in the Ukraine, went wrong. An explosion blew open the reactor and released radiation into the atmosphere. Over 6 tonnes of nuclear material escaped. The world-wide release of radiation may have been as high as 10 000 000 000 000 000 000 Bq.

Radiation absorbed by the body can cause damage. The **dose** is the energy absorbed per kilogram of body mass. A dose of 1 gray or 1 Gy means each kilogram has absorbed 1 joule of energy. As some types of radiation are absorbed more than others the dose is multiplied by a 'quality factor'. This number is 1 for beta and gamma radiation, and 20 for alpha radiation. The dose multiplied by the quality factor produces the **dose equivalent**, which is measured in sieverts (Sv).

The background radiation to which we are all exposed, on average, is 2 mSv per year. Doses of 3 Sv (1500 times the average background radiation) will cause radiation

burns to the skin. Radiation workers have to keep their dose levels low as effects can build up with time. A weekly dose of 1 mSv might be considered safe. Clean-up workers at Chernobyl received an average dose of 170 mSv. 134 workers received doses high enough to give **acute radiation sickness** and 28 died.

A large amount of radioactive iodine 131 was released in the Chernobyl accident. Food chains were contaminated. Iodine collects in the thyroid gland and young children are particularly affected. For the people living near Chernobyl, thyroid radiation doses reached several sieverts in some cases. Already, 800 thyroid cancer cases have been reported and it is expected that this may reach a few thousand.

Eventually, the remaining reactors at Chernobyl will be shut down. The damaged reactor still holds 200 tonnes of nuclear fuel and its ten-year-old protective covering of concrete still contains 700 000 000 000 000 000 Bq of activity.

Cancer cells can be killed by radiation

Radiation can kill living cells because it ionises molecules inside them. Cancer cells can be killed in this way. Cobalt 60 is often used. It produces gamma rays which can penetrate deep inside the body. For skin cancers, phosphorus 32 or strontium 90 may be used instead. These produce beta radiation. The dose of radiation has to be carefully controlled. Otherwise the radiation could do more damage than help.

Patients undergoing radiation treatment often feel ill, because the radiation also damages other cells.

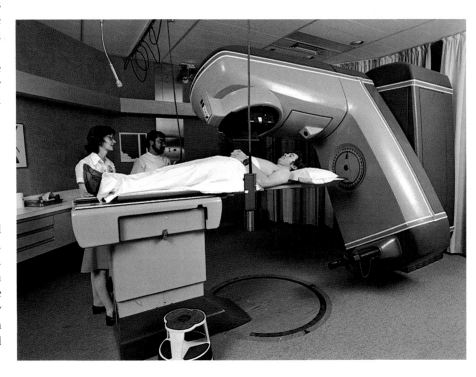

Fig. 27.1 A patient receiving radiotherapy treatment using a linear accelerator.

Radiation can be used to sterilise surgical instruments

Gamma rays are often used to kill bacteria and viruses on dressings, syringes, and other medical equipment. This is called **sterilisation**. Sterilisation means killing all living things. These items used to be sterilised using very high temperatures, or steam. Gamma radiation is a more convenient and more effective method.

Radioactive tracers show what happens during biological processes

Inside a living organism, a radioactive isotope of a particular substance behaves in just the same way as the normal isotope. So, if a plant is given carbon 14, for example, it will use it in exactly the same way as it always uses carbon 12. But the carbon 14 produces beta radiation. By measuring the radioactivity in different areas of the plant, the path taken by the carbon atoms can be followed.

In a similar way radioactive iodine can be used to check that a person's thyroid gland is working properly. The thyroid gland uses iodine. If a person is given a tablet containing iodine 131, the thyroid takes it up as though it was normal iodine. The amount of radioac-tivity emitted by the thyroid gland can then show how much iodine has been taken up.

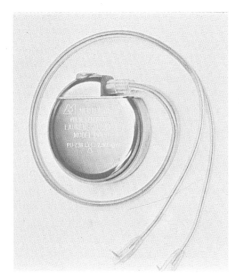

Radiation can provide energy

When an atom decays it gives out energy. This energy can be used to provide electricity. **Nuclear batteries** use this process. A nuclear battery lasts for a very long time. Nuclear batteries are often used to power heart pacemakers. The heat energy produced by radioactivity can be used to generate electricity on a very large scale. In a **nuclear reactor** special nuclear reactions are encouraged that release large amounts of energy.

Fig. 27.2 This pacemaker can be inserted into a human heart whose own pacemaker is faulty. It emits regular pulses, which stimulate the heart to contract rhythmically. If powered by a nuclear battery, rather than a conventional one, it can run for much longer before the battery needs to be replaced.

Radiation can be used to check the thickness of metal sheets

Gamma radiation is used to make sure that steel sheets are made to the correct thickness. Figure 27.3 shows how this is done. The steel is pressed between rollers to produce a sheet of a particular thickness. A source of gamma radiation is then positioned on one side of the steel sheet. A detector is positioned opposite it, with the steel in between. Gamma radiation can pass through steel, but the thicker the steel the less radiation gets through. If the sheet comes through thicker than usual, the radiation picked up by the detector falls. This causes the pressure on the roller to be increased, until the radiation detected increases to its normal level. What do you think happens if the detector picks up *more* radiation than usual? A similar method can be used to check the thickness of sheets of paper. This time, though, alpha radiation is used.

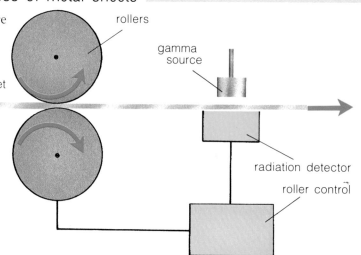

Fig. 27.3 Using radiation to check metal sheet thickness.

Questions

1 a How can radioactivity help in the treatment of cancer?

b Why do you think that gamma radiation is used to treat cancers inside the body, but beta radiation is used to treat skin cancers?

2 a What is meant by sterilisation?

b Why do you think it is important that surgical dressings should be sterilised?

c Why is gamma radiation, not beta radiation, used for sterilising dressings?

3 Why are nuclear batteries, rather than ordinary batteries, used to power heart pacemakers?

---

4 ^{14}C makes up about $\frac{1}{10\,000\,000}$ of the carbon in the air.
The half-life of ^{14}C is 5600 years.

a In what substance is carbon present in the air?

b Which is the commonest isotope of carbon?

c ^{14}C is constantly decaying. So why does the amount of ^{14}C in the air not decrease?

d Carbon is taken in by plants. In what substance is the carbon? What is the process by which the plants take it in?

e What happens to the amount of ^{14}C in a plant, or in something made from the plant, as time goes by?

f The amount of ^{14}C in a piece of linen is analysed. It is found to make up $\frac{1}{20\,000\,000}$ of the carbon in the cloth. How old is the cloth?

EXTENSION

Radiocarbon dating

Air contains 0.04 % carbon dioxide. Most of the carbon atoms in the carbon dioxide are carbon 12 atoms. But a small proportion are carbon 14 atoms. These carbon 14 atoms decay to nitrogen, emitting beta particles. But new carbon 14 atoms are always being produced by the action of cosmic rays. So the amount of carbon 14 atoms in the carbon dioxide in the air stays the same.

When plants photosynthesise they take in carbon dioxide from the air. The carbon atoms become part of molecules in the plant. The carbon 14 atoms in these molecules slowly decay to nitrogen. But, unlike the carbon 14 in the air, they will not be replaced.

So the amount of carbon 14 in the plant gradually falls.

Many things might happen to the plant. It might be eaten by an animal. It might be made into material such as cotton or linen. It might form coal. But whatever happens the carbon 14 in it gradually decays. After 5600 years there will only be half as many carbon 14 atoms as there were when they first entered the plant from the air.

So, by finding out how much carbon 14 there is in an object, we can work out how long ago the plant from which it was made was alive. The less carbon 14 compared with carbon 12, the older the object is.

Fig. 27.4 Part of the Turin shroud. This ancient piece of cloth shows marks which some people believe to have been made by Christ's body after crucifixion. In this detail, you can see the image of hands, and a mark which could have been made by a nail. In 1988 three small pieces of the shroud were dated using the radio-carbon technique. Three different laboratories all showed the shroud to be about 500–600 years old. Although this shows that it cannot really have been Christ's shroud, the way in which the marks were made is still a mystery.

EXTENSION

Questions

1 a Name the three types of radiation emitted by radioisotopes.

b Which is negatively charged?

c Which is made up of positively charged helium nuclei?

d Which can penetrate the farthest in air?

e All three types are ionising radiation. What does this mean?

2 Film badges are worn by people who may be exposed to radiation.

a Why do they wear the badges?

b Why must no light be allowed to fall on the badges?

3 Find out what pitchblende ore is and where it comes from.

4 To measure engine wear an engine is run with piston rings that have been made radioactive by bombarding them with neutrons in a nuclear reactor. Radioactive material goes into the oil when the rings wear against the sides of the cylinders. The radioactivity in the oil is used to estimate the amount of metal that has worn off the rings. Why is it important to know the half-life of the radioisotope in the piston rings?

5 The level of radiation from a radioactive source was measured for just over one minute. The results are shown below.

Time (s)	0	10	20	30	40	50	60	70
Level of radioactivity	112	103	109	111	116	117	109	107

a What causes the change in the readings?

b What was the average reading?

c When the source was removed from the room, the average count fell by 80 counts. If the source has a half-life of two years, estimate the count (with the source back in the room) in six years' time.

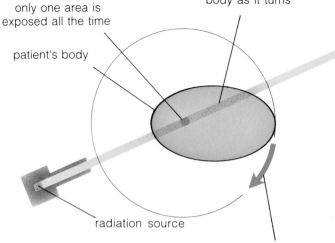

beam of radiation sweeps through the body as it turns

only one area is exposed all the time

patient's body

radiation source

body is rotated on an axis passing through the cancer

6 The diagram above shows how gamma rays may be used to kill cancer cells.

a Why are gamma rays used, rather than alpha or beta radiation?

b During treatment the person is rotated in a circle with the tumour at the centre. Why is this done?

7 The table shows the radiation count from a source over a period of 40 min.

Time (min)	0	5	10	15	20	25	30	35	40
Count	152	115	87	66	50	38	29	22	17

a Plot a line graph to show these results.

b What is the time taken for the count to drop from 110 to 55 counts?

c What is the time taken for the count to drop from 80 to 40 counts?

d What is the half-life of this radioactive element?

e How long would it take for the count to reach 10 counts?

EXTENSION

8 The energy from nuclear reactions is used in power stations. Uranium 235 is encouraged to split into two similar-sized nuclei by bombarding it with neutrons. One possible reaction is:

$$^{235}_{92}U + ^{1}_{0}n \longrightarrow ^{141}_{56}Ba + ^{92}_{36}Kr + 3^{1}_{0}n$$

Neutrons are released, which could go on to start other reactions.

If one of the daughter nuclei was barium 144 and the other was krypton 90, what difference would this make? (Hint: rewrite the equation, putting in the values that you know and recalculate the nucleon and proton numbers – how many neutrons are released now?)

9 Cobalt 60 is commonly used as a source of gamma rays for medical and industrial uses. It can be used for imaging, thickness measurements and killing cells in cancer treatment. It decays to nickel by beta particle emission and the nickel nucleus then releases gamma radiation. Write a nuclear equation for this decay.

10 A Geiger–Müller tube was used to measure the radiation at different distances from a source emitting alpha and gamma radiation. As the Geiger–Müller tube was gradually moved away from the source, it was found that the level of radiation fell very rapidly over the first few centimetres. After that, the radiation fell more slowly. Why was this?

11 Smoke detectors contain radioactive Americium 241, which is an alpha and gamma source with a half-life of 460 years. If the distance travelled by the alpha particles is decreased significantly, the smoke detector is activated.

a Why is an isotope with a long half-life used in smoke detectors?

b What is the approximate range of alpha particles in air?

c What effect would the presence of smoke particles have on this range? What does this do to the smoke detector?

12 A manufacturer intends to check that cereal packets are full to the top as they pass down a conveyor belt. If you were provided with a selection of radioactive sources and a detector, how could you do this?

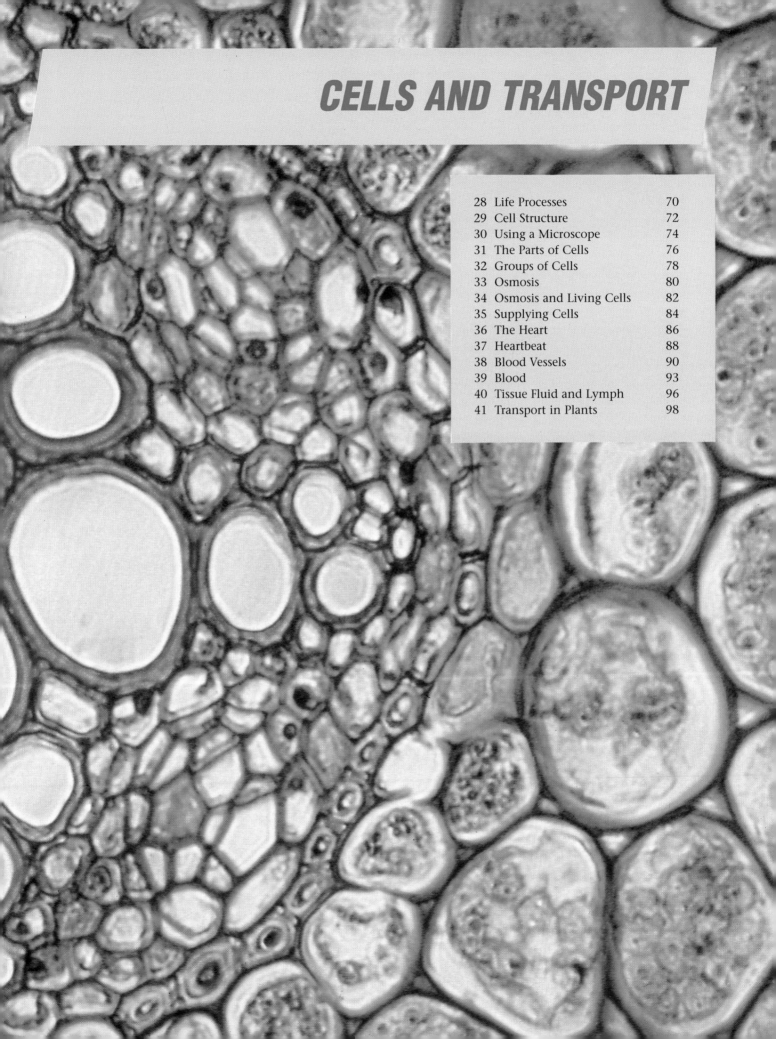

CELLS AND TRANSPORT

28 LIFE PROCESSES

All living things carry out seven processes. Non-living things do not carry out all of these processes. Living things are also known as organisms.

Fig. 28.1 What can you see here that is alive? What is not alive? Do any of the non-living things share some of the 'life processes' of the living organisms?

Living things include plants and animals

If you look around you, you will see many things which are alive and many which are not alive. Living things are often known as **organisms**. The living things you can see might include people and other **animals**, and also **plants**. There may also be other, smaller living things around you. For example, there are **bacteria** almost everywhere. And there are probably some **fungi** (such as toadstools) not too far away.

All of these living things carry out seven vital processes. These processes are special to living things – they are not all carried out by non-living things. They are called **life processes**. The seven life processes are **nutrition, respiration, excretion, sensitivity, movement, reproduction** and **growth.**

Nutrition includes feeding and photosynthesis

All living things need to take in substances from around them. They use the atoms in these substances to build the molecules which make up their bodies. Taking in substances for this purpose is called **nutrition**.

For animals (including humans), nutrition involves **feeding**. We eat substances which we call 'food'. Food is material containing substances such as proteins, fats and carbohydrates. These substances have been made by plants. You can read more about feeding in humans in Topics 98 to 101.

For plants, nutrition involves **photosynthesis**. Plants do not need to eat food – they make what they need from carbon dioxide and water, plus some minerals. They use energy from sunlight to help them to convert these substances into the molecules they need. You can read more about photosynthesis in Topic 88.

Respiration releases energy from food

Living things do not only need atoms to build up their bodies. They also need **energy**. All living organisms get the energy they need by breaking down certain molecules, such as glucose molecules. The glucose molecules contain energy, and when they are broken down, in a series of chemical reactions, the energy is released. This process is called **respiration**. It takes place in every cell in every living organism. You can read more about respiration in Topic 92.

Movement may involve all or part of the body

Living things can **move**. You can see this most clearly in animals, most of which can move around from place to place. Plants do not move their whole bodies around like this, but parts of their bodies can move. For example, the petals of many flowers close up at night. You can read more about movement in Topic 249.

Excretion removes waste products of metabolism from the body

Respiration is one example of a series of chemical reactions that takes place inside living cells. The chemical reactions which happen in living organisms are called **metabolic reactions**. Some metabolic reactions produce harmful substances, which the living organism does not want and must get rid of. For example, respiration produces carbon dioxide. The removal of waste products of metabolism, such as carbon dioxide, from the body, is called **excretion**. You can read more about excretion in animals in Topic 252.

Reproduction is the formation of new organisms

Living organisms are able to produce new organisms from themselves. This is called **reproduction**. Sometimes, a single organism just grows a new organism from its body. This is called **asexual reproduction**. For example, a spider plant grows new spider plants. Sometimes, the organisms produce sex cells called **gametes**, which fuse together in a process called **fertilisation**. This produces a new cell, which then divides to form a new organism. This is called **sexual reproduction**. You can read more about reproduction in Topics 128 to 137.

Sensitivity involves responding to stimuli

Living things are able to detect changes in their environment. This is called **sensitivity**. For example, plants and animals are sensitive to light. The light is a **stimulus**, which may cause the plant or animal to **respond** to it in some way. In an animal, the response might be a movement. In a plant, the response might be by growing in a particular way. You can read more about sensitivity and responses in Topics 244 to 247.

Growth is a permanent increase in size

When a new organism is first formed, it is usually small. As it gets older, it gets bigger. This is called **growth**. Growth happens as the individual cells in an organism get bigger, and also as these cells divide to produce more cells. You can read more about growth in Topic 133.

Questions

1   Match each term with its description.
   terms:
   nutrition, growth, sensitivity, reproduction, excretion, movement, respiration.
   definitions:
   - a permanent increase in size, brought about by an increase in the size of individual cells, and also an increase in the number of cells.
   - the production of new individuals from one or more parents; it may be asexual or sexual.
   - the release of energy from food, which takes place in every living cell.
   - the removal of waste products of metabolism from the body; some of these products would be toxic (poisonous) if allowed to accumulate inside the body.
   - the ability to detect changes in the environment, called stimuli, and to respond to them.
   - changing the positions of parts of the body, or of the whole body.
   - taking in substances containing atoms and molecules which can be used to build new molecules that make up the structure of the body; and which can also be used to provide energy.

2   Which characteristics of living things does a computer have? Why isn't a computer a living thing?

All cells are made of cytoplasm surrounded by a cell surface membrane. Most have a nucleus. Plant cells also have a cell wall around them.

Cells are the building blocks for living things

All living things are made up of cells. A cell is a very small piece of transparent jelly-like substance, surrounded by a thin covering. The jelly is called **cytoplasm**. The covering is called a **membrane**. Most cells also have a dark area inside the cytoplasm, called a **nucleus**.

An average size for a cell is about $\frac{1}{100}$ mm across. Some cells, for example bacterial cells, are a lot smaller than this. Some may be much larger. Some of the largest cells of all are egg cells. As cells are so small, it takes many of them to make up a large organism such as yourself. Your body contains several million cells. Other organisms are so small that their whole body is made of just one cell.

Plant cells always have cell walls

The cells in every kind of living organism are very similar. But plant and animal cells do have some important differences between them. One of these differences is that all plant cells have a **cell wall** outside their cell surface membrane. The cell wall is made of fibres which criss-cross over one another. This makes a strong, protective covering around the plant cell. The fibres are made of a carbohydrate called **cellulose.**

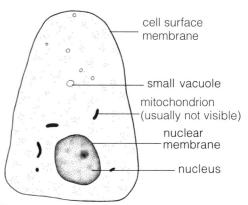

Fig. 29.1 An animal cell as seen through a light microscope.

cell surface membrane

small vacuole

mitochondrion (usually not visible)

nuclear membrane

nucleus

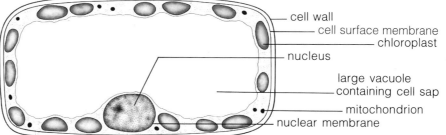

cell wall
cell surface membrane
chloroplast
nucleus
large vacuole containing cell sap
mitochondrion
nuclear membrane

Fig. 29.2 A plant cell as seen through a light microscope.

Fig. 29.3 A 3-D view of a plant cell.

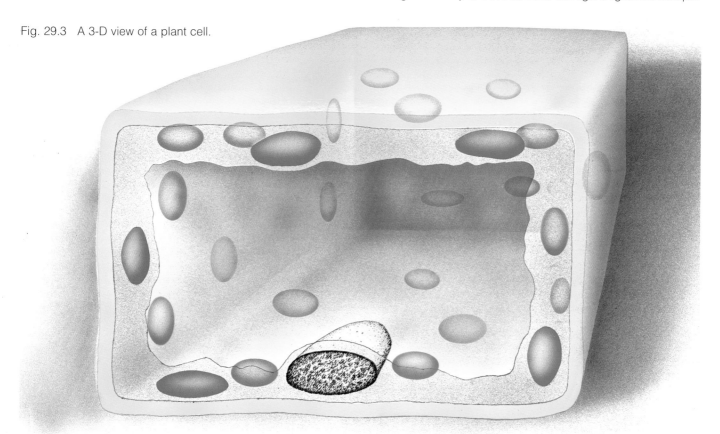

Plant cells may have chloroplasts and large vacuoles

Many plant cells contain small, green objects inside their cytoplasm. These are **chloroplasts**. They are green because they contain a green colour or pigment called **chlorophyll**. It is chlorophyll which makes plants look green. Only the parts of the plants which are above ground contain chlorophyll. Chlorophyll absorbs light energy, which is used in photosynthesis.

Most plant cells also contain a large fluid-filled area called a **vacuole**. The vacuole contains a sugary liquid called **cell sap**.

Animal cells never have cell walls, never have chloroplasts and never have vacuoles as large as those in plant cells.

Question

Copy and complete this table.

Structures contained in		
all cells	all plant cells but not animal cells	some plant cells but not animal cells

Question

Which of these groups of cells are animal cells, and which are plant cells? Give a reason for each of your answers.

Fig. 29.4   Cells from plants and animals.

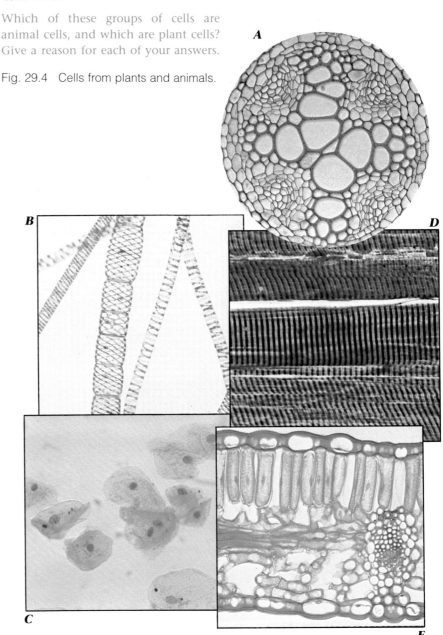

Sizes of cells

As cells are so small, they are usually measured in micrometres. A micrometre is one thousandth of a millimetre. The shorthand way of writing 'micrometre' is μm.

Questions

1   Which is the smallest of the cells shown?
2   How many times larger is the cheek cell than the bacillus?
3   How many micrometres are there in 1 cm?
4   How many micrometres are there in 1 m?

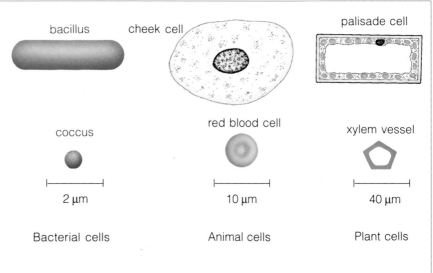

30 USING A MICROSCOPE

As cells are very small, a microscope is needed to see them clearly. Using a microscope is an important skill which all scientists should have.

Using a microscope

These instructions may look long and complicated, but you **must** follow them every time you use a microscope if you want to be sure of seeing things quickly and clearly. As you work through them, use the labels on the large diagram to find the different parts of the microscope.

1 Place the microscope so that it is facing towards a source of light. This could be a window, or a lamp.

2 Swivel the objectives until the smallest one (the lowest power) is over the centre of the stage.

Fig. 30.1 How to use a microscope

3 Look down through the eyepiece, and move the mirror until everything looks bright.

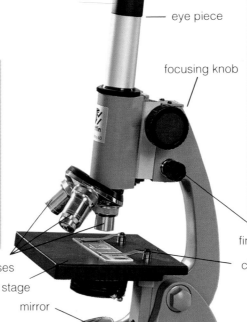

objective lenses

stage

mirror

eye piece

focusing knob

fine focusing knob

clips to hold slide in position

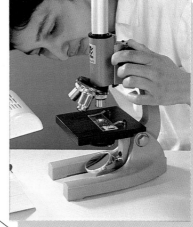

4 Put your slide on the stage. Put the part you want to look at in the middle of the stage.

5 Looking from the *side* of the microscope, *not* down the eyepiece, turn the focusing knob to bring the stage close to the objective lens.

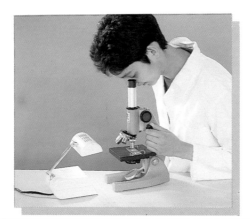

6 Now look down the eyepiece, and gradually move the stage *away* from the objective until the slide is in focus.

7 If you want to use a higher magnification, make sure that the piece of the object you want to look at is right in the middle of your field of view. Then swing a higher powered objective lens into place over the stage. Focus as before. You will need to bring the objective lens so close to the slide that it almost touches it.

Making a microscope slide

1 Make sure that the microscope slide you are going to use is clean.
2 Put the object you are going to look at in the centre of your slide. The object must be so thin that light can pass right through it. Thick objects cannot be seen clearly through an ordinary microscope.
3 If the object is not already liquid, it must have liquid added to it. This could be water or it could be a stain. Add the liquid carefully, using a pipette. Add just enough to cover the object.
4 Now, gently, lower a cover slip over the liquid on the slide. The diagram shows how to do this without trapping air bubbles.
   The cover slip is important for several reasons. It holds the object flat, making it easier to focus on it. It stops the liquid evaporating from the slide. It also stops the liquid getting on to the lenses of the microscope. *Always* use a cover slip when you make a slide.

5 Finally, use filter paper to mop up any excess liquid on the slide, or on top of the cover slip. This is important, because if liquid gets on to the objective lenses on your microscope, you will not be able to see through tnem properly.

Fig. 30.2   Lowering a coverslip

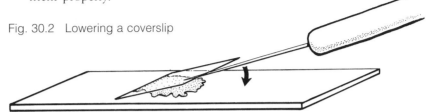

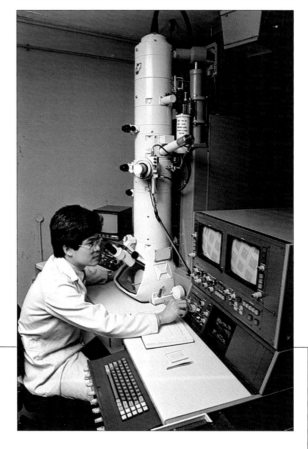

Fig. 30.3   A microscopist using a scanning electron microscope. Electron microscopes work in a similar way to light microscopes, but use electron beams instead of light beams. The beams are focused with electromagnets instead of glass lenses. The image on these screens shows a grid of known size, used to calibrate the microscope.

Looking at plant cells

1 Take a clean microscope slide, and put a drop of water into the centre of it.
2 Cut a piece about 1 cm² from a section of onion bulb. Using fine forceps, carefully pull away the very thin inner lining, or **epidermis**, from your piece of bulb. Working quickly, put the piece of epidermis into the drop of water on the slide, trying not to let it curl up.
3 Take a cover slip, and gradually lower it over the epidermis, as shown in the diagram. Doing it this way allows any air bubbles to escape.

4 Use filter paper to soak up any water which has seeped out from under the cover slip.
5 Set up a microscope and look at your slide on low power.
6 Make a labelled drawing of what you can see.
7 Now take a piece of pond weed, and put this on to a slide in the same way as you did for the onion cells. Again, cover with a cover slip, and look at it under the microscope. Draw and label what you can see.

Questions

1 Which were the larger – onion epidermis cells or pond weed cells?
2 What colour were
   a the onion cells
   b the pond weed cells?
3 What is the reason for this difference in colour?

THE PARTS OF CELLS

Every part of a cell has its own part to play in the smooth running of the cell's activities.

The cell surface membrane controls what enters and leaves the cell

The cell surface membrane is one of the most important parts of a cell. It makes sure that the contents of the cell stay inside. It also stops unwanted substances outside the cell from entering. But it must allow some substances into the cell. These substances include water, oxygen and food materials. It must also allow waste substances, such as carbon dioxide, to get out. So the membrane lets some substances through, but not others. It is called a **partially permeable** membrane.

Chemical reactions take place in the cytoplasm

The cytoplasm is a jelly made of proteins dissolved in water. Many chemical reactions go on in the cytoplasm. These reactions are called **metabolic reactions**.

The nucleus contains inherited information

Most cells also have a nucleus. The nucleus contains thin threads called **chromosomes**. Chromosomes contain information, inherited from the parent cell or cells, which give information to the cell about what it does. Usually, the chromosomes in the nucleus are so long and thin that they cannot be seen

Mitochondria release energy from food

The cytoplasm contains small objects called **mitochondria**. Inside mitochondria, sugar and oxygen react together to release energy. The number of mitochondria in a cell can often give you a clue about what that particular cell does. Muscle cells, for instance, contain huge numbers of mitochondria, because they need so much energy.

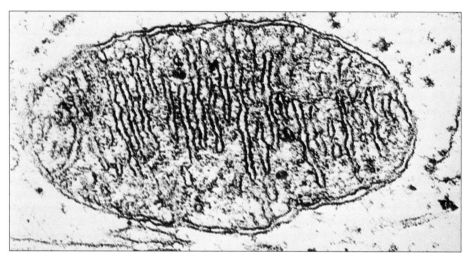

Fig. 31.1 A photograph of a mitochondrion, taken with an electron microscope. The reactions of aerobic respiration take place in the space inside it, and on the folded membranes. The actual size of a mitochondrion can be about 2 micrometres long. By how much has this one been magnified?

except with an electron microscope. But when the cell divides, they thicken. You can then see them with a light microscope.

Chloroplasts make food by photosynthesis

All the structures described so far are found in every cell. But some plant cells also contain very important structures called **chloroplasts**. Chloroplasts are quite large – often as big as the nucleus. They are always green because they contain the green pigment **chlorophyll**. Their function is to make sugar and other types of food, using energy from sunlight, water and carbon dioxide. Chlorophyll is essential for this, because it traps energy from sunlight. The process is called **photosynthesis**.

Many plant cells contain starch grains

Plant cells which live underground have no use for chloroplasts, as there is no sunlight for them to absorb. These cells may have other structures, similar to chloroplasts, but with no chlorophyll in them. These structures contain food stores. Potato cells, for example, contain many **starch grains**. The cells in the potato plant's leaves, which do contain chloroplasts, make sugar by photosynthesis. Some of this sugar is carried down to the potato underground. It is then converted to starch and stored there to be used later on. Chloroplasts can also store starch grains.

Specialised cells

Many cells have their own special jobs to do. Their sizes, shapes and contents are often different from other cells, to enable them to do their job really efficiently.

Palisade cells from a leaf are rectangular in shape. They contain large numbers of chloroplasts, often arranged around the edge of the cell where they can get the most sunlight for photosynthesis.

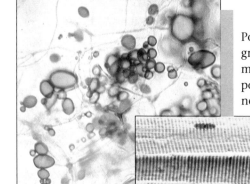

Potato cells store starch. The starch grains fill most of the cell. The starch may be used to provide energy for the potato to begin to grow new shoots next year.

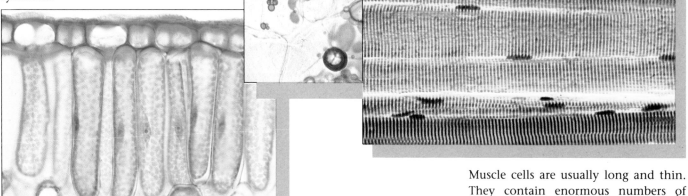

Muscle cells are usually long and thin. They contain enormous numbers of mitochondria, to provide the energy they need for movement.

INVESTIGATION 31.1

The effect of heat on plant cell membranes

Beetroot cells contain a red pigment (colour). The cell surface membrane keeps this pigment inside the cell. High temperatures damage the cell surface membrane, allowing the pigment to leak out.

1 Take a raw beetroot, and cut a cylinder out of it using a cork borer. Wash the cylinder very thoroughly, until no more red colour comes out of it.

2 Set up a Bunsen burner, tripod and gauze. Put some sand into a sand tray, and put this on the gauze. Half fill a beaker with water, and stand this in the sand in the sand tray.

3 Put the piece of washed beetroot into the beaker of water. Light the Bunsen burner. Hold a thermometer in the water, and watch the beetroot carefully as the temperature rises. As soon as any red colour begins to come out of the piece of beetroot, stop heating, and note the temperature of the water.

Questions

1 Why must the piece of beetroot be washed thoroughly?
2 Why does some red colour come out of the piece of beetroot *before* it is heated?
3 Why is the beaker of water heated in sand?
4 At what temperature did red colour begin to leak from the beetroot?
5 What was happening to the beetroot cells at this temperature?
6 Vegetables which have been boiled for a long time contain fewer nutrients than raw or lightly cooked vegetables. Can you explain why?

Questions

1 Which part of a cell performs each of the following functions:
  a releases energy from sugar and oxygen?
  b controls what enters and leaves the cell?
  c makes food by photosynthesis?
  d contains information inherited from the cell or cells that produced it?

2 Below is a list of cells.

A cheek cell   **B** leaf cell
C muscle cell   **D** onion epidermis cell
E carrot root cell   **F** human liver cell

Which of the cells from this list would you expect to contain each of the structures below? Give a reason for each of your answers.
  a cell membrane
  b chloroplasts
  c mitochondria
  d nucleus
  e cell wall

32

GROUPS OF CELLS

Large organisms are made of many different kinds of cells, each specialising in a different job. Cells doing the same job are often grouped together into a tissue.

A group of similar cells is called a tissue

In a single-celled organism, like *Amoeba*, the one cell has to carry out all the jobs which need doing. But in a large organism, such as yourself, there are millions of cells. Most of these cells are specialised to carry out just a few functions really efficiently. Muscle cells, for example, are specialised to produce movement. Nerve cells carry messages.

The cells lining your digestive system digest food.

Cells which specialise in the same function are usually grouped together. A group of similar cells is called a **tissue**. The onion epidermis which you looked at in Investigation 30.3 is an example of a tissue. The cells lining your stomach are another example.

A group of tissues can form an organ

Tissues themselves are often grouped together to form even larger structures. A group of tissues working together to carry out a particular function is called an **organ**. An eye is an example of an organ. If you look at Figure 246.2, you can see some of the tissues that it contains. The sclera, the choroid, the conjunctiva and most of the other structures labelled on the diagram are all tissues.

Plants have organs too. A leaf is an organ. The different tissues – for example, the epidermis and palisade layer – are labelled in the diagram on page 217.

A group of organs which all work together to perform a particular function is called an **organ system**, or just a system. For example, you have a digestive system, containing organs that are shown on page 244. Other systems in the human body include the nervous system, excretory system, circulatory (blood) system and breathing system.

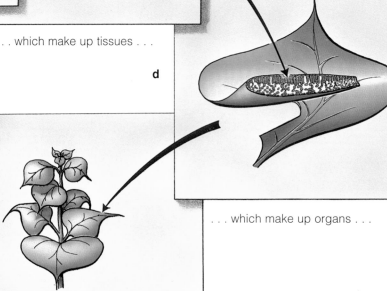

a Organelles . . .

b . . . make up cells . . .

c . . . which make up tissues . . .

d . . . which make up organs . . .

e . . . which make up organisms.

Fig. 32.1 Organelles to organisms. The small structures found inside cells are called organelles.

Can you name:
a the organelle shown in diagram a?
b the cell shown in diagram b?
c the tissue shown in diagram c?
d the organ shown in diagram d?

Questions

1 Copy and complete this paragraph.

All cells contain a jelly-like substance called .... which is surrounded by a .....
......Within the cell are smaller structures called organelles. The largest organelle is usually the ..... This contains ....., which hold inherited information about what the cell should do.

In addition, plant cells contain green organelles called ..... These contain a green pigment called ..... which traps energy from ..... This energy is used to make food, in the process known as .....

Unlike animal cells, plant cells always have a .... outside their cell surface membrane. This is made of ....

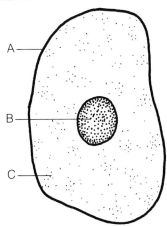

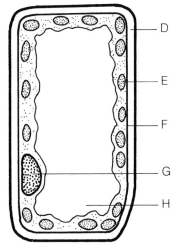

2 Name the structures labelled on the diagrams above.

3 The following instructions for looking at a slide under a microscope on high power, are in the wrong order. Write down the letters in the correct sequence. One of them needs to be used twice.

a Put the slide on to the stage.

b Turn to the highest power objective lens.

c Focus the microscope.

d Position the microscope facing a light source.

e Look from the side, and bring the stage and the objective lens close together.

f Adjust the mirror to fill the field of view with light.

g Turn to the lowest power objective lens.

4 Below is a photograph of some human cheek cells taken through a light microscope.

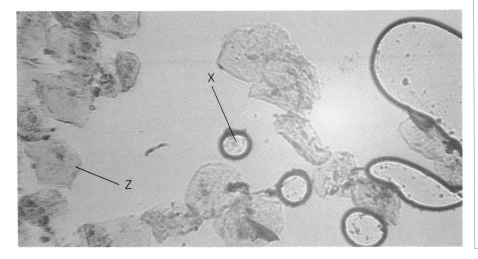

4 a Give two reasons why you could tell that these were animal and not plant cells.

b What are the structures labelled X?

c Describe how the slide should be made to make sure that there are not too many of these structures.

d Make a drawing of Cell Z, and label its nucleus, cytoplasm and cell surface membrane.

e A human cheek cell is about 10 μm or 0.01 mm across. By how much has Cell Z been magnified in this photograph?

5 The following list contains two organelles, two cells, two tissues, two organs and two organisms. Which is which?

onion epidermis	oak tree
heart	chloroplast
mitochondrion	cheek cell
pig	conjunctiva
leaf	red blood cell

EXTENSION

6 Identify the structures labelled A to E on this electron micrograph of a plant cell. (An electron micrograph is a photograph taken using an electron microscope. Electron microscopes use beams of electrons instead of beams of light, and can show clearly much smaller objects than light microscopes.)

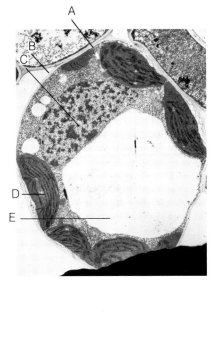

33 OSMOSIS

Water molecules are small. The molecules and ions of solutes are often much larger. This means that water molecules can get through holes through which solute molecules are too large to pass.

Water molecules are smaller than most solute molecules

A water molecule is made up of two hydrogen atoms and one oxygen atom. Hydrogen atoms are the smallest atoms which exist. Oxygen atoms are not particularly large, either. So water molecules are quite small as molecules go.

All sorts of substances can dissolve in water. Many of these substances have quite large molecules. Sugar is one example. A molecule of cane sugar (the sort you put in tea or coffee) has 12 carbon atoms, 22 hydrogen atoms and 11 oxygen atoms. So you can see that a sugar molecule is quite enormous compared with a water molecule.

Water molecules can diffuse through very small holes

Figure 33.1 shows a sugar solution. The sugar molecules are spread amongst the water molecules. The sugar solution is separated from some pure water by a thin piece of material called a **membrane**. This particular membrane is a piece of **visking tubing**. Visking tubing has extremely small holes in it. The holes are big enough to let water molecules through. But sugar molecules are much too big to pass through the holes. A membrane like this, which will let some molecules through but not others, is called a **partially permeable membrane**.

What will each kind of molecule do? The sugar molecules cannot do very much at all. They move around on their side of the membrane, bumping into each other, into water molecules, and into the membrane. But they cannot cross on to the other side of the membrane. The water molecules also move around, bumping into other molecules and the membrane. Some of them will 'bump into' a hole in the membrane, and go through to the other side. Water molecules from both sides of the membrane will cross on to the other side. There is a two-way traffic of water molecules from one side of the membrane to the other.

But there are far more water molecules on side A than on side B. On average, more water molecules on this side will bump into holes in the membrane, because there are more of them. So more water molecules will go from side A to side B than will go the other way. The water molecules **diffuse** from the pure water into the sugar solution.

This is the same process as a gas diffusing across a room. The water molecules diffuse from where there are a lot of them to where there are not so many. The sugar molecules would do the same if they could. But they cannot.

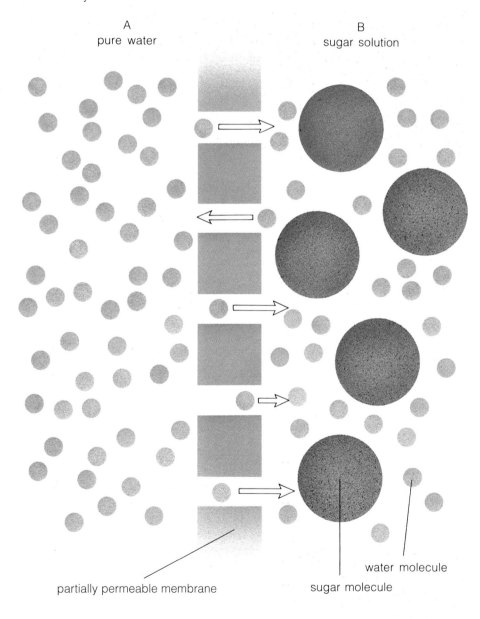

A
pure water

B
sugar solution

partially permeable membrane

sugar molecule

water molecule

Fig. 33.1 Osmosis

The net movement of water molecules into the sugar solution is called osmosis

The overall result of all this bumping around of molecules is that water molecules move from the pure water into the sugar solution. This process is called **osmosis**.

Osmosis can be defined as *the net movement of water molecules, through a partially permeable membrane, from a place where there is a high concentration of water molecules to a place where there is a lower concentration of water molecules.*

Osmosis is really just a special sort of diffusion, where only water molecules can diffuse through a membrane.

INVESTIGATION 33.1

Osmosis and visking tubing

Fig. 33.2

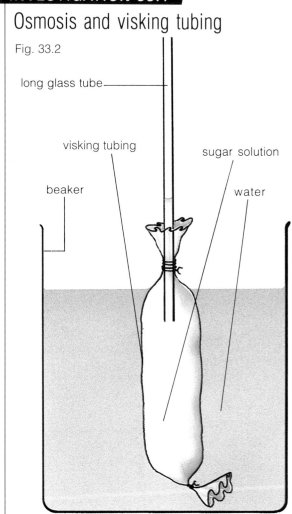

long glass tube

visking tubing

sugar solution

beaker

water

1 Set up the apparatus as in Figure 33.2. You must take great care to tie your knots very tightly, so that there is no chance that any liquid can leak out. You must also wash the outside of the visking tube after filling it and before putting it into the beaker, in case any sugar solution has got on to the outside of it.

2 Mark the level of the liquid in the tube. At 2 min intervals, measure any increase in height above this initial height. Record your results in a chart.

3 When you have about 15 readings, or when the liquid has stopped rising, stop collecting results. Draw a line graph of your results.

Questions

1 Explain, as fully as you can, why the liquid rises up the glass tube.

2 Did the liquid rise at a steady rate? If not, suggest some reasons for any variations.

3 What would you expect to happen, and why, if you set up this experiment:
   a with pure water in the tubing, and a concentrated sugar solution in the beaker?
   b with equal concentrations of sugar solution in the tubing and the beaker?
   c with a concentrated sugar solution in the tubing and a dilute sugar solution in the beaker?

Questions

1 Write down definitions of:
   a diffusion   b osmosis

— E X T E N S I O N —

2 A piece of visking tubing is filled with starch solution. The tubing is tied tightly at both ends, and then put into a beaker. The beaker is filled with iodine solution. The concentration of water molecules in the two solutions is the same.
   a Draw a labelled diagram of what the apparatus would look like at this stage.
   Starch molecules are too big to get through visking tubing. Iodine molecules can get through.

   b Explain what you think will happen to:
      i the water molecules
      ii the starch molecules
      iii the iodine molecules
   When starch and iodine are mixed together, a blue-black colour is produced. Iodine solution is brown.
   c Draw and colour the apparatus as it would look at the end of the experiment.

34 OSMOSIS AND LIVING CELLS

All living cells are surrounded by a partially permeable membrane. Osmosis can occur through this membrane.

Living cells are surrounded by a partially permeable membrane

All living cells are surrounded by a **cell surface membrane**. This membrane is partially permeable. It will let water molecules through, but many other molecules are not allowed to pass through freely. The contents of a living cell are a fairly concentrated solution of proteins, sugars and other substances in water. These solute molecules are not allowed out of the cell. They cannot get through the cell surface membrane.

Fig. 34.2 A plant cell in pure water. The cell swells and becomes turgid. The strong cell wall stops it from bursting.

Fig. 34.1 Plant and animal cells. The cell surface membrane is a partially permeable membrane.

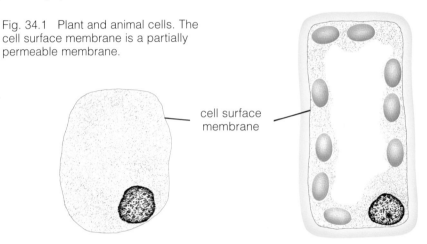

cell surface membrane

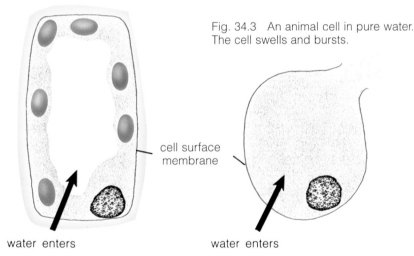

cell surface membrane

water enters

Fig. 34.3 An animal cell in pure water. The cell swells and bursts.

water enters

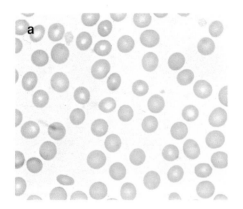

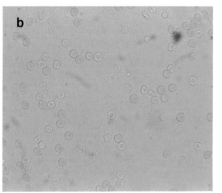

Fig. 34.4 Red blood cells in (a) a solution of the same concentration as their cytoplasm, and (b) in pure water. The cells in (a) look normal. Because the solutions on either side of their cell surface membranes are of equal concentration, they neither gain nor lose water. The cells in (b) are bursting. Water is entering them by osmosis, making them swell and break their membranes.

Cells take up water if put into pure water

What might happen if you put a cell into pure water? The situation is like that in Figure 7.1. Water molecules will tend to move from where they are in high concentration – outside the cell – to where they are in a lower concentration – inside the cell. Water moves by osmosis into the cell. As the cell surface membrane will not let other molecules out, the cell gets fuller and fuller. If this goes on for long, it may even burst. This does actually happen if you put red blood cells in pure water. Water goes into them by osmosis and they burst.

A red blood cell is an animal cell. Plant cells also have partially permeable membranes. So plant cells also take up water by osmosis if you put them into pure water. But they do not burst. This is because every plant cell has a tough, outer covering called a **cell wall**. The cell wall is strong enough to stop the cell bursting, even if it takes up a lot of extra water. Plant cells in pure water simply get very full. The extra water which goes into them makes the contents of the cell push out against the cell wall. The cell becomes firm and rigid. It is said to be **turgid**.

Cells in a concentrated solution tend to lose water

Imagine a cell in a solution which is more concentrated than itself. Now the higher concentration of water molecules is *inside* the cell. So water molecules tend to diffuse out of the cell. Remember, many molecules dissolved in water, either inside the cell or outside, cannot get through the cell surface membrane. Only the water molecules are allowed through. As water leaves the cell, the cell starts to shrink. A red blood cell in a concentrated sugar solution looks very shrivelled. Other animal cells behave in the same way.

But, once again, the cell wall of a plant cell makes it behave rather differently. Just as in an animal cell, water leaves the plant through its partially permeable membrane. The contents of the cell shrink. But the cell wall stays quite firm. It collapses inwards a bit, but not much. As the cell loses water, it becomes less rigid. It gets floppy, or **flaccid**. If the plant cell loses a lot of water, its contents may shrink so much that the inside parts pull away from the cell wall. This is called **plasmolysis**.

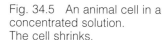
Fig. 34.5 An animal cell in a concentrated solution. The cell shrinks.

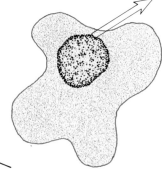
water leaves the cell

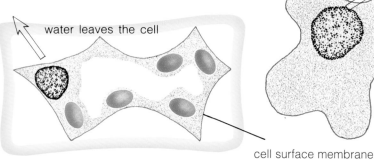
water leaves the cell

cell surface membrane

Fig. 34.6 A plant cell in a concentrated solution. The cell contents (but not the cell wall) shrink. The cell becomes flaccid. If the contents shrink so much that the cell surface membrane is pulled away from the cell wall, the cell is said to be plasmolysed.

Fig. 34.7 Red blood cells in a concentrated solution. Water has moved out of their cytoplasm, by osmosis, through their cell surface membranes into the solution surrounding them. The cells have become shrunken.

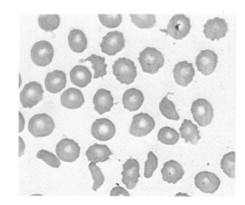

INVESTIGATION 34.1

Osmosis and raisins

Raisins are dried grapes. They are made up of plant cells. The grapes are left in the sun to dry. A lot of the water inside their cells evaporates, leaving each cell full of a very concentrated sugar solution.

1   Draw a raisin.
2   Put some raisins into a petri dish and cover them with water. Leave them until your next lesson.
3   Draw a raisin after soaking it in water.

Questions

1   Where were the following when you put the raisins into water:
   a a concentrated solution
   b an extremely dilute solution
   c partially permeable membranes?
2   Explain, as fully as you can, why the raisins became swollen after soaking in water.

Questions

1   The diagram below shows a plant cell which has just been put into a concentrated sugar solution.

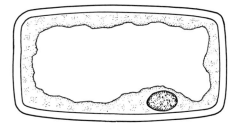

   a Copy the diagram and label: concentrated sugar solution; cell wall; cell surface membrane; fairly concentrated solution.
   b The diagram below shows the same cell after being in the solution for 30 minutes. Copy the diagram and label: concentrated sugar solution (in two places); cell wall; cell surface membrane.

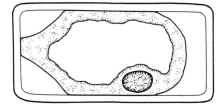

   c Which part is the partially permeable membrane?
   d Which cell is the flaccid cell?
   e How could you make the flaccid cell turgid?
2   *Amoeba* is a single-celled organism made up of a cell about 0.1 mm across. It lives in fresh-water ponds. Like all cells, it is surrounded by a partially permeable cell surface membrane, surrounding a fairly concentrated solution of proteins and other substances in water. *Amoeba* has a small, water-filled space inside it called a contractile vacuole. The contractile vacuole keeps filling up with water, and then emptying this water outside the cell. It does this non-stop, at frequent intervals.
   a Explain why you think that *Amoeba* needs a contractile vacuole, and what might happen to it if it did not have one.
   b A different kind of *Amoeba* lives in the sea. It does not have a contractile vacuole. Why do you think this is?

35 SUPPLYING CELLS

Cells need water, food and oxygen. In small organisms cells can get these things by diffusion. In large organisms a transport system is needed.

Cells need a supply of water, food and oxygen

All cells need water and food. Most cells need oxygen. These substances get into the cell through its cell surface membrane. A cell on its own, such as *Amoeba*, gets its food, water and oxygen from the water in which it lives. The molecules go into the cell by **diffusion**. Because cells are very small it takes hardly any time at all for the molecules to diffuse right into the middle of the cell.

Large organisms need transport systems

In a large organism, such as fish, the cells in the middle of its body are a long way from the water around it. It would take far too long for oxygen to diffuse into them from the water. So large animals and plants need a **transport system**. The transport system carries substances around their bodies and supplies every cell with whatever it needs. In many animals, including humans, this transport system is the **blood system** or **circulatory system**.

The human blood system is made up of the heart and blood vessels

The blood system is made up of thousands of tubes which carry blood to every cell in the body. These tubes are called **blood vessels**. There are several different sorts of blood vessels. There is a pump (the heart) which keeps the blood moving swiftly through these vessels. It works continuously throughout life.

Humans have a double circulatory system

The heart is really two pumps side by side. One side pumps blood to the head and body. The other side pumps blood to the lungs.

The two pumps are closely joined together, and pump with exactly the same rhythm. But blood cannot get directly from one side to the other.

To understand how the system works, look at Figure 35.1. Begin in the right-hand side of the heart (the left-hand side of the diagram). The heart pumps blood out of its right-hand side, along a vessel to the lungs.

In the lungs, the blood collects oxygen from the air you breathe in. The blood goes back along another vessel to the heart. This time it goes into the left-hand side of the heart.

This side pumps the blood out along another vessel, this time to the head or some other part of the body. Here, cells will take oxygen from the blood. The blood which has had some oxygen removed flows along yet another vessel which carries it back to the right-hand side of the heart. The whole journey now begins again.

So, on one complete journey round the body, the blood goes through the heart twice – first one side and then the other. This arrangement is called a **double circulatory system**. All mammals have a blood system like this.

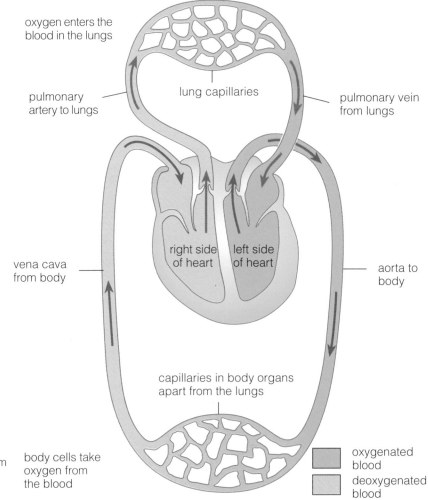

Fig. 35.1 The double circulatory system of a mammal.

oxygen enters the blood in the lungs

pulmonary artery to lungs

lung capillaries

pulmonary vein from lungs

vena cava from body

right side of heart

left side of heart

aorta to body

capillaries in body organs apart from the lungs

body cells take oxygen from the blood

oxygenated blood

deoxygenated blood

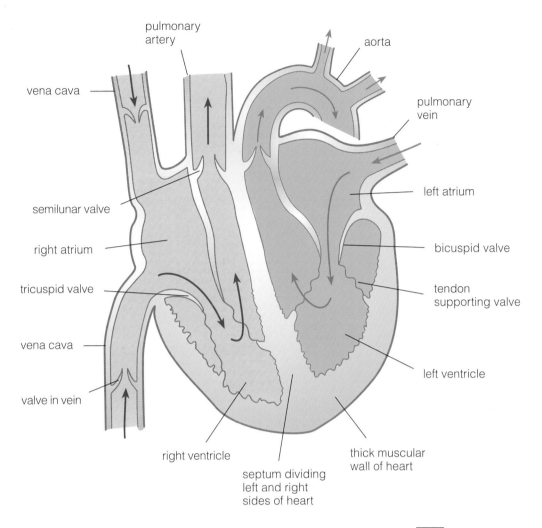

Fig. 35.2   A vertical section through a mammal's heart.

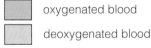   oxygenated blood

deoxygenated blood

Double circulatory systems keep blood moving fast

Fish do not have a double circulatory system. Their blood goes only once through the heart on one complete journey round the body. The heart pumps the blood to the gills. It picks up oxygen from the gills. From the gills, the blood continues around the body, without going back to the heart first.

A double circulatory system is a really good arrangement for keeping blood moving at a good pace around the body. Blood has to go to the lungs to collect oxygen before it goes to the other cells in the body, or it would not have any oxygen to give them. But if it went straight to these cells from the lungs, it would lose a lot of the speed

and pressure which the heart had given it. By going back to the heart again after going to the lungs, the blood is given another boost to push it round the body quickly. This enables the blood to carry its oxygen swiftly to all the cells which need it. It is no coincidence that it is the most active, fast-moving animals such as birds and mammals, which have double circulatory systems. The quick, efficient supply of oxygen to all their cells means that these cells can respire rapidly, producing energy for movement and keeping warm.

Questions

1   Besides speed, what other advantage does a double circulatory system have over the type found in fish?

2   Suggest how the structure of a fish's heart might differ from the structure of a mammal's heart. Find out if you are right.

36 THE HEART

Your heart beats all your life, pumping up to 25 litres of blood every minute. Like all pumps, it has a power source – the heart muscle – and valves to ensure flow in one direction.

The most reliable pump ever!

Mass:
  between 250 g and 400 g
Flow rate:
  up to 25 dm³/min
Pressure developed:
  right side up to 5 kPa
  left side up to 26 kPa
Rate of beating:
  65 beats per minute on average
Total number of beats in a 70 year lifetime:
  2 391 480 000
Fuel:
  glucose and oxygen
Fuel consumption:
  26 cm³ oxygen per minute at cruising speed
  150 cm³ oxygen per minute at high speed
Servicing intervals:
  with care, should last a lifetime
Maintenance instructions:
  use regularly at higher beat rates;
  keep supply vessels clear by avoiding saturated fats in diet; do not smoke;
  keep body weight at reasonable level.

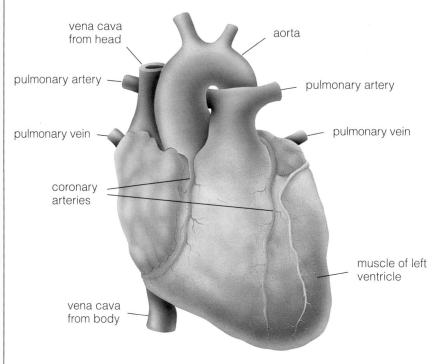

Fig. 36.1   External view of the human heart

The heart has four chambers

The heart contains four spaces called chambers. The two upper ones are small, with thin walls. These are the right and left **atria**. These are the parts of the heart into which blood flows first.

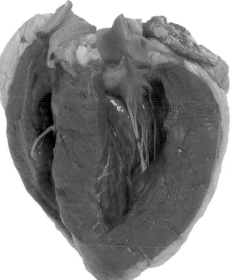

Fig. 36.2   A sheep's heart, cut open to show the chambers and valves. Try to identify: left and right ventricles; left atrium (this is very small, and has not been cut open); openings to the aorta and the pulmonary artery; semi-lunar valve in the aorta; bicuspid or mitral valve at the top of the left ventricle; tendons supporting this valve. Notice how thick the wall of the left ventricle is, and the septum which separates the two sides of the heart.

Beneath the atria are the two **ventricles**. These are the real power-houses of the heart. Their thick walls are made of cardiac muscle. When this muscle shortens·or contracts, it squeezes inwards on the blood inside the ventricles. This pushes the blood out of the heart. The left ventricle is quite a bit larger than the right ventricle. The main reason for this is that the left ventricle has to be able to produce a much bigger force, as it has to push blood all around the body. The right ventricle has to push blood only to the lungs, which are very close to the heart. To produce a big force, a lot of muscle is needed, so the left ventricle has extra thick walls.

The atria need only thin walls, as they have to produce only enough force to push blood into the ventricles below them.

Valves keep blood moving in the right direction

All pumps need valves, to keep the fluid they pump moving in one direction. The heart has four sets of valves.

Two of these sets are between the atria and the ventricles. On the left side is the **mitral** or **bicuspid** valve. The one on the right is the **tricuspid** valve. These valves make sure that blood can flow easily from the atria into the ventricles, but not in the opposite direction. Figure 36.3 shows how they work.

The other two sets of valves are in the entrances to the two large arteries through which the blood leaves the heart. They are called **semi-lunar** (half-moon) valves because of their shape. They allow blood to flow out of the ventricles into the arteries, but not back the other way.

Fig. 36.3 How the bicuspid valve works. Only the left-hand side of the heart is shown.

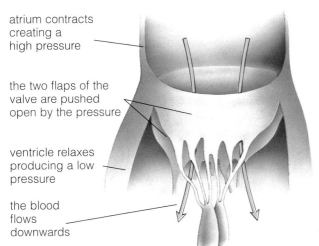

atrium contracts creating a high pressure

the two flaps of the valve are pushed open by the pressure

ventricle relaxes producing a low pressure

the blood flows downwards

atrium relaxes

blood is forced up against the valve flaps pushing them together

ventricle contracts producing a high pressure

the tendons attached to the valves will hold them, so that they cannot swing too far upwards

a When the ventricle relaxes and the atrium contracts, blood is squeezed against the flaps of the valve. This pushes the flaps downwards, and the blood flows through from the atrium to the ventricle.

b When the ventricle contracts, it pushes the blood up against the valves. They will be pushed upwards. The two flaps will be pushed tightly together, so no blood will be able to flow upwards.

Tendons hold valves in position

The valves between the atria and the ventricles have strong cords attaching them to the wall of the ventricle. These cords are called **tendons**. The tendons stop the valves from swinging too far up into the atria. When the ventricles contract, blood is squeezed up against the valves. The valves are pushed upwards. They swing up until the tendons are pulled taut. The tendons hold them in just the right position to make a barrier so that blood cannot get through.

INVESTIGATION 36.1

Looking at a mammal's heart

Sheep and pig hearts are sold in super-markets and butcher's shops.

1 Describe the shape and size of the heart.

2 The heart is made almost entirely of dark red muscle. What is the name of this kind of muscle?

3 Running over the surface of the heart are branching blood vessels. What are these called?

4 If the heart has not been too badly damaged, you will be able to find the two atria at the top of it. How thick are their walls compared with the thickness of the ventricle walls? Why is there a difference between them?

5 Look at the large tubes coming out of the top of the heart. These are the **aorta** and the **pulmonary artery**. Push a pencil – or better still, a finger – down through these tubes to find out which part of the heart they are coming from. Which part is it? Where would each of these arteries take blood to in the living animal?

6 Using scissors, cut through the walls of the aorta, continuing the cut to cut right through the wall of the left ventricle. Look for the thin, floppy **semi-lunar valves** in the wall of the aorta. What is their function?

7 The inside of the ventricle has little bumps on it. Attached to these are white cords, called **tendons**. What is their function? Attached to the top of the tendons, at the top of the ventricle, are more valves. What is the function of these valves?

8 Cut into the right ventricle. Is it bigger or smaller than the left ventricle? Why?

9 Notice the thick wall of muscle which separates the right and left ventricles. What is this dividing wall called? There is no way for blood to get through it. So how does blood get from one side of the heart to the other?

37 HEART-BEAT

Your heart beats at around 65 times a minute. It speeds up or slows down according to how much work you are doing. A blockage in the vessels supplying the heart muscle with oxygen can cause a heart attack.

Heart rate is controlled by a pacemaker

The rate at which your heart beats is controlled by a patch of muscle in the right atrium, called the **pacemaker**. When you are resting, the pacemaker beats at a speed of about 65 beats per minute (yours might be slower or faster than this). When you exercise, the brain sends messages along nerves to the pacemaker to make it beat faster.

Sometimes, the heart's pacemaker does not work properly, perhaps because it has become diseased. A person with this problem can be fitted with an artificial pacemaker. This is a small, battery-powered device which gives out electrical signals at regular intervals. The heart muscle responds to the signals by contracting at the same rate.

Electrocardiograms

Electrocardiograms, or ECGs, give information about the way the heart is beating. Electrodes attached to the skin of the chest, near the heart, pick up electrical signals from it as messages travel from the pacemaker to the rest of the muscles. ECG A is a normal one, showing a normal steady heartbeat. ECG B shows a heart which is contracting in an un-coordinated, haphazard way. This is called **fibrillation**, and is fatal if not treated immediately.

Fig. 37.1 Electrocardiograms

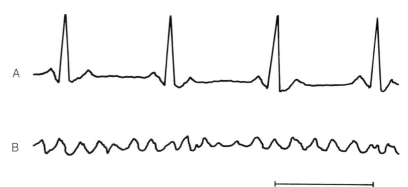

1 second

EXTENSION

How does exercise affect the rate at which your heart beats?

The more exercise you do, the more oxygen your muscles need. To supply them with this extra oxygen, your heart beats faster than usual. This pushes blood quickly to the lungs to collect oxygen, and then to the muscle cells to deliver it.

Fig. 37.2 Finding your pulse. Let your left wrist flop downwards. Using two or three fingers of your right hand – not your thumb – find the tendon near the outside of your wrist. Feel down in the hollow beside it.

1 Rest quietly, until you are really relaxed. Find your pulse and count how many times your heart beats in 1 min. Record this in a results chart.
2 Still resting, record your pulse rate again, and fill it in on your chart.
3 Carry out some exercise until you feel slightly out of breath.

You could do some press-ups, or step up and down on to a chair.
4 As soon as you finish exercising, take your pulse rate again. Keep taking it every minute, filling in each result on your chart. Keep taking the rate until you have two readings which are nearly the same as the ones you began with.

Questions

1 Draw a line graph to show your results.
2 What is your normal pulse rate at rest?
3 What was your maximum pulse rate during this investigation?
4 After you stopped exercising, how long did it take your pulse rate to return to normal?
5 Explain why your heart beats faster then normal when you exercise.

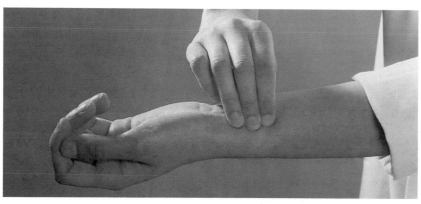

Heart sounds give information about how the valves are working

When a person has a medical checkup, the doctor will usually listen to their heartbeat. A normal heartbeat has two sounds when heard through a stethoscope. The first is made by the valves between the atria and ventricles as they snap shut. The second is a similar, but slightly quieter, sound made by the closure of the valves in the aorta and pulmonary artery.

Sometimes, an extra sound can be heard. This is called a **heart murmur**. Many people who have heart murmurs are perfectly healthy. But sometimes a murmur may be caused by turbulence as blood flows past a faulty valve. If the problem is very bad, then artificial heart valves can be put into the heart to replace the faulty ones.

Fig. 37.3 Heart surgery being carried out.

Heart attacks are caused by damaged coronary arteries

The muscle in the heart needs a continuous supply of oxygen. If it does not get it, it may stop beating, or begin to beat very irregularly. This is a **heart attack**, or **cardiac arrest**.

Oxygen is supplied to the heart muscle by the coronary arteries (see Figure 36.1). If these are blocked, then the amount of blood flowing through them is reduced, and the heart muscle runs out of oxygen. The most usual cause of a blockage in a coronary artery is the build-up of a fatty substance called **cholesterol**, which narrows the artery and stiffens its wall. This is most likely to happen in people who smoke, or eat a lot of animal fat (such as butter), and who do little exercise. If the problem is realised in time, a badly damaged coronary artery can be by-passed. A healthy piece of blood vessel is taken from another part of the patient's body. This is often a leg. It is then attached to the heart in such a way that it carries blood past the damaged artery to the heart muscle which needs it. Coronary by-pass operations have saved many lives.

Heart attacks are still one of the major causes of death in Britain. You can cut down your chances of having a heart attack when you are older by the way you live now. Do not smoke; do not eat too much animal fat; and do some regular exercise.

Questions

Read the following passage. Use the information in the passage, and your own knowledge, to answer the questions.

The benefits of by-pass surgery

The coronary artery supplies the muscles of the heart with oxygenated blood. Several smaller coronary vessels branch off from the coronary artery. These vessels sometimes become partially blocked. The narrowing of the coronary vessels reduces blood flow to the heart muscles. One symptom of coronary artery problems is pain in the region of the heart and left arm. This pain is called angina. Patients with poor coronary circulation are unable to perform vigorous exercise, because the supply of oxygen to the heart muscle is inadequate. They run a high risk of a heart attack, which could be fatal.

It has been accepted for some time that coronary by-pass surgery can relieve the pain of angina and improve the quality of life of sufferers from coronary artery problems. Comparative studies have been carried out to see what happened to patients who had this surgery, or were just treated with drugs. The studies found that, five years after treatment, 92% of patients who had had coronary by-pass surgery were still alive. Only 83% of those treated with drugs were still alive. Twelve years after treatment, 71% of surgery patients and 67% of those treated with drugs were still alive.

This suggests that coronary by-pass surgery is a better treatment for sufferers from coronary problems than the use of drugs alone. However, twelve years after treatment the difference between the two groups is not very great. This is partly because the by-pass grafts tend to deteriorate with time. This may mean that a second by-pass operation is needed, which is much riskier than the first. The implication of these studies is that mild coronary problems are probably better treated at first with drugs rather than surgery. This means that if a by-pass operation does turn out to be needed later, there is less likelihood that a second, risky, one will have to be used. But severe coronary artery disease should be treated immediately with by-pass surgery.

1 a Why does coronary artery disease limit a person's ability to perform vigorous exercise?
b Give one symptom of coronary artery disease.
c Give two factors which may increase a person's risk of suffering from coronary artery problems.
d What is 'coronary by-pass surgery'?
e What evidence is there from this study that coronary by-pass surgery is more successful than the use of drugs in treating coronary artery disease?
f Explain why it is recommended that mild coronary artery disease should initially be treated with drugs rather than surgery.

38 BLOOD VESSELS

Blood is carried around the body in tubes called blood vessels. Arteries take blood away from the heart. Veins return blood to the heart. Capillaries, which link arteries and veins, deliver blood to the body tissues.

Blood vessels take blood all over the body

In humans, the blood is contained in tubes called **blood vessels**. There are three types – arteries, veins and capillaries.

Blood leaves the heart by one of two large **arteries**. These are the **pulmonary artery**, which goes to the lungs, and the **aorta**, which goes to the rest of the body. The pulmonary artery divides into two, one branch going to each lung. But the aorta splits into many smaller arteries, each delivering blood to one of the many body organs. Figure 38.7 shows just some of these many branches.

On reaching its destination, each artery divides into many tiny vessels called **capillaries**. These penetrate right inside every tissue, forming a network which takes blood close to every individual cell. A network of capillaries is sometimes called a **capillary bed**.

The capillaries then gradually join up with each other to form larger vessels called **veins**. Veins carry blood back to the heart. The veins from the body empty into one of the two large veins called **venae cavae**, which empty into the right hand side of the heart. The veins from the lungs are called **pulmonary veins**, and these empty into the left-hand side of the heart.

Arteries have muscular walls

Arteries carry blood away from the heart. This blood is at high pressure, travelling fast, so the walls of arteries must be strong. Artery walls are also elastic, as this allows them to give a little as the blood surges through. You can feel this happening when you feel your pulse. Each surge of blood from the heart pushes outwards on the artery wall, and in between surges the wall recoils inwards again. This helps to keep the blood flowing smoothly, as the elastic recoil of the wall gives the blood an extra 'push' in between the pushes from the heart.

Capillaries have very thin walls

The job of capillaries is to deliver oxygen, food and other substances to body tissues, and to collect waste materials from them. So their walls must be really thin, to let these substances move into and out of them. Often, these walls are only one cell thick. They have small gaps in them, too, to make it even easier for substances to pass through. Capillaries are very tiny, in many cases only 7 μm (0.007 mm) in diameter.

Fig. 38.1 An artery

thick outer wall

thick layer of muscles and elastic fibres

small lumen

smooth lining

Fig. 38.2 A capillary

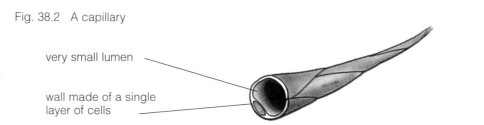

very small lumen

wall made of a single layer of cells

Fig. 38.3 A vein

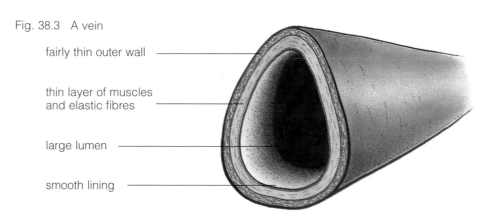

fairly thin outer wall

thin layer of muscles and elastic fibres

large lumen

smooth lining

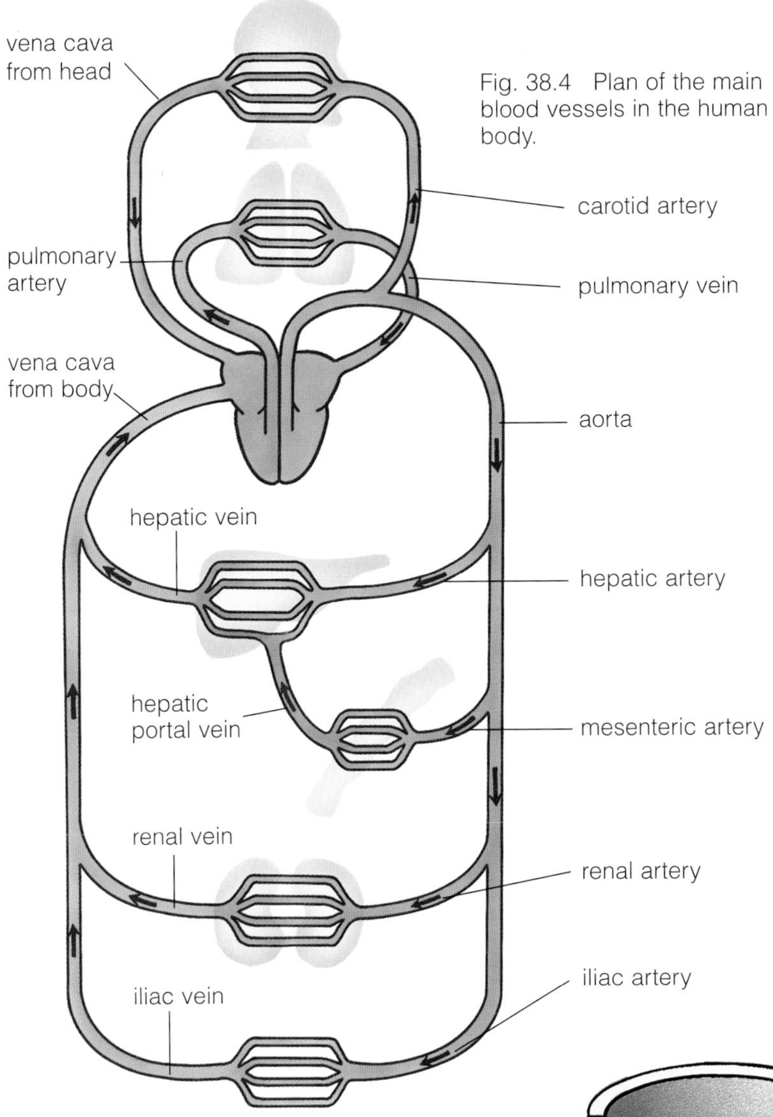

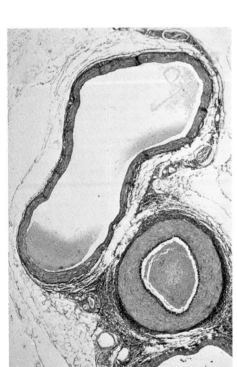

vena cava from head

pulmonary artery

vena cava from body

hepatic vein

hepatic portal vein

renal vein

iliac vein

Fig. 38.4 Plan of the main blood vessels in the human body.

carotid artery

pulmonary vein

aorta

hepatic artery

mesenteric artery

renal artery

iliac artery

Veins have thin walls, and valves

Veins carry blood back to the heart from the tissues. By the time it enters the veins, the blood has lost most of its impetus, and is travelling quite slowly at low pressure. So veins do not need to have strong, elastic walls like arteries. Instead, they are wide and have valves to help the blood to flow through them easily. The valves stop blood going backwards, and make sure that blood flow is always towards the heart. Many veins lie between muscles and this also helps to keep the blood flowing in them. The big veins in your legs lie between the leg muscles. When you walk, the movements of these muscles squeeze in on the veins, pushing the blood along inside them.

Fig. 38.5 Valves in a vein

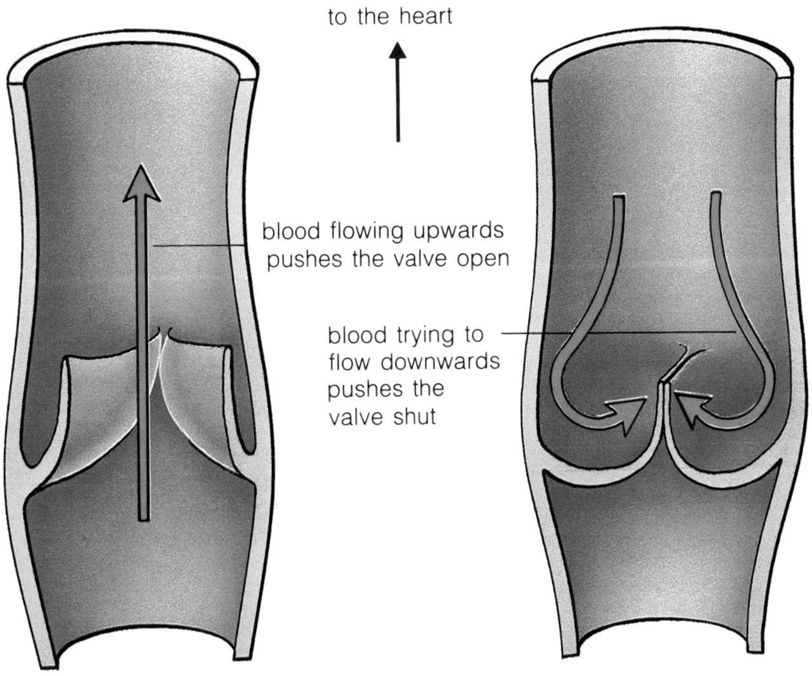

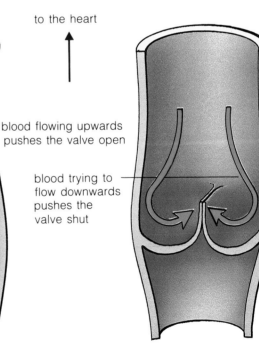

to the heart

blood flowing upwards pushes the valve open

blood trying to flow downwards pushes the valve shut

Question

1 The photograph in Figure 38.6 shows a section through two blood vessels. One is an artery and one a vein. Which is which? Give as many reasons as you can for your choice.

Fig. 38.6

Problems with blood vessels

Problems with blood vessels are one of the commonest medical complaints, especially in people who are overweight and unfit. However, you do not need to be overweight or unfit to suffer from a **bruise**. A bruise is caused when blood capillaries in or just under the skin are broken, and blood leaks from them. The dark colour is the blood seen through the skin. As the blood is gradually broken down the bruise changes colour.

Varicose veins are raised veins near the surface of the skin, usually on the legs. They are caused when the valves in the leg veins stop working properly, allowing blood to flow the wrong way. The blood collects in these veins instead of flowing back to the heart, and stretches their walls.

Hardening of the arteries is medically known as **atherosclerosis**. It is a stiffening of the artery walls so that they cannot stretch and recoil as the blood pulses through. It can be caused by a build-up of cholesterol. If this happens in the coronary arteries a heart attack may result. The artery walls become weaker, and are more likely to burst if the blood inside them is at high pressure. If this happens in the brain, blood spills out and damages brain cells. This is called a **stroke**.

A **thrombosis** is a blood clot. Blood clots inside blood vessels can be dangerous, because they may block important vessels such as the coronary arteries, causing a heart attack.

Questions

1 For each of the following, state whether they are associated with arteries, veins or capillaries:

a valves

b pulsating, muscular walls

c blood taken away from the heart

d walls made of a single cell

e blood taken towards the heart

f blood taken very close to every cell

g leaky walls

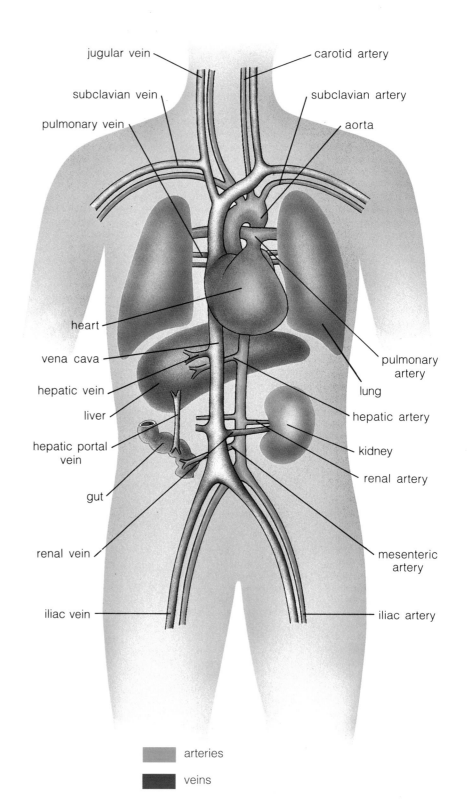

Fig. 38.7   The positions of some of the main vessels in the human body.

arteries

veins

39 BLOOD

Blood is made of a liquid in which float several kinds of cells. It transports all sorts of substances around the body, and defends you against diseases.

The liquid part of blood is called plasma

You might be forgiven for thinking that blood is a red liquid. But, in fact, if you see blood under a microscope you realise this is not quite true. The red part of blood is not actually liquid at all. Blood looks red because it contains red cells, which float in a pale yellowish liquid. The liquid part of the blood is called **plasma**. Plasma is mostly water. Many different substances are dissolved in the water. They include glucose, amino acids, vitamins, ions such as sodium and chloride, and blood proteins. These blood proteins have some important roles to play in helping the blood to clot, and in fighting disease.

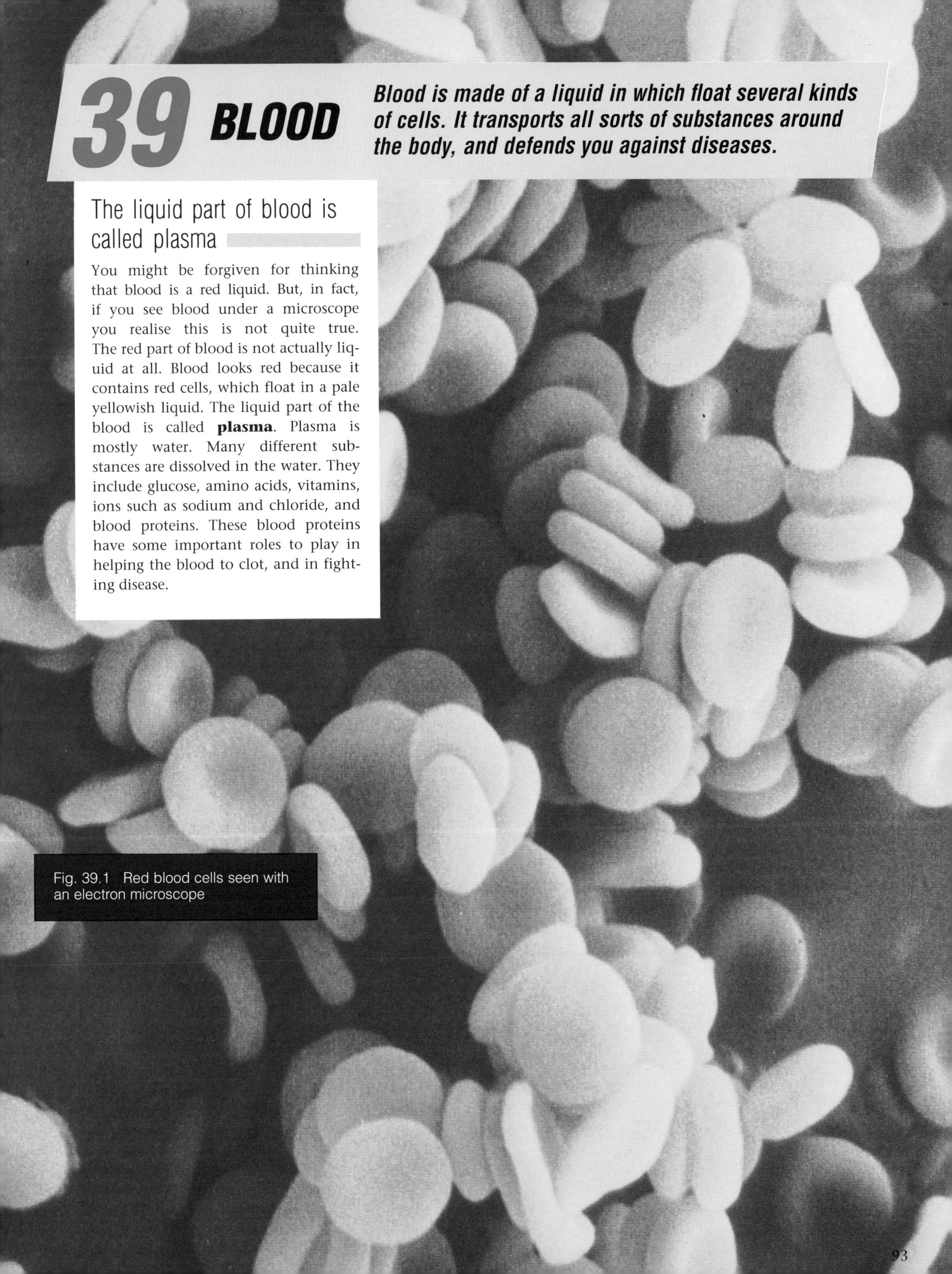

Fig. 39.1   Red blood cells seen with an electron microscope

Red blood cells transport oxygen

Red blood cells are much the most common cells in your blood. They are very small cells. You have about 5 million of them in 1 ml of blood!

Red cells are red because they are full of a protein called **haemoglobin**. Haemoglobin is sometimes called **Hb** for short. Hb combines with oxygen as the blood flows through the lungs, becoming **oxyhaemoglobin** or **oxyHb**. The blood flows out of the lungs, back to the heart, and then round the body. As it passes through the capillaries, the oxyHb gives up its oxygen to the body cells, becoming Hb again. OxyHb is bright red, and Hb is purplish red, so your blood actually changes colour as it goes round and round the body.

Red cells are very unusual because they have no nucleus. This is thought to be so that there is more room to pack in as much Hb as possible. Their shape gives them a large surface area, so that a lot of oxygen can get in or out of them very quickly. Their small size, and the fact that they are quite flexible, makes it possible for them to squeeze along even the smallest capillaries. They may have to go in single file if the capillaries are very tiny, but this is useful because it means that every red cell gets very close to the cells to which it is delivering oxygen.

Blood transports carbon dioxide

The blood also transports carbon dioxide. Body cells produce carbon dioxide when they respire. The carbon dioxide diffuses into the blood. Most of the carbon dioxide is transported in solution in the blood plasma. Some of this is in the form of carbon dioxide molecules, but most of it is as hydrogencarbonate ions. A small amount of the carbon dioxide combines with the haemoglobin inside the red blood cells. So red cells are not the main way of transporting carbon dioxide, but they do help. When the blood reaches the lungs, the carbon dioxide leaves the blood and diffuses into the air spaces in the lungs, before being breathed out.

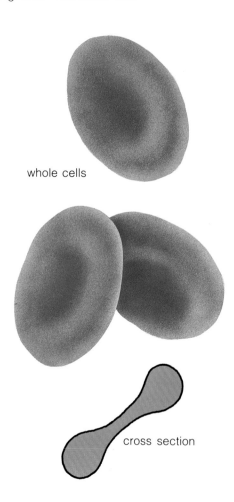

Fig. 39.2   Red blood cells

whole cells

cross section

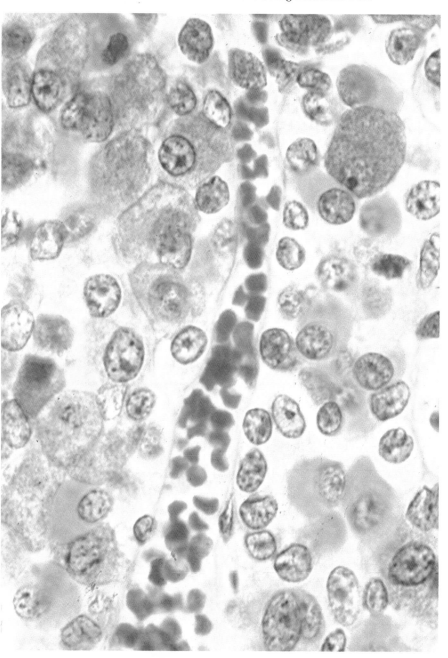

Fig. 39.3   Red blood cells in a capillary. The capillary runs from top to bottom through the centre. On either side are cells making up the tissue through which the capillary runs. Notice how small the red blood cells are, in comparison with the tissue cells.

White cells fight disease

There are far fewer white cells than red cells in your blood. White cells are slightly larger than red cells and they always have a nucleus. There are several different sorts, which can be divided into two main groups.

Phagocytes are irregularly shaped white cells, often with a lobed nucleus. They crawl actively around the body, squeezing in and out of capillary walls, and finding their way into every bit of you. Their job is to find and 'eat' any invading bacteria, or any of your own cells which have become damaged or worn-out. They do this by flowing around the bacterium, enclosing it in a vacuole, and then secreting enzymes on to it to digest it.

Platelets help in blood clotting

As well as red and white cells, the blood contains little cell fragments called **platelets**. These are smaller than red or white cells, and they do not have a nucleus – just cytoplasm surrounded by a cell surface membrane. Platelets come into action when a blood vessel is damaged, for example if you cut your skin. They react by secreting chemicals which stimulate the formation of a blood clot, and which encourage white cells to attack any bacteria that might invade through the broken skin. The platelets also help to block the wound by sticking to each other, to other blood cells, and to the walls of the damaged blood vessel.

Blood diseases

Some people don't have enough red cells in their blood. This makes them look pale, and they feel tired because not enough oxygen is being carried around their body. The disease is called **anaemia**. One very common cause of anaemia is not having enough iron in the diet. Iron is needed to make haemoglobin, the major component of red blood cells.

Leukaemia is a type of cancer which affects the white cells. So many white cells are made that there is not enough room in the blood for red cells. Leukaemia can be fatal, but is often treated successfully.

Lymphocytes also attack bacteria, but in a different way. They make proteins called **antibodies**. These are carried around the body in the blood and destroy foreign invading cells.

You can read more about how white cells help you to fight disease in Topic 217.

Fig. 39.4   White blood cells

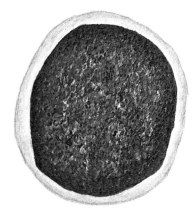

b. a lymphocyte

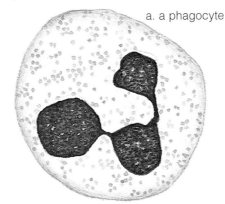

a. a phagocyte

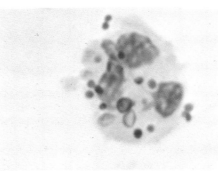

Fig. 39.5   A phagocytic white blood cell. The purple areas are the lobed nucleus. The smaller, redder objects are organisms which have been engulfed by the cell.

AIDS is another disease which affects the white cells. The AIDS virus reproduces inside lymphocytes and destroys them. This means that the body has lost its defence system, and is open to attack from any germs which come along. A person with AIDS eventually dies from one of these infections. You can find more about AIDS on pages 213–214.

Haemophilia is an inherited disease in which the blood does not clot properly. It can cause all sorts of problems. One of the most painful problems is caused by bleeding into the joints. Haemophilia is a sex-linked disease – only men suffer from it. It can be kept under control by giving regular doses of a substance called Factor 8, which is obtained from the blood of unaffected people. Haemophilia cannot be cured.

Fig. 39.6   Blood collected from donors is carefully labelled before storage. What information do you think needs to go on the label?

Questions

1   Briefly list the function of:
  a blood plasma
  b red blood cells
  c white blood cells
2   Explain why:
  a blood is red
  b blood in arteries is bright red, whereas blood in veins is purplish red
  c people feel tired if they do not have enough iron in their diet
  d someone with AIDS is likely to suffer from many different infections.

40 TISSUE FLUID AND LYMPH

Blood plasma leaks out of capillaries and forms tissue fluid. This is collected in lymphatic capillaries and returned to the blood.

Spaces between your cells are filled with tissue fluid

Capillaries leak. Tiny gaps in their walls allow blood plasma to leak out as the blood flows through them. The leaked blood plasma fills all the spaces between your cells. It is called **tissue fluid**.

Tissue fluid helps to carry substances between the blood and the body cells. Oxygen and glucose, needed by the cells for respiration, move from the blood to the cells. Carbon dioxide and other waste substances move from the cells to the blood.

The concentration of substances like glucose and water in the blood plasma is kept just right, in the processes known as **homeostasis**. As tissue fluid is made from blood plasma, it too has the correct concentrations of these important substances. So the cells are bathed in a fluid containing the correct concentrations of the substances that they need, allowing them to work really efficiently.

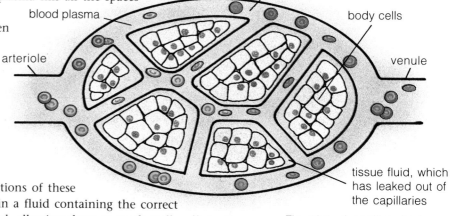

Fig. 40.1 A capillary bed

Fig. 40.2 Gas exchange through tissue fluid

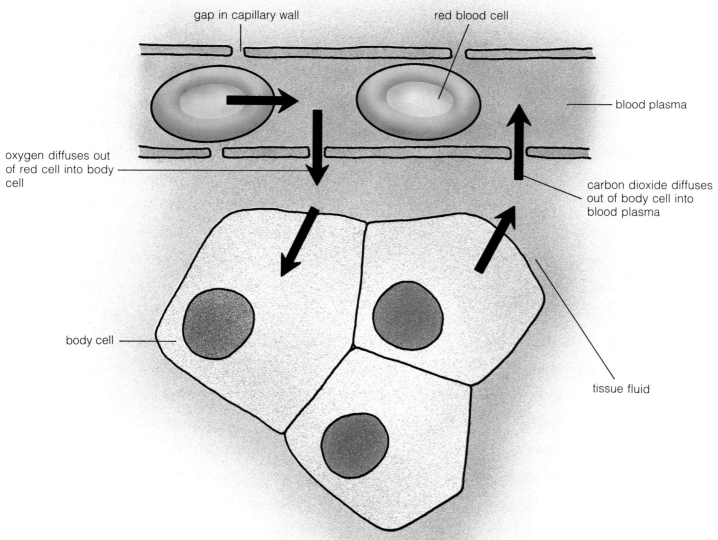

Tissue fluid is returned to the blood in lymph vessels

Blood capillaries are not the only tiny vessels running in amongst the body cells. There are also **lymphatic capillaries**. These collect up the tissue fluid from between the cells and drain into the lymphatic capillaries. These carry it to the neck. Here, they empty the fluid back into the blood. Lymphatic vessels have no pump to keep the fluid moving in them. They use the same method as veins. They lie between muscles, so that when the body moves, the muscles squeeze in on them. Valves in the lymphatic vessels make sure that the fluid only goes in one direction.

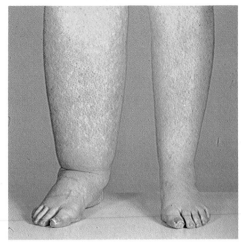

Fig. 40.3 This person has oedema. The tissue fluid between the cells in their right leg is not being carried away in the lymphatic vessels. The fluid builds up, making the leg swollen and puffy.

EXTENSION

Questions

The chart shows the changes in pressure inside the left atrium and left ventricle of a human heart during one heart beat.

a How long does one heart beat last?

b How many beats will there be in 1 min?

c For what proportion of one heart beat is the ventricle contracting?

d What is the maximum pressure reached in the ventricle in this heart beat?

e What is the maximum pressure reached in the atrium in this heart beat?

f Make a simple copy of this graph. On it, show the time in which you think the atrium is contracting. Show it in the same way as the ventricular contraction has been shown.

g Draw a vertical line on the graph to mark the time at which you think the valve between the atrium and ventricle will close.

h Draw another vertical line on the graph to mark the time at which you think this valve will open.

i Explain why you have drawn the two vertical lines in these positions.

j This graph shows what is happening in the left-hand side of the heart. Will the events in the right-hand side of the heart show a similar pattern at the same time?

k Will the pressures in the right-hand side of the heart be the same, higher or lower than shown on this graph? Explain your answer.

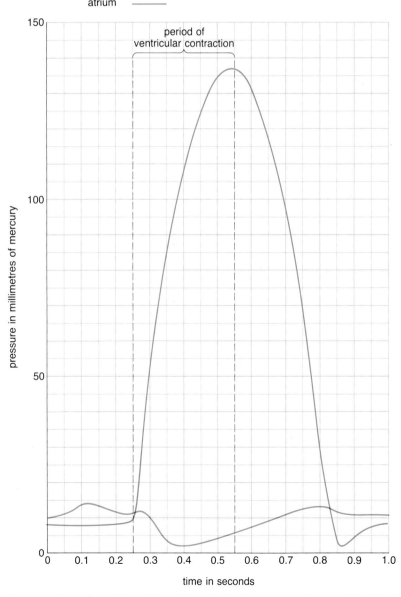

41 TRANSPORT IN PLANTS

Plants have two transport systems. Xylem transports water and minerals. Phloem transports sugars and amino acids. Transpiration pulls water up through the plant.

Plants have two transport systems

Plants have two separate transport systems. A network of **xylem vessels** transports water and mineral ions from the roots to all the other parts of the plant. **Phloem tubes** transport food made in the leaves to all other parts of the plant. Neither of these systems has a pump. Plants can manage without a heart to pump substances around their bodies, because they are not so active as animals. Muscle cells in an animal need rapid supplies of food and oxygen, but plant cells do not.

Neither xylem nor phloem transports oxygen. Plants do not have a special oxygen transport system. Oxygen gets to a plant's cells by diffusion. Both stems and roots contain xylem vessels and phloem tubes. In a stem, these are grouped into **vascular bundles** arranged in a ring. In a root, they are in the centre, forming a structure called the **stele**.

The **epidermis** is a layer of protective cells on the outside of the stem and root. The cells of the **cortex** are quite large, and often store starch. **Cambium** cells can divide, so that the root or the stem can grow wider.

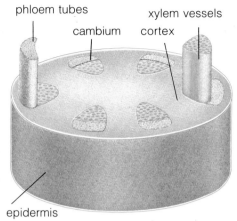

Fig. 41.1 Transverse section through a stem

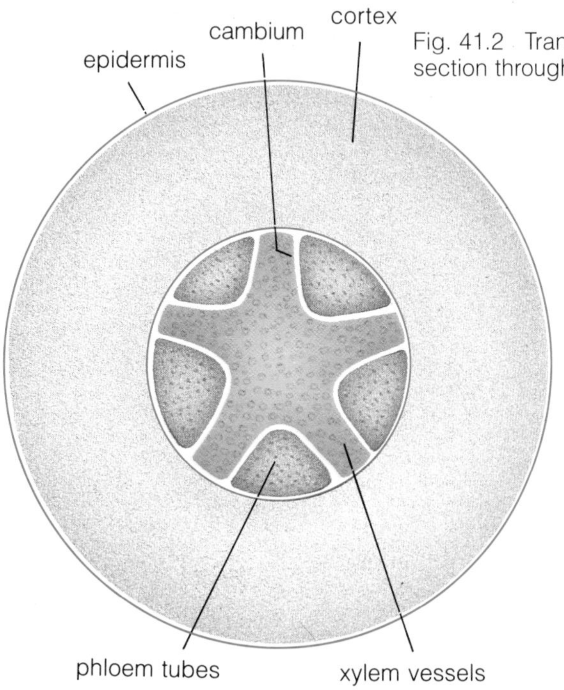

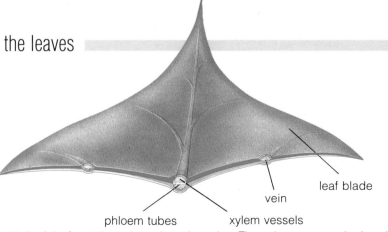

Fig. 41.2 Transverse section through a root

Xylem vessels carry water and minerals from the roots

Water gets into a plant through its roots. Near the tip of each branch of a root, there are thousands of tiny **root hairs**. Water from the soil moves into these hairs by osmosis. The small size and large number of root hairs gives them an enormous surface area. So a lot of water can get into the plant very quickly.

Mineral ions are dissolved in the water. These might include nitrate, phosphate, calcium and magnesium ions. These also enter the root hairs. They travel into the plant dissolved in water.

Once inside the plant's root, the water and ions move into the xylem vessels. The xylem vessels are like long drainpipes reaching all the way from the root to the tip of every leaf. They are made of dead cells, joined end to end.

Phloem vessels carry sugars from the leaves

The leaves of a plant make sugars by photosynthesis. These sugars are carried to other parts of the plant by the phloem vessels. Phloem vessels run from the leaves to every part of the plant. They also carry other substances made by the plant's cells, such as amino acids.

Phloem and xylem vessels often run side by side. A group of phloem and xylem vessels is called a **vascular bundle**. The veins in a leaf are vascular bundles.

Fig. 41.3 A leaf, cut through to show the veins. The veins are vascular bundles. They branch all over the leaf, taking water and collecting food from the leaf cells. The rigid, dead xylem vessels also act as a skeleton supporting the leaf blade.

Transpiration pulls water up through xylem vessels

Plants have no heart to pump water through their xylem vessels. Yet water travels upwards very fast through them. Try standing a piece of freshly-cut leafy celery in a beaker of ink. If your celery is healthy, you can actually watch the ink moving up through the vascular bundles.

What makes water move up through xylem vessels? When the water reaches the top of the xylem vessels it goes into the leaves. Leaves contain large air

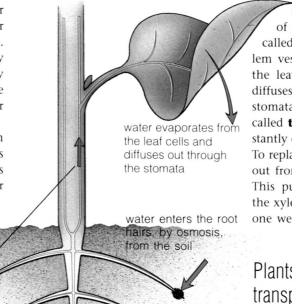

water evaporates from the leaf cells and diffuses out through the stomata

Fig. 41.4 The transpiration stream

the water travels up through the xylem vessels in the root, stem and leaf

water enters the root hairs, by osmosis, from the soil

spaces. They also have hundreds of small holes on their undersides, called **stomata**. Water from the xylem vessels evaporates when it gets to the leaves. It turns into gas. The gas diffuses out through the air spaces and stomata, into the air. This process is called **transpiration**. So water is constantly evaporating from a plant's leaves. To replace this water, more water moves out from the top of the xylem vessels. This pulls water in at the bottom of the xylem vessels. It is as though someone were 'sucking' water up the plant.

Plants may need to reduce transpiration

Transpiration is useful to a plant. It helps to draw water into the roots, and through the plant. Another useful effect of transpiration is that the evaporation of water from the leaves has a cooling effect. This may be very important to plants growing in hot conditions.

But if water is lost from a plant by transpiration faster than it can be taken up through the roots, then the plant's cells may become short of water. The cells become flaccid (Topic 34). When the cells are full of water, they are firm and rigid and help to support soft parts of the plant, such as its leaves. But when the cells are flaccid, the leaves become soft and floppy. The plant wilts. If the plant is short of water for too long, it may die.

Most plants are able to reduce the rate of transpiration if water is in short supply. They can close their stomata, so water vapour cannot escape through them from the leaves. They usually have a layer of wax on their leaves, so water vapour cannot escape through the leaf surface. Plants which live in dry places, such as deserts, may have a very thick layer of wax, and also other special features to help to cut down the loss of water from their leaves.

INVESTIGATION 41.1

Using a potometer to compare rates of water uptake

A potometer measures water uptake. There are many different types of potometer. But all of them have a tube into which a plant shoot can fit tightly. The tube is full of water. It is connected to a capillary tube. By watching how fast the water moves along the capillary tube, you can see how fast the shoot is taking up water.

1 Fill your potometer with water. There must be no air bubbles, and any joints must be completely airtight. Put Vaseline around any that you are not sure about.
2 Cut a leafy shoot from a plant. (Wash any Vaseline off your fingers first.) Try to choose a shoot of the right thickness to fit into your potometer. A slanting cut is often more successful than a straight one.
3 Push your shoot firmly into the potometer. It must make a really tight fit. If it does not fit, cut another shoot, or ask for help.
4 Leave the potometer for a few minutes to settle down. If everything is airtight, you should see the water meniscus moving along the capillary tube towards the plant. While you are waiting, draw up a results chart.

5 When the meniscus is moving smoothly, begin to record its position every minute for about 10 min. Record your results in a table and then draw a line graph of them.
6 If you have time, try changing the conditions around your shoot and collecting a new set of results. You could try the experiment in a cooler or warmer place, a lighter or darker place, blowing a fan on to it, or removing some of its leaves. Draw graphs to show any other results that you manage to obtain.

Questions

1 Draw a large, labelled diagram of the potometer which you used.
2 Why does the meniscus move towards the plant?
3 Why does the potometer not work properly if you fail to make everything airtight?
4 Even if you get everything airtight, the potometer will not work if a large air bubble gets trapped in the tube. Why?
5 If you managed to get more than one set of results, describe and explain any differences between them.

Questions

1 The photograph shows a blood smear from a healthy person. The blood has been stained and photographed through a microscope.

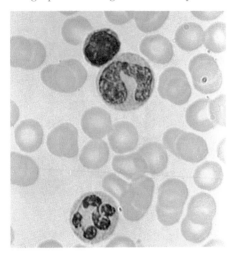

a How many white cells are visible in this photograph?

b Draw one of the white cells. Label its cell surface membrane, cytoplasm, and nucleus.

c What is the function of the type of white cell you have drawn?

d Give three ways in which the structure of red blood cells appears different from that of the white cells in this photograph.

e What is the function of red blood cells?

2 The diagram shows a vertical section through a mammalian heart.

a Copy the diagram. Add labels to each labelling line.

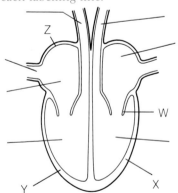

b On your diagram draw arrows to show how the blood enters, flows through, and leaves the heart. You will need to draw arrows on both sides of the heart.

c Why is the wall labelled X thicker than the wall labelled Y?

d Why is the wall labelled Z thinner than walls X and Y?

e What is the function of part W?

	Wall	Diameter	Valves	Explanation
Artery				
Vein				
Capillary				

3 Copy and complete this comparison table (above). The 'explanation' column needs plenty of space, because you may need to write quite a lot in it.

4 The graph shows how the rate of transpiration of an oak tree changed over a period of 24 hours.

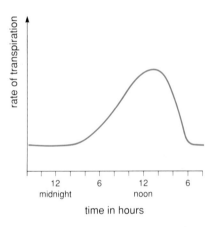

a At what time of day was the transpiration rate greatest?

b Why do you think the transpiration was highest at that time?

c The day on which these data were collected was a warm, dry day in summer. Make a copy of the graph. On your copy draw another line to show what you might expect to find on a cool, moist day in winter.

d Explain, as fully as possible, why you have chosen to draw your line in this way.

5 The photograph shows a micrograph of a transverse section through a plant stem.

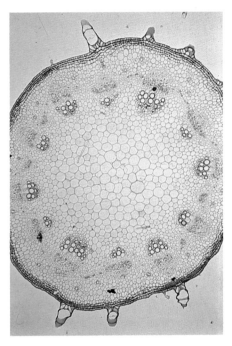

a Make a careful drawing of this photograph. You do *not* need to draw all the individual cells!

b On your diagram label the following:

epidermis   xylem
phloem   cortex

c Which of the four structures that you have labelled is made up of dead cells joined end-to-end to form long tubes?

d Which of the structures that you have labelled transports sugars from the leaves to the roots?

e Are the structures you have labelled:
A organelles
B organs
C tissues
D organisms?

FORCES

42 FORCES

The most simple forces are pushes and pulls. If we push or pull on an object, it often moves. Sometimes the force makes the shape of the object change.

There are many different types of force

A force can start an object moving. It can also slow down or speed up a moving object. A force can change the shape of an object. Sometimes a force seems to be doing nothing. This might be because it is cancelling out the effects of another force.

There are many types of force. They include elastic, magnetic, electrostatic, compressive, tensile, gravitational, turning, squashing, squeezing, twisting, stretching... you can probably think of many more. Often, these are just different names for the same thing. A compressive force is a squashing force. A tensile force is a stretching force.

Fig. 42.1 A pile driver uses gravitational force to pull the large mass downwards.

Large electric currents in superconductors levitate the disc with magnetic forces.

Forces are measured in newtons

To measure forces we use a newton meter. This is a spring and a scale. A large force stretches the spring. The scale is calibrated in **newtons**. The newton, N, is the unit of force. It is named after Sir Isaac Newton.

INVESTIGATION 42.1

Measuring the extension of a spring

1 Read through the experiment and design a results chart.
2 Set up the apparatus as shown in the diagram. Measure the length of the spring. Add a load of 1 N to the spring. Measure how much longer the spring now is. You should record by how much the spring has **stretched** – not its actual length.
3 Repeat the measurements, increasing the load a little each time.
4 Plot your results as a line graph. Put 'load' on the horizontal axis, and 'extension' – how much the spring has stretched – on the vertical axis.

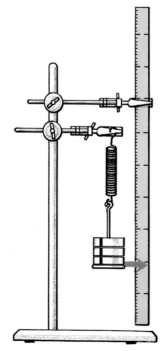

Fig. 42.2

You may have loaded your spring so much that it became permanently stretched. If you did this, your graph will change shape at this point. The point where this happens is called the **elastic limit** of the spring.

Questions

1 What is the shape of your graph?
2 If your spring extended by 25 mm, what force was being applied to it?
3 If the load on your spring is doubled, what happens to the extension?
4 If you reached the elastic limit of your spring, at what load did this happen?
5 What would happen if the spring in a newton meter went beyond its elastic limit? How is this prevented in a newton meter?

Forces change the way things move

Sir Isaac Newton spent a lot of time thinking about forces. He stated some important laws about them.

Newton's First Law says that **an object keeps on going as it is, unless an unbalanced force acts on it.**

This helps to describe what a force is. A force is something that changes the way in which something moves. The force will either speed it up or slow it down.

Imagine a book on a table. If you give it a push, it starts moving. Newton's First Law says that when you stop pushing, the book will keep on going – unless it is acted on by an unbalanced force. You know that when you stop pushing, the book very quickly stops moving.

450N

500N

So there must be an unbalanced force acting on it!

What exactly is an **'unbalanced force'**? Think about a seesaw. If the people on either side push down with the same force, they cancel one another out. The seesaw is balanced and it does not move. The two forces are balanced. But if one person is heavier then the forces are unbalanced. In Figure 42.3 there is an unbalanced force of 50 N. This moves the seesaw down at one end. So if an unbalanced force acts on a stationary object it will start moving. If *all* the forces acting on a moving object are balanced it will keep on moving.

Fig. 42.3 Unbalanced forces. The right-hand end of the seesaw will move downwards.

Friction slows things down

Now think about the book on the table again. When it slides across the table the unbalanced force is friction. The surface of the book and table are not perfectly smooth. The two surfaces catch on one another, and stop the book moving.

Does this mean that if the surfaces were perfectly smooth the book would keep on going for ever? Newton's First Law says that it would. But the book also has to push the air aside as it moves. Air resistance is another frictional force, and it will slow the book down. But if you were out in space and you pushed the book away from you, it really would go on moving away from you for ever. There is no surface and no air to slow it down.

One force always produces a reaction to itself

If a book is placed on a table, the force of its weight acts downwards. So why does it not *move* downwards? A simple answer would be that the table gets in the way!

Fig. 42.4

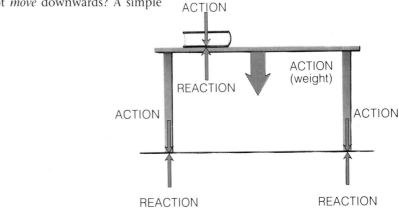

ACTION

ACTION (weight)

REACTION

ACTION

ACTION

REACTION

REACTION

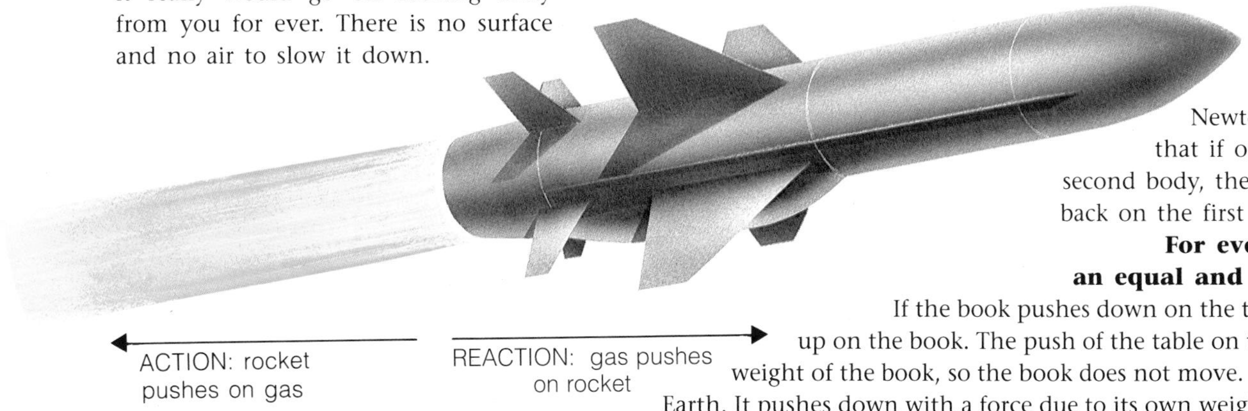

Newton's Third Law states that if one body pushes on a second body, the second body pushes back on the first with the same force. **For every action there is an equal and opposite reaction.** If the book pushes down on the table, the table pushes up on the book. The push of the table on the book balances the weight of the book, so the book does not move. The table rests on the Earth. It pushes down with a force due to its own weight, and the book. The Earth pushes back with an equal and opposite force. So the table stays where it is.

ACTION: rocket pushes on gas

REACTION: gas pushes on rocket

Fig. 42.5

Forces can cause objects to change shape

Balanced forces will not make an object move – but they might make it change shape. Think of the book on the table again. The book and the table stay where they are, because the push of the book on the table is balanced by the push of the Earth on the table. But the table is 'squashed' between the force of the book pushing down on it, and the force of the Earth pushing up on it. This could change the shape of the table. If the book were very heavy, and the table legs very thin, they could bend. You can see this happening when you squeeze a balloon. If you push hard on either side of a balloon, the forces of each hand are balanced. The balloon does not go anywhere (unless you suddenly let go!) but it does change shape.

Fig. 42.7 Balanced forces squashing a balloon.

Fig. 42.6 Even when forces are balanced, things can happen.

force of the Earth on the table legs

force of the book on the table

Questions

1 If an object is not moving, then either no forces are acting on it at all (which is unlikely), or all the forces acting on it are balanced.
For each of the diagrams shown:
  a Say whether the forces are balanced or unbalanced
  b Describe any 'missing' forces
  c If the forces are unbalanced, say how big the resulting force is

2 A spring is stretched from 2.5 cm to 5 cm when it is loaded with a weight of 5 N.
  a What would be the new length of the spring if a weight of 2 N was added to the 5 N weight?
  b Can you make a sensible prediction for the length of the spring if a weight of 100 N was hung on it?

3 A body builder uses a chest expander. It has five springs. It takes 300 N to pull one spring out by 20 cm.
  a What force is needed to pull all five springs out by 20 cm?
  b How hard will the body builder have to pull to extend all five springs by 40 cm?

4 The weight of a car produces a downward force of 10 000 N. This pushes each of the four springs in its suspension down by 20 cm. If

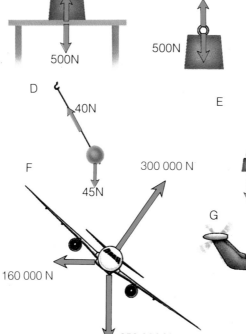

five adults get into the car, the downwards force increases by 4000 N. If the force is shared equally by the four springs, how much more will they be pushed down?

5 A set of bathroom weighing scales contain a spring. As the top is pushed down, the movement of the spring turns a pointer on the scale.
  a When person A stands on the scales, they depress the top by 10 mm. Person B weighs 2/3 as

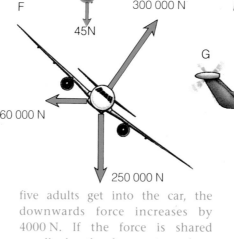

much as person A. By how much would person B depress the top of the scales?
  b What would happen if 10 people, all weighing the same as person A, balanced on the top of the scales?

6 An aircraft in flight has four forces acting on it. They are weight, lift, thrust and drag.
  a Draw a diagram with arrows showing these forces.
  b Which forces must balance each other?

EXTENSION

104

43 GRAVITY, MASS AND WEIGHT

Gravity attracts objects towards each other, producing a force called weight. Weight and mass are not the same.

Mass is one way of measuring how much of something there is

Mass is a quantity of matter. The mass of an object tells you how much of it there is. Mass is measured by finding out the force needed to change the way the object moves. The greater the force needed, then the greater the mass of the object.

If you push on a book, it moves faster than if you push with the same force on a car. This is because the car has more mass than the book. If you had two identical tins, one containing lead and the other full of feathers, you could find out which was which by pushing them. The lead-filled tin has more mass, so you need more force to move it. We can say that the car and the lead-filled tin are more 'reluctant' to move than the book and the tin of feathers. We call this reluctance to move **inertia**. The larger the mass of an object, the larger its inertia.

Moving objects have inertia, too. A moving object needs force to make it stop. A moving car has more inertia than a moving book. It needs more force to make it stop.

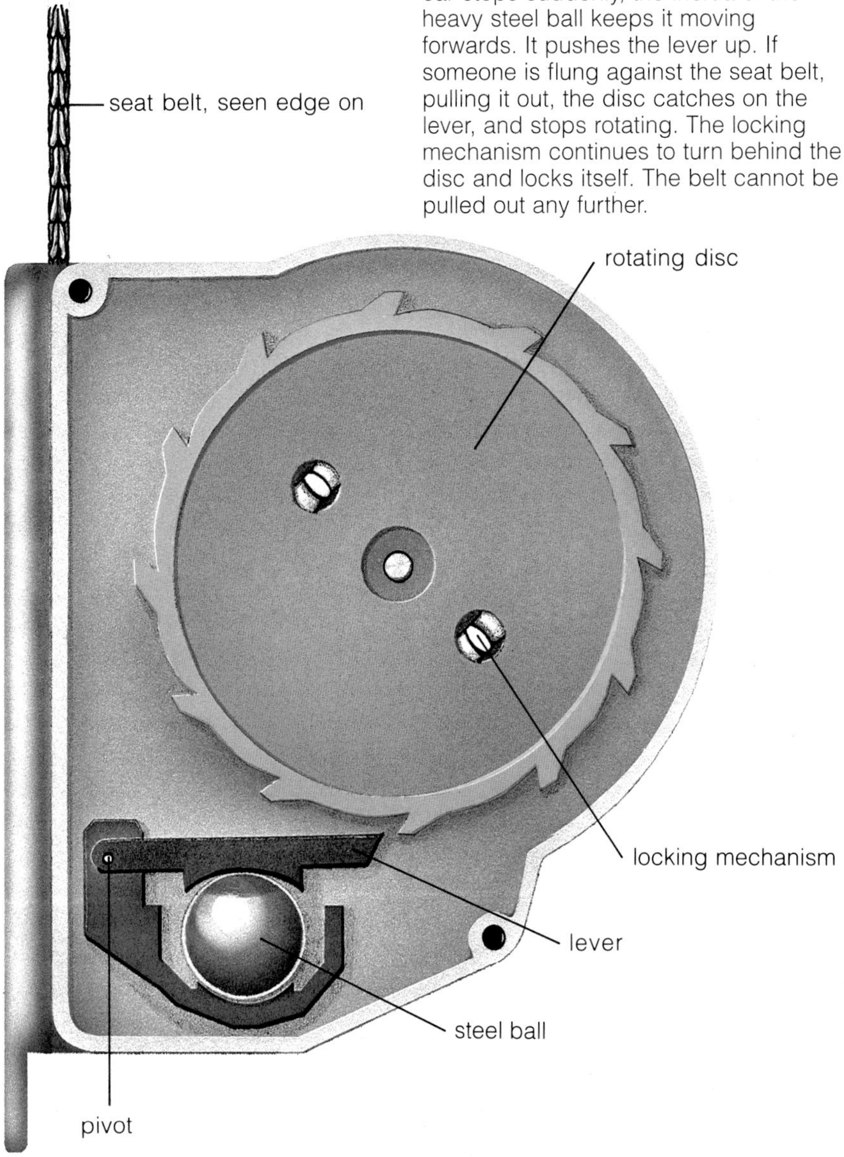

Mass is measured in kilograms

The kilogram (1000 g) is the usual scientific unit of mass. The standard kilogram is the mass of a particular cylinder of platinum-iridium alloy kept near Paris in France. All masses that are measured are compared (usually rather indirectly) with this.

Large masses are measured in **tonnes** (t). One tonne is 1000 kg.

Fig. 43.1 An inertia reel seat belt; the locking mechanism and disc rotate together when the belt is pulled out. If the car stops suddenly, the inertia of the heavy steel ball keeps it moving forwards. It pushes the lever up. If someone is flung against the seat belt, pulling it out, the disc catches on the lever, and stops rotating. The locking mechanism continues to turn behind the disc and locks itself. The belt cannot be pulled out any further.

— seat belt, seen edge on

rotating disc

locking mechanism

lever

steel ball

pivot

Questions

1 a When a car stops suddenly, you appear to be 'thrown' forwards. What actually happens, in terms of mass, forces and inertia?

  b If children who are not wearing seat belts sit in the back space in an estate car, what will happen to them if the car is suddenly knocked forwards? Why?

2 If a car suddenly speeds away from traffic lights, it exerts a backwards force on the road. Why does the road not move backwards?

Weight is a force

A mass is pulled down towards the Earth. This is because all mass is attracted together. The force which pulls masses together is called **gravity**. Normally, the force is too small to notice. But with a mass as big as the Earth, the force becomes quite large.

You are attracted towards the Earth, and the Earth is attracted towards you, with the same force. But the Earth has a much greater mass, and much greater inertia, than you have. If you jump up in the air, you and the Earth pull on each other with the same force. But the Earth is much more reluctant to move than you are, because it has more mass. So the Earth does not seem to come up to you – you drop down to the Earth.

A large mass is pulled towards the Earth more than a small one. So we can compare masses by measuring the force that pulls them down. The size of the force pulling an object towards the Earth is called its **weight**. Like all forces it is measured in newtons.

Weight = mass x g

A kilogram mass is pulled towards the Earth by a force of 9.81 N. So the weight of 1 kg is 9.81 N.

9.81 is very close to 10. To make calculations much easier, we can use 10 instead of 9.81 most of the time. We say that the strength of gravity (on Earth) is 10 newtons per kg, or 10 N/kg. The strength of gravity is given the symbol g. So, on Earth, g is 10 N/kg.

The apparent weight of an object is **the force it exerts on its support**.

The weight of a book is the force it exerts on the table it is resting on, or the force it exerts on the spring balance it is hanging from. If the book has a mass of 2 kg, gravity pulls down with a force of 10 N per kg. So the total force of the book on the table is 2 x 10 N, which is 20 N.

So, if you know the mass of an object, you can find its weight by multiplying its mass in kilograms by g. **Weight = mass x g**.

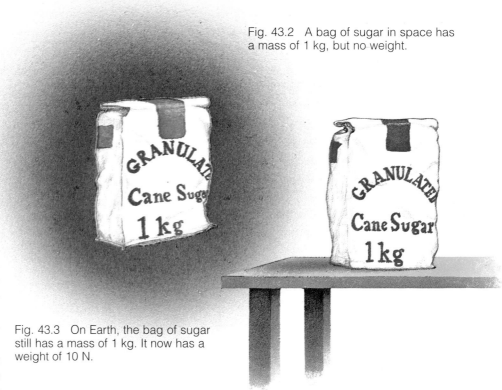

Fig. 43.2 A bag of sugar in space has a mass of 1 kg, but no weight.

Fig. 43.3 On Earth, the bag of sugar still has a mass of 1 kg. It now has a weight of 10 N.

If the scale of a balance is in kg, you are not weighing things – you are massing them

What happens when you weigh something? If you are measuring the force due to gravity on an object, then you are measuring its weight. You are weighing it. A spring balance does this. Its scale is in newtons.

But if you use a lever arm balance, you are comparing the force pulling on the object's mass with the force pulling on a known mass. You are measuring the mass of your object. The scale will be in kilograms. You can easily multiply the mass in kilograms by 9.81 to find the weight in newtons. Many balances have scales where this has already been done.

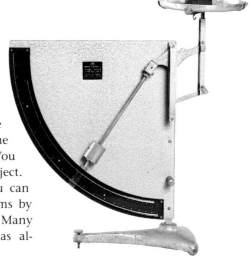

Fig. 43.4 A lever arm balance. Gravity acting on the mass on the pan produces a force which rotates the arm clockwise, while gravity acting on the mass near the pointer produces a force which rotates the arm counterclockwise. If gravity was different – say on the Moon – it would be different for both masses, so the reading would be the same as on Earth. So a lever arm balance measures mass, rather than weight.

Gravity is not the same everywhere

Gravity is not the same all over the Earth's surface. The accepted value for the force due to gravity is 9.81 newtons per kilogram of mass. But this varies, depending where you are.

On the Moon the pull downwards is much less than on the Earth. This is because the Moon is many times smaller than the Earth. The force due to gravity on the Moon is one-sixth that on the Earth. The force of gravity on the Moon is 1.67 newtons per kilogram. So the weight of a 1 kg mass on the Moon is 1.67 N.

Weightless objects still have mass

Out in space you would be weightless. You are too far from the Earth, or any other large body, to be pulled towards it by gravity.

You would be able to pick up a very heavy object – because the object would not be heavy! It would have no weight because there would be no gravity. You could hold it up with no effort at all. But both you and the object would still have mass. And you would both have inertia. To start a 1000 kg mass moving through space, you would still have to push very hard. It would only begin to move very slowly. No matter where it is, a 1000 kg mass has a mass of 1000 kg. Because it has a large mass, its inertia is large too. Even out in space, if the mass was moving towards you fast, it could still crush you against your spacecraft. It has a lot of inertia so it would take a lot of force to stop it moving.

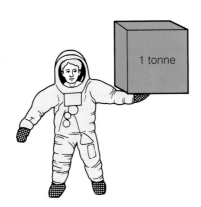

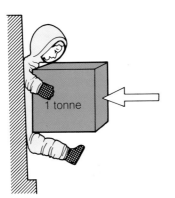

Gravity produces acceleration

There are two ways of thinking about gravity. So far, we have thought about it as causing a force on an object, which we call its weight. The force of gravity acting on a 1 kg mass produces a force of 9.81 N, which is the object's weight.

But gravity can also pull on an object and make it move. The force of gravity starts the object moving, and makes it go faster and faster. Gravity causes the object to **accelerate**. The acceleration which gravity causes is 9.81 m/s^2.

So we can either think of gravity as causing weight, or causing acceleration.

Summary

A lot of very important ideas have been covered in these last few pages. Learn and try to understand them!

An unbalanced force changes the motion of an object.

Every force has an equal and opposite reaction.

Mass is a quantity of matter measured in kg.

Mass has inertia.

Inertia is the reluctance of an object to have its motion changed. Inertia increases with mass.

Gravity acts on mass and gives it weight.

Weight is a force and is measured in newtons.

Weight is found by multiplying the local gravitational field strength (g) by the mass of the object:
weight = m x g.

Questions

1 On Earth, the force of gravity is about 10 N/kg.
Complete the following table.

Mass (kg)	Weight (N)
1	10
2	
4.6	
	85

2 An empty space shuttle has a mass of 68 t. It can carry a cargo of 29 t.

a What is the total mass of the full shuttle in kilograms?

b What is the weight of the shuttle before launching?

c What force, or **thrust**, would the rockets have to provide to just balance this weight?

d Draw a diagram showing all the forces acting on the space shuttle at take-off.

3 The force due to gravity is given the symbol *g*. On Earth *g* is about 10 N/kg. To find the weight of an object, you multiply its mass in kilograms by *g*. So on Earth the weight of an object is found by multiplying its mass in kilograms by 10.

Complete the following table:

Place	Object	Mass	Local value of g	Weight
Earth	1 kg mass	1 kg	10 N/kg	
Jupiter	1 kg mass			24.9 N
Earth	bag of coal			250 N
Sun	bag of sugar	1 kg	274 N/kg	
Moon	car	2000 kg	1.67 N/kg	

44 CENTRES OF GRAVITY

All objects have a point at which we can consider all their mass to be located. The position of this point affects the stability of the object.

We can say that gravity acts at a single point

The weight of an object is the force due to gravity, when the object is at rest.

Imagine a stone resting on the ground. It is pulled down towards the Earth by gravity. We can think of the stone as many particles, all pulled towards the Earth by many little forces. We can also think of a single force pulling the stone down. This single force acts on the **centre of gravity** or **centre of mass** of the stone.

All objects can be thought of as behaving as though all their mass is concentrated at a single point. If the object is supported under that point, it will balance.

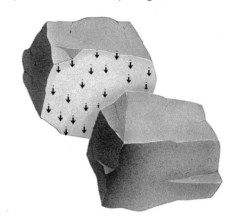

Fig. 44.1a Gravity acts on all the particles in a stone, pulling it down.

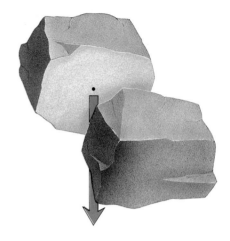

Fig. 44.1b All the individual forces can be represented by a single force.

Finding centres of gravity

If you support an object under its centre of gravity, it balances. Half of its weight tries to topple it one way. The other half tries to topple it the other way. The two forces cancel one another. The same thing happens if it is hung so that the centre of gravity is directly below the support. We can use this to find the centre of gravity of an object.

1 Take your first shape. Hang it from a pin as shown in the diagram.

Fig. 44.2a

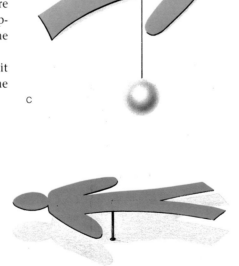

2 Now hang a thread, with a weight on the end, from the same pin. The thread will hang vertically downwards. Wait until the shape and the thread come to rest.

3 The thread is now passing through the centre of gravity of your shape. Draw this line down your object.

4 Now hang your object from the same pin, but in a different position. Again, the vertical line of the thread passes through the centre of gravity of your object. Draw the line on it.

The point at which the two lines cross is the centre of gravity of the object. But it is probably a good idea to repeat with the object in different positions, to improve your accuracy.

5 You can check that you really have found the centre of gravity by trying to balance your object on the point of a pin at exactly the point you have marked.

The centre of gravity does not always lie inside an object

The centre of gravity of a ruler is in the middle of it. The centre of gravity of a sphere is at its centre. But the centre of gravity of some objects is *outside* them. A boomerang is a good example of an object whose centre of gravity is not inside it. Figure 44.3a shows where it is. You could certainly not support the boomerang at this point! But the boomerang does behave as though all its mass was concentrated at this point. When the boomerang spins, it spins about its centre of gravity.

centre of gravity

a

Fig. 44.3a The centre of gravity of a boomerang lies outside its shape. It rotates about this point when thrown.

b When thrown, the two spheres of the bolas spin about their centre of gravity, which is halfway along the rope which joins them.

b

If two equal masses are attached to each other by a string, and thrown through the air, they spin round the central point of the string. This is how a South American bolas works. The centre of gravity of the two masses is half way between them. If one mass is larger than the other, then the centre of gravity is nearer to the larger one.

The Earth and the Sun are rather like the two masses joined by a string. Gravitational attraction stops them flying apart. The Sun is 330 000 times the mass of the Earth. The centre of mass of the Earth-Sun system is very close to the centre of the Sun. The Sun and Earth orbit around this point. The orbit is the path of one body in space around another. (This path is an ellipse.)

Many stars also orbit around each other in pairs. A pair of stars like this is called a **binary system**. The two stars often appear to be very close together. The centre of mass of the binary system lies somewhere between the two stars. The two stars rotate about this point. Some binary stars take centuries to complete one orbit! Some orbit very fast.

The way in which the stars move can tell us how far apart the stars are, and how heavy they are. Sometimes a star may seem to be 'wobbling' around in space. This gives astronomers a clue that there is another undiscovered star nearby, spinning around with it.

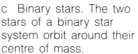
c

c Binary stars. The two stars of a binary star system orbit around their centre of mass.

The position of the centre of gravity helps us to balance an object

An object is balanced when its centre of gravity and its point of support lie on a vertical line. The forces on each side are balanced. The object is in **equilibrium**. There are two possible ways of balancing the object. The supporting point can be placed either *above* or *below* the centre of gravity. One of these is much easier to find than the other. Which is the tricky one?

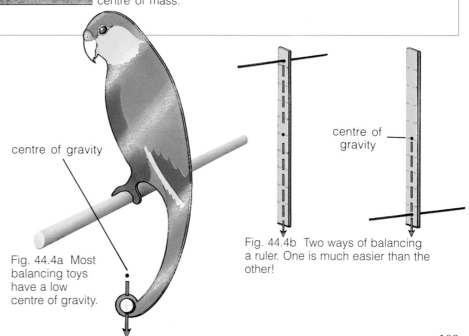

centre of gravity

Fig. 44.4a Most balancing toys have a low centre of gravity.

centre of gravity

Fig. 44.4b Two ways of balancing a ruler. One is much easier than the other!

45 EQUILIBRIUM

An object is said to be in equilibrium when all the forces acting on it are balanced. There are three types of equilibrium – stable, unstable and neutral.

There are three types of equilibrium

If you tried to balance a cone like the one in the diagrams, you could find three different positions in which it would stay. Two are quite easy. One is so difficult as to be almost impossible!

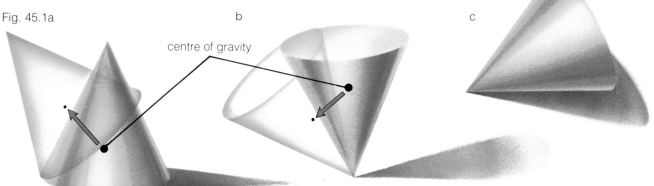

Fig. 45.1a   b   c

centre of gravity

This cone is in a position of **stable equilibrium**. If you tried to push it over you would *raise* its centre of gravity. It is as though you are lifting the cone. When you let it go it falls back to where it was.

A ball on a horizontal table is in neutral equilibrium all the time. This is because, however it rolls, its centre of gravity stays at the *same height*. It will only become unstable or stable if you change the shape of the surface on which it is resting. (See question 1 below.)

This cone is in a position of **unstable equilibrium**. If you tilt it you will *lower* its centre of gravity. When you let go it carries on falling. Even the slightest tilt will make it fall.

d

This cone is in a position of **neutral equilibrium**. If you tilt it the centre of gravity is still at the same height as it was before. In neutral equilibrium, a push does not change the height of the centre of gravity. When you push the cone it just stays in its new position.

Stable objects have a low centre of gravity

An object is stable if its centre of gravity lies above its base. An object is unstable when its centre of gravity lies outside its base. In other words, an object is unstable if a line drawn between its centre of gravity and the centre of the Earth does not pass through its base.

A stable object becomes unstable when it has been tilted so far that any more tilt starts to lower the centre of gravity. The critical point is reached when the centre of gravity is vertically above the edge of the base.

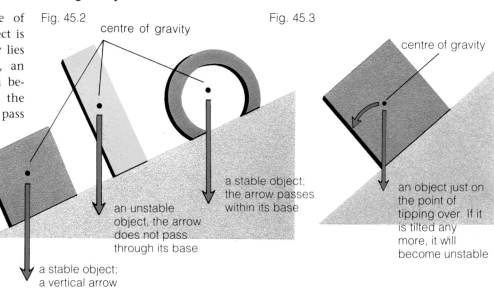

Fig. 45.2

centre of gravity

Fig. 45.3

centre of gravity

an unstable object, the arrow does not pass through its base

a stable object; the arrow passes within its base

a stable object; a vertical arrow from its centre of gravity passes through its base

an object just on the point of tipping over. If it is tilted any more, it will become unstable

There are two ways of making an object more stable. One way is to *lower its centre of gravity*. Racing cars have very low centres of gravity, so that they are less likely to roll over even when cornering at high speeds.

The other way is to *make the base of the object wider*. Racing cars have very wide bases with the wheels far apart from each other.

So even if the car tips, the centre of gravity still lies above its base, and it is unlikely to turn over.

Fig. 45.4 A low centre of gravity and wide wheelbase prevents a racing car from turning over as it corners at speed.

INVESTIGATION 45.1

Equilibrium in animals

You will be provided with some plasticine and cocktail sticks. Use them to make models of animals, and to answer the following questions. Plan your investigation carefully before you begin. Make diagrams of your models. Describe how you make your measurements of stability, and the evidence you have for each of your answers.

1 Are two legs more stable than one leg?
2 Are two legs more stable than four legs?
3 Which is more stable – a hippopotamus or a giraffe?
4 How could a giraffe make itself more stable?

Fig. 45.5 Tractors are designed with low centres of gravity, as they may have to work on steep slopes.

Questions

1 The diagrams show a ball bearing on a smooth surface. In which case is the ball bearing:

a in neutral equilibrium?
b in stable equilibrium?
c in unstable equilibrium?

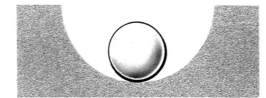

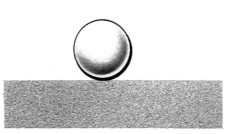

2 A boat has a high mast and sails. The force of the wind on the sails could turn the boat over.
  a Where should the centre of gravity be to make the boat as stable as possible?
  b How does a heavy keel help stability?

EXTENSION

3 The diagram shows a double decker bus on a slope. At what angle will the bus become unstable?

centre of gravity

4 a Some fast boats have keel fins. Find out about these. What are their advantages?
  b Fish have fins. Find out about the different functions of the fins of fish. Do fins help with stability? What else do fish use their fins for?

111

46 FLOATING AND SINKING

When something is placed in water its weight is reduced. The object experiences an upthrust from the water. If the weight is reduced enough, the object floats.

An object displaces fluid

When you climb into a bath the water level rises. You **displace** some water. The volume of water you displace is the same as your volume.

You may have used this method to measure the volume of complicated shapes.

If you put a stone into water it weighs less than it does in air. If you collect and weigh the water which the stone displaces, you will find that its weight is the same as the lost weight of the stone. If the stone weighs less then something must be lifting it up. The lifting force comes from the water and is called **upthrust**.

The force of the upthrust is the same as the weight of the water displaced.

This was discovered by Archimedes more than 2000 years ago.

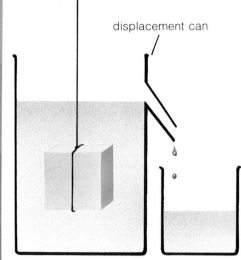

displacement can

an object displaces a volume of water equal to its own volume

Fig. 46.1   The force of upthrust is the same as the weight of water displaced.

| If you weigh a stone in air... | ...and then in water, you find it weighs less. | The weight the stone lost is the same as the weight of water displaced. |

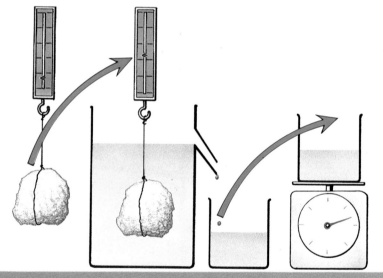

Any liquid or gas produces upthrust

It is not only water which produces upthrust. Any liquid or gas will do it. For example, the upthrust of air keeps a hot air balloon up. The balloon displaces a volume of air equal to its own volume. The air displaced is *colder* than the hot air inside the balloon. A certain volume of cold air is heavier than the same volume of hot air. So the air displaced is *heavier* than the air in the balloon. And, remember, the force of the upthrust produced is the same as the weight of air displaced. So the force of the upthrust is greater than the weight of the balloon. There is enough extra force to lift not only the balloon, but also the passengers and their basket.

Fig. 46.2   Hot air displaces cold air in a balloon, producing upthrust. This upthrust is balanced by the weight of the balloon.

Why does water produce upthrust?

When an object is immersed in water the water pushes in on it. This force is called water pressure. Water pressure is greater the deeper you go. So the water pressure on the bottom of an object is greater than the water pressure on the top. This tends to push the object upwards.

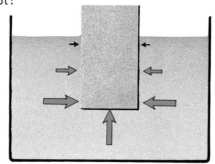

Fig. 46.3 Upthrust. The sizes of the arrows represent the sizes of the forces. The difference between the water pressure on the bottom and top of an object results in an overall upwards force, called upthrust.

EXTENSION

Does it float or sink?

Block A is floating. The upthrust of the water must be balancing the weight of the block. So the weight of water displaced must equal the weight of the block.

Not all the block is under water. The volume of water displaced is less than the volume of the block. But the weight of water displaced equals the weight of the block. So equal volumes of block and water cannot weigh the same. The block would be lighter. We say that the **density** of the block is less than the density of water.

A

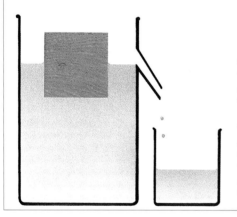

B

Block B is just floating. The upthrust of the water must be just balancing the weight of the block. So the weight of water displaced must equal the weight of the block.

All of the block is under water. The volume of water displaced is the same as the volume of the block. So both weight and volume of the block and the displaced water are equal. The density of the block and the water are the same.

Block C has sunk. The upthrust from the water is not enough to make the block float. So the weight of the displaced water must be less than the weight of the block.

So equal volumes of water and the block do not weigh the same. The block is denser than water.

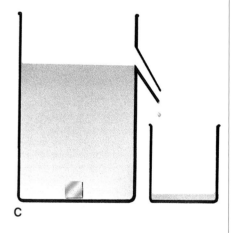

C

Density is the mass of a certain volume of a substance

Density is a way of comparing the masses of equal volumes of materials. It is meaningless to say that lead is heavier than feathers. It is more correct to say that a certain volume of lead is heavier than the same volume of feathers. We can say that lead has a greater density than feathers.

If an object has the same density as water, it just floats. If its density is less than water, it floats well. If its density is greater than water, it sinks.

EXTENSION

Questions

1 A balloon weighs 2000 N. It is filled with 5000 N of helium and it displaces 10 000 N of air.
   a What is the upthrust on the balloon?
   b What load could the balloon carry?
2 A block of copper is weighed in air, hydrogen, water and salt water. Which would give:
   a the highest reading?
   b the lowest reading?
   c If the copper block was massed in air, hydrogen, water and salt water, what would be the result?

47 DENSITY

Density is mass per unit volume. It can be measured in g/cm³ or kg/m³.

Density is the mass per unit volume

The density of something is the mass of a particular volume of it.

One cubic centimetre of water has a mass of one gram. So the density of water is one gram per cubic centimetre. In shorthand this is 1 g/cm^3.

Densities can also be written in terms of metres and kilograms. One cubic metre of water has a mass of 1000 kilograms. The density of water can therefore be written as 1000 kg/m^3.

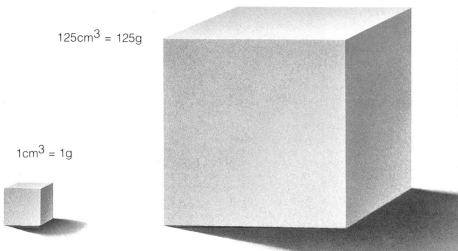

$125\text{cm}^3 = 125\text{g}$

$1\text{cm}^3 = 1\text{g}$

Mass = volume x density

If you know the volume and density of a substance, you can work out its mass. For example, the density of sand is 1600 kg/m^3. If you ordered 3 m^3 of sand, you would get:

mass = volume x density
$= 3 \text{ m}^3 \times 1600 \text{ kg/m}^3$
$= 4800 \text{ kg}$

Fig. 47.1 Density. Different sizes of the same material have different masses, but the mass for a particular volume is the same. The mass of 1 cm³ of a material is called its *density*. The density of this material is 1 g per cm³.

Rearrange the formula to find volume or density

The formula can be rearranged:
mass = volume x density
volume = mass / density
density = mass / volume

For example, the density of 22 carat gold is 17.5 g/cm^3. The density of 9 carat gold is 11.3 g/cm^3.
If you have a piece of gold jewellery, you can find out whether it is 22 carat or 9 carat by weighing it, and finding its volume. If it weighs 5 g, and has a volume of 0.286 cm³, then:

density = mass / volume
= 5 g / 0.286 cm³
= 17.48 g/cm³
So it must be 22 carat gold.

Archimedes used this method to find out if the King's jeweller had made a crown out of pure gold. Pure gold has a density of 19.3 g/cm³. So a 386 g crown should have a volume of 386 g / 19.3 g/cm³
This works out as 20 cm³.
When the crown was put into water, it should displace 20 cm³ of water. This should make it weigh 20 g less in water than in air. But it did not. Archimedes proved that the gold had been mixed with cheaper metal.

Fig. 47.2 A formula triangle can be used to rearrange a formula. If you cover up the quantity you want to find, the arrangement for the other two is shown.

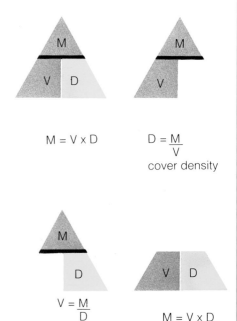

$M = V \times D$

$D = \dfrac{M}{V}$
cover density

$V = \dfrac{M}{D}$
cover volume

$M = V \times D$
cover mass

Finding the density of an object

You are going to find out the volume of an object by measuring the volume of water it displaces. If you also measure its mass, you can calculate its density.

1   Choose an object, and a beaker large enough for it to fit into easily.
2   Tie a thread around the object. Weigh it. Calculate its mass. Remember:

weight = mass x g
or mass = $\dfrac{\text{weight}}{g}$ g is 10 N/kg

3   Either: *If you have a displacement can available:*
    Fill the displacement can. Put your object into it, until it is completely immersed. Catch all the water which overflows, and measure its volume.

Or: *If you have a beaker:*
Fill the beaker with water until it overflows, using a measuring cylinder. Record the volume of water it took to fill your beaker. Immerse the object in the beaker, letting the water that it displaces overflow. Measure how much water is left in the beaker. Work out how much water the object displaced.

4   Record the volume of your object.
5   Calculate the density of your object by using the formula:
    density = $\dfrac{\text{mass in grams}}{\text{volume in cm}^3}$
6   Suspend the object from the thread again. Measure its weight when it is immersed in water.

Questions

1   What weight has the object lost in water, compared with its weight in air?
2   What weight of water must have been displaced?
3   What mass would this water have? (Use the formula in step **2**.)
4   What volume would this mass of water have? (The volume of 1 g of water is 1 cm³.)
5   Does this volume agree with the volume you measured in step **3**?
6   Suggest any possible causes of error in this experiment. How could you improve its accuracy?

Finding the density of sand

1   Measure out 200 g of dry sand. Find out, and record, the volume of your dry sand.
2   Calculate the density of your dry sand.
3   Measure out 200 g of water. Find out, and record, the volume of your water.
4   Add the sand to the water. Stir, and allow to settle. Then measure and record the volume of the sand and water together.
5   The volume of sand and water together will be less than the two separate volumes added together. Why is this?
6   Work out the actual volume of the sand particles.
7   Calculate the density of sand if it contains no air spaces.
8   Glass is made by heating sand, so that the particles fuse together. The density of glass is around 2.5 g/cm³. Can you link this information to the results you have obtained in this experiment?

Questions

1   Marble has a density of 2.7 g/cm³. What is the mass of 200 cm³ of marble?
2   The density of butter is 0.9 g/cm³. What is the volume of 800 g of butter?
3   A cast iron pair of kitchen scales has a volume of 500 cm³ and weighs 3.5 kg.
    a What is the density of cast iron?
    b If the scales had been made in mild steel, which has a density of 7.9 g/cm³, how much heavier or lighter would they be?
4   Milk has a density of 1.03 g/cm³. One pint is the same as 568 cm³. How much heavier is a pint of milk than a pint of water?

5   A floating buoy weighs 2000 N. When its cable snaps, it drifts away, but stays at the same height in the water.
    a What weight of water must the buoy displace?
    b If g is 10 N/kg, what mass of water does the buoy displace?
    c If the density of water is 1000 kg/m³, what is the volume of the water displaced by the buoy?
6   Coal has a density of 4.0 g/cm³. Paraffin has a density of 0.8 g/cm³. A piece of coal is lowered into paraffin. Its weight is reduced by 1 N. Assume that g = 10 N/kg.
    a What mass of paraffin is displaced?
    b What volume of paraffin is displaced?
    c What is the volume of the piece of coal?
    d What is the mass of the piece of coal?

— *E X T E N S I O N* —

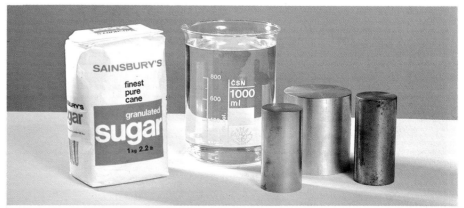

Fig. 47.3   1 kg masses of different substances. From left, they are sugar, water, brass, aluminium and steel. Which has the greatest density? Which has the lowest density?

Some animals use water to support their bodies

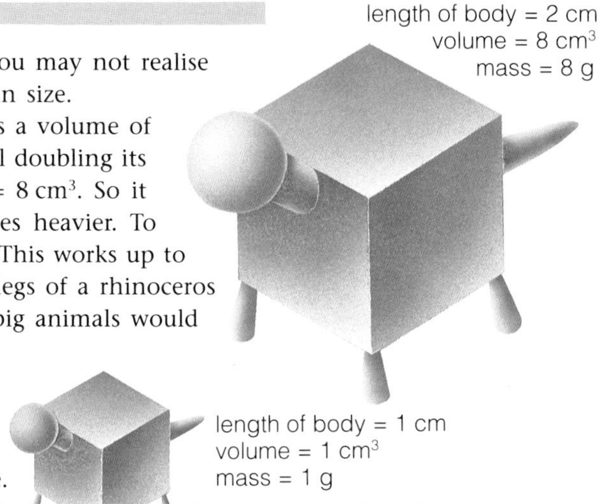

length of body = 2 cm
volume = 8 cm³
mass = 8 g

It is fairly obvious that a large animal is heavier than a small one! But you may not realise just how great an increase in weight is caused by quite a small increase in size. Imagine a cube-shaped animal. If it is 1 cm wide in each direction, it has a volume of 1 cm³. If its density is 1 g/cm³, then it weighs 1 g. Now imagine the animal doubling its size. It now measures 2 cm in each direction. Its volume is now 2x2x2 = 8 cm³. So it now weighs 8 g. So, if an animal doubles its size, it becomes eight times heavier. To support its weight, its legs would need to have an area eight times larger! This works up to a point. If you think of the relative size of the legs of a mouse and the legs of a rhinoceros you can see that legs do get relatively larger in bigger animals. But *very* big animals would need legs so big that they could not move them.

The biggest animal which has ever lived on Earth is the blue whale. It may have a mass of over 100 t. It would be impossible for it to live on land. But in water, the water it displaces reduces its weight. The water helps to support it. It can manage with quite a small skeleton for its size. If a blue whale gets stranded on a beach, it dies, because its body weight crushes its ribs and stops the lungs working.

length of body = 1 cm
volume = 1 cm³
mass = 1 g

Fig. 47.4 An animal that is twice as big is eight times heavier.

Fish and submarines can alter their density

Fish and submarines have a similar problem. They need to be able to stay at a particular depth in the water, without using unnecessary energy.

If the density of a fish is more than the density of water, it will sink. If it is less than the density of water, it will float. To float at a particular depth in the sea, a fish must be able to control its density. Most fish have a **swim bladder**. The swim bladder contains air. The more air there is in the swim bladder, the lower the density of the fish. If the fish tends to sink, it can add a little more air to its swim bladder until it is the same density as the water. The fish then has **neutral buoyancy**. It will neither sink nor rise in the water. Since the density of water changes with temperature, depth and saltiness, fish are always making small adjustments to keep their position in the water. The air needed comes either from their gut, or from their blood through capillaries. Submarines work on the same principle. They have **ballast tanks**. Compressed air is blown into the tanks to make the average density of the submarine less. This makes the submarine go up. To make it sink, air is released from the tanks.

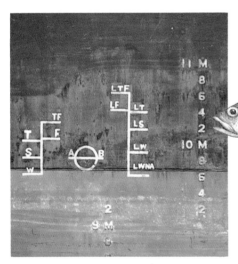

Fig. 47.5 Plimsoll lines. These lines on the side of a ship show the level at which it may safely lie in the water. The line AB, across the central disc, shows the standard position. The lines on the left show safe levels in different circumstances. TF stands for tropical fresh water, F for fresh water, T for tropical, S for summer and W for winter. Can you explain their relative positions?

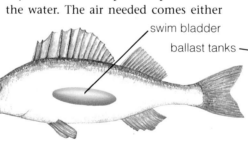

swim bladder
ballast tanks

Fig. 47.6 The amount of air in the swim bladder controls the average density of the fish.

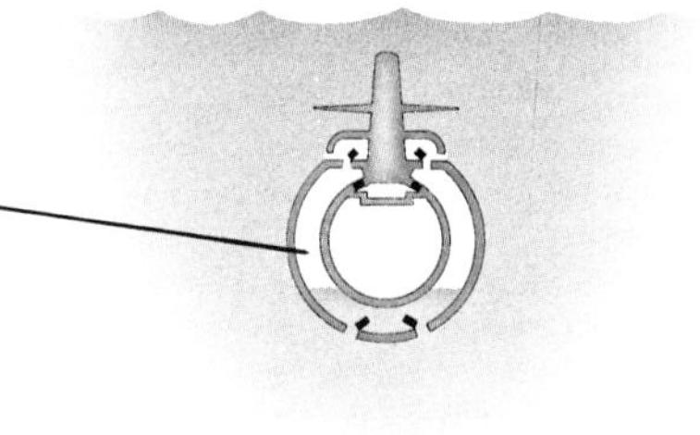

Fig. 47.7 The amount of air in the ballast tanks controls the average density of the submarine.

Ships float because their average density is less than water

Steel is denser than water. A lump of steel sinks in water. But a steel ship contains a lot of air. Its average density is less than water. So a steel ship floats.

A ship displaces water. The weight of the displaced water gives the amount of weight that can be supported. To take more weight, more water must be displaced. This means that the ship sits lower in the water. In the early days of shipping, owners were often tempted to put too much cargo on ships. This made the ships float so low in the water that they were likely to sink. Now, to show the safe maximum load, a Plimsoll line is marked on the side of a ship. This was enforced in England by act of Parliament in 1785. The marks show the safe water level. They are different for summer and winter, and for fresh and salt water. This is because the density of water changes with temperature and saltiness.

48 FIELDS

Some forces act only when objects are in contact. Others can act at a distance. A field is a region in which one object can exert a force on another, even when they are not in contact.

There are forces resulting from contact and forces acting at a distance

There are two types of force that we meet in everyday life. One type comes from contact with something. When you push on a door, the surface of your hand and the surface of the door touch. The door pushes back on you just as hard as you push on it. The force only seems to act when your hand touches the door. This sort of force results from contact. The other type of force does not need any contact. If you jump up in the air, there is no contact between you and the Earth. But a force acts on you. The attraction between you and the Earth is still there. The force of gravity still acts even when there is no contact.

A field is a region in space

The force of gravity acts between you and the Earth, even when you and the Earth are not in contact. There is a certain region all around the Earth, stretching out into space, where gravity acts on objects, pulling them towards the Earth. This region is called the Earth's **gravitational field**. In science, a field is a space where one object can exert a force on another object without touching it.

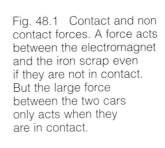

Fig. 48.1 Contact and non contact forces. A force acts between the electromagnet and the iron scrap even if they are not in contact. But the large force between the two cars only acts when they are in contact.

There are three main types of field

There are three types of field that you are likely to meet.

1 **Gravitational field** In a gravitational field, matter is attracted to other matter. All matter produces a gravitational field. But only very large masses, such as the Earth, have strong gravitational fields which produce large forces.

Fig. 48.2 The Earth's gravitational field.

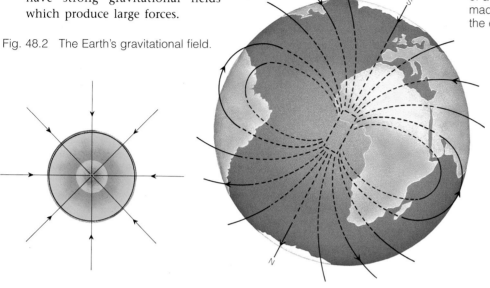

2 **Magnetic field** All magnetised materials produce magnetic fields. These can attract or repel, depending on what is placed in the field.

Fig. 48.3 The Earth's magnetic field.

3 **Electric field** All charged objects produce electric fields. Like magnetic fields, these can attract or repel, depending on what is placed in the field.

Fig. 48.4 This girl is touching the dome of a Van de Graaff generator. The pattern made by her hair suggests the pattern of the electric field around her head.

Fig. 48.5 This magnetically levitated linear railroad car (MLV-002) is suspended in a magnetic and a gravitational field. The forces from each field cancel each other and the train floats above the track.

49 MAGNETS

Magnets can attract or repel one another. Permanent magnets keep their magnetism. Magnetism results from small magnetic regions called domains.

The pole of a magnet is where the magnetism is strongest

A magnet always has two poles. These are the points where the magnetism is strongest. The two poles are always different. If the magnet is freely suspended, it will swing until one end points towards the North Pole of the Earth. This end of the magnet is the **north seeking pole**, or **north pole** of the magnet. The other pole is the **south pole**, which points towards the South Pole of the Earth.

Fig. 49.1   A magnet attracts a paper clip. The magnet induces magnetism in the paper clip. The end of the paper clip nearest to the magnet's north pole becomes a south pole. The north and south poles are attracted to each other – which is why the paper clip is attracted to the magnet.

Fig. 49.2a   Opposite poles attract

Fig. 49.2b   Like poles repel

Fig. 49.3   A magnet induces magnetism in a piece of iron.

Like poles repel, unlike poles attract

A north pole of a magnet attracts south poles. It repels north poles. A south pole attracts north poles, and repels south poles.

Some materials, such as iron, are easily magnetised. If a magnet is brought near to a piece of iron, the magnet causes the iron to become a magnet with poles at either end. The iron is said to be an **induced magnet**. If the piece of iron is approached by the north pole of the magnet, then the nearest end becomes a south pole, and is attracted to the magnet. The furthest end becomes a north pole.

Iron is a **magnetic** material. This means that it can be made into a magnet. You can find out if a material is magnetic by seeing if it is attracted to a magnet. But this does not prove that it is a magnet! To find out if a piece of iron is a magnet, you must try approaching it with both ends of a magnet. If one of your magnet's poles *repels* the piece of iron, then the piece of iron is a magnet. If it is only a weak magnet, though, this repulsion can be very hard to detect.

A compass needle is a magnet

A compass needle is a small magnet. It is supported so that it can swing round freely. Its north pole points towards the North Pole of the Earth.

If a compass needle is put near a magnet, its north pole will point towards the south pole of the magnet. What does this tell you about the North Pole of the Earth?

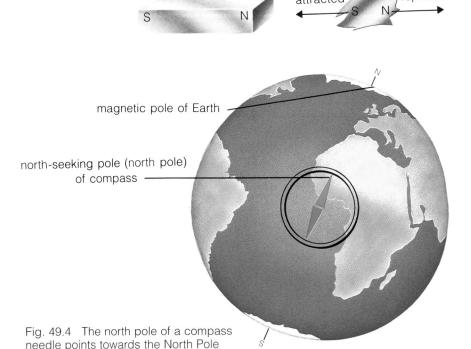

magnetic pole of Earth

north-seeking pole (north pole) of compass

Fig. 49.4   The north pole of a compass needle points towards the North Pole of the Earth.

Plotting magnetic fields

The space around a magnet in which it can affect other objects is called its **magnetic field**.

1 Put a bar magnet on a sheet of paper, and draw round it. Label the N and S poles.
2 Leave the bar magnet in position on the paper. Place a plotting compass near one end. Note which way the needle is pointing. Mark a dot on the paper against the plotting compass to show the direction that it points in. Move the compass forward so that the back end of the needle points at the spot you have just marked. Mark a new spot on the paper against the edge of the plotting compass to show the direction that it points in now.
3 Repeat this until the compass reaches the edge of the paper or the magnet. Draw a smooth line to join all the points. Mark arrows on the line to show the direction of the force on the compass.
4 Repeat 2 and 3 at several positions around the bar magnet.

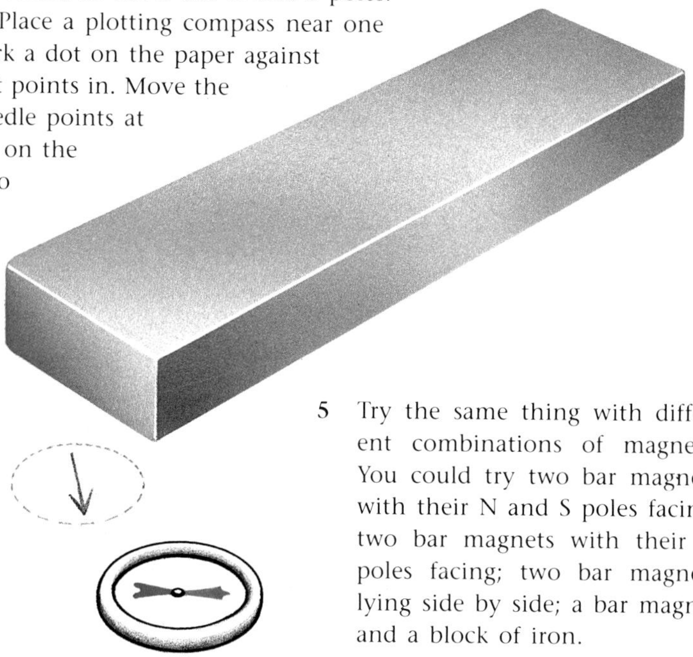

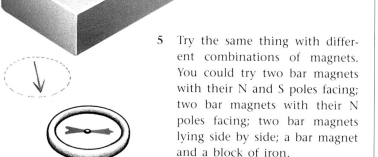

5 Try the same thing with different combinations of magnets. You could try two bar magnets with their N and S poles facing; two bar magnets with their N poles facing; two bar magnets lying side by side; a bar magnet and a block of iron.

Fig. 23.5 Plotting field lines. The direction of the compass needle is drawn for several positions around the magnet.

Comparing the strength and permanence of iron and steel magnets

1 Take a strong bar magnet. Use it to pick up a piece of iron wire, and a piece of steel wire. The iron and steel pieces should be of the same size.
2 Hold the magnet so that the iron and steel touch a heap of iron filings. Lift the magnet. Which picks up the most iron filings - the iron or the steel?
3 Carefully detach the iron and steel from the magnet, and put them near the iron filings again. Do either of them still pick up any iron filings? Have either of them kept their magnetism?
4 Take fresh, unmagnetised pieces of iron and steel. Magnetise each of them by stroking with a strong magnet in one direction only. Make sure that you do it fairly. Design a way to find out:
  a which is the most strongly magnetised
  b which keeps its magnetism best
  and then carry out experiments to find out.

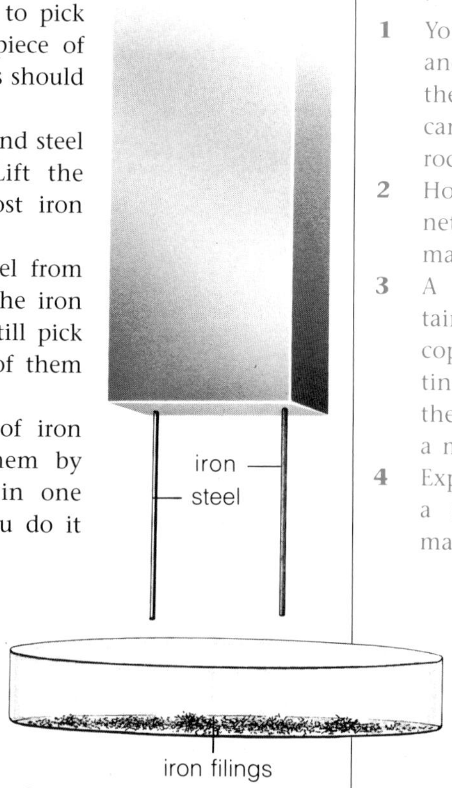

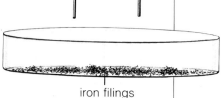

iron —
— steel

iron filings

Fig. 49.6 Comparing iron and steel magnets

Questions

1 You are given a piece of thread and three iron rods. Two of the rods are magnetised. How can you find out which of the rods is not magnetised?
2 How can you show the magnetic field pattern around a magnet?
3 A heap of scrap metal contains iron, steel, nickel, zinc, copper, aluminium and some tin cans. Find out which of these could be removed using a magnet.
4 Explain the difference between a magnetic material and a magnet.

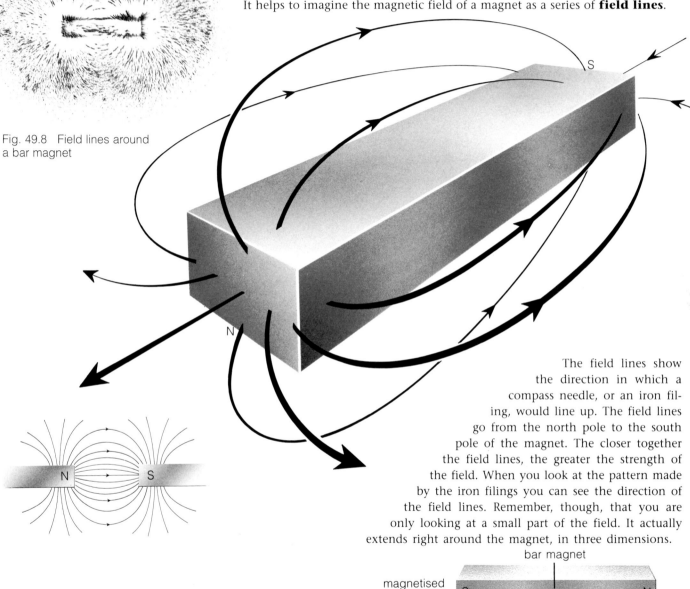

Fig. 49.8 Field lines around a bar magnet

The magnetic field is the space around a magnet in which it can affect other objects

Every magnet has a space around it in which it can affect some objects. This space is called its **magnetic field**.

You cannot see a magnetic field. But you can show its effects. If you scatter iron filings on a piece of paper, and place it over a bar magnet, each iron filing acts like a tiny compass needle. Figure 49.7 shows the patterns they make.

It helps to imagine the magnetic field of a magnet as a series of **field lines**.

The field lines show the direction in which a compass needle, or an iron filing, would line up. The field lines go from the north pole to the south pole of the magnet. The closer together the field lines, the greater the strength of the field. When you look at the pattern made by the iron filings you can see the direction of the field lines. Remember, though, that you are only looking at a small part of the field. It actually extends right around the magnet, in three dimensions.

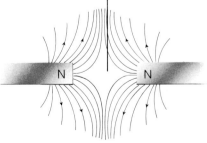

neutral point; the magnetic forces are balanced in this region, so there is no magnetic field effect

Fig. 49.9 Magnetic field lines between two bar magnets

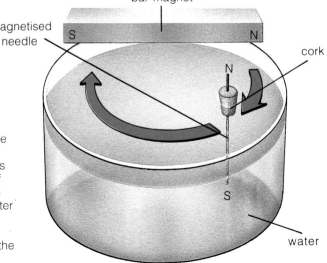

Fig. 49.10 Field lines show the direction in which a north pole would move. You can show this by setting up this apparatus. If the needle is long enough, the south pole is too low in the water to be affected by the magnet. The north pole moves along a curved path from the north to the south pole of the bar magnet.

The Earth behaves like a giant magnet

The Earth is like an enormous magnet. Its North Pole is actually a magnetic south pole – the north poles of magnets are attracted towards it. Its South Pole is a magnetic north pole – the south poles of magnets are attracted towards it.

The field lines of the Earth's magnetic field extend far out into space.

Magnetism is caused by domains

A magnetic material is made up of tiny magnetic regions, or **domains**. These regions are about 20 millionths of a metre in size.

An electron spinning round the nucleus of an atom produces a tiny magnetic field. In the domain of a magnet, the spinning electrons are lined up in a way that lines up their magnetic fields. This produces a tiny magnetic region in the material.

If the material is unmagnetised, the domains point in lots of different directions. The different field directions cancel each other out so there is no overall magnetic effect.

If the unmagnetised material is put into a magnetic field, the domains which are pointing in the same direction as the magnetic field grow bigger. They gradually 'take over' the piece of material, until all the domains have fields pointing in the same direction as the magnetic field. The material has now become a magnet.

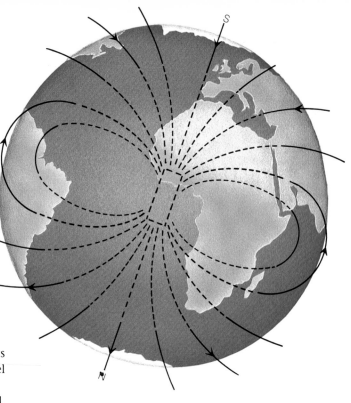

Fig. 49.11   The Earth's magnetic field. The Earth behaves like a giant bar magnet, with a south-seeking pole at the North Pole, and a north-seeking pole at the South Pole. In fact, the poles of the magnet are not exactly at the geographical North and South poles, but just a few degrees away.

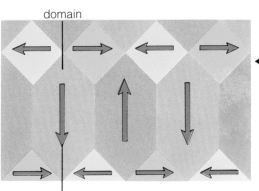

domain

direction of magnetic field

Fig. 49.12   Magnetising a substance changes the size of some of the domains. In an unmagnetised sample of iron, the domains' magnetic fields point in all directions. When the iron is magnetised, the domains in the direction of the outside field get bigger, at the expense of the domains not in the same direction.

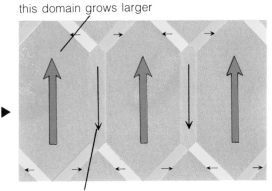

this domain grows larger

this domain grows smaller

Iron is easier to magnetise than steel

When a magnet is held close to iron filings, each iron filing becomes a tiny magnet. Its domains easily change position. But as soon as the magnet is taken away, the filings lose their magnetism. Its domains do not stay in their new position.

We say that iron is a **soft** magnetic material. It is easy to magnetise and makes a strong magnet, but it easily loses its magnetism. This is useful if you want a magnet that you can switch on and off. This type of magnet is called a **temporary magnet**. Soft magnetic materials are used in electromagnets.

Paper-clips are made of steel. If you stroke a paper-clip in one direction with the pole of a magnet, you can make its domains line up, so that it becomes a magnet. When you take the magnet away, the paper-clip stays magnetised. We say that steel is a **hard** magnetic material. Once its domains are lined up, they stay lined up. The steel becomes a **permanent magnet**. Permanent magnets are used in door catches, motors, cassette and video tape and for computer discs. But even so-called permanent magnets can be demagnetised. Anything which can knock

domains out of alignment can demagnetise a magnet. Dropping or hitting or heating the magnet can remove the magnetism. So can exposure to a strong magnetic field.

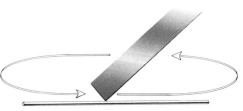

50 ELECTRIC FIELDS

Atoms contain electrons and protons. If these are separated, a charge results. This charge can be positive or negative.

Some materials can be given an electric charge

Atoms contain electrons and protons. If the number of protons equals the number of electrons, the atom is uncharged.

If you rub two materials together, it is possible that electrons might be rubbed off one and on to the other. The electrons and protons in each material are now no longer balanced. One material has extra electrons and the other is missing some electrons. The materials become **charged**. The Greeks discovered this in the 6th century B.C. When a piece of amber is rubbed, it becomes charged. It attracts dust and small pieces of paper. The Greek word for amber is 'elektron'. We now call the charge **static** (stationary) **electricity**.

Some charged objects, such as perspex and polythene, attract one another. But two pieces of charged polythene repel each other. This is because

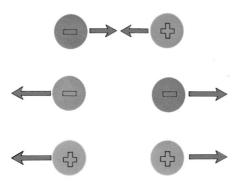

Fig. 50.1 Like charges repel. Unlike charges attract.

there are two kinds of charge. If the charges are the same, they repel each other. If the two charges are different, they attract each other.

Charged materials may attract other objects

If either a positively or negatively charged rod is brought close to tiny pieces of foil, the foil is attracted to the rod. But if they touch the rod, they acquire the same charge as the rod, and are repelled.

Before touching, when the charged rod is brought close, a charge separation is **induced** in the foil. If the rod is negatively charged, for example, it has extra electrons. These repel electrons in the foil, which move away from its surface. The part of the foil nearest to the rod becomes positively charged, and is attracted towards the rod.

But when the foil touches the rod, some of the extra electrons from the rod jump on to the foil. Now the foil is negatively charged too. So it is repelled from the rod.

Extra electrons produce a negative charge; too few electrons produce a positive charge

If a polythene or amber rod is rubbed with a cloth, electrons are transferred from the cloth to the rod. The rod now has extra electrons. Electrons have a negative charge, so the rod has an overall **negative** charge. The cloth now has too few electrons and too many protons. Protons have a positive charge, so the cloth has an overall **positive** charge.

Rubbing the rod has not produced the charge. The energy used in rubbing has only separated the positive and negative charges already in the atoms.

If a perspex rod is rubbed with wool, it loses electrons to the wool. The perspex rod becomes positively charged. The wool becomes negatively charged.

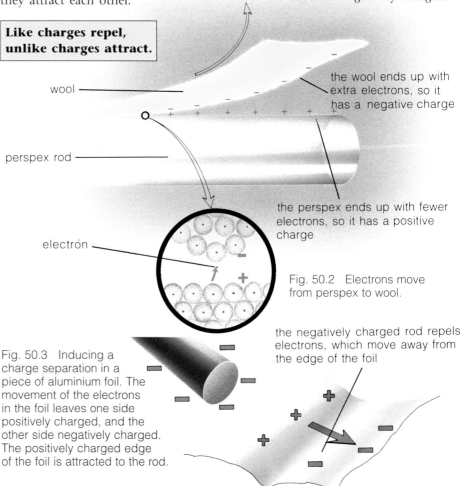

Like charges repel, unlike charges attract.

wool

perspex rod

electron

the wool ends up with extra electrons, so it has a negative charge

the perspex ends up with fewer electrons, so it has a positive charge

Fig. 50.2 Electrons move from perspex to wool.

the negatively charged rod repels electrons, which move away from the edge of the foil

Fig. 50.3 Inducing a charge separation in a piece of aluminium foil. The movement of the electrons in the foil leaves one side positively charged, and the other side negatively charged. The positively charged edge of the foil is attracted to the rod.

Static electricity stays where it is, until discharged

When two materials are rubbed together and become charged, electrons are transferred from one to the other. If the material does not conduct electricity, the charge stays where it is. A material which does not conduct electricity is called an **insulator**. If the charge stays where it is, it is called static electricity. Static means 'not moving'.

If you walk around on a nylon carpet, electrons can be rubbed off the carpet on to you. You can build up quite a large negative charge of static electricity. But you do not realise it, because the charge stays where it is. If you now touch a metal post connected to the ground, the electrons can escape. The metal post is a **conductor**. It will let the electrons rush away from you into the ground. And you can certainly feel that! You get a small electric shock. The escape of the charge from your body is called a **discharge**. You can get a similar effect when you pull off a nylon jumper. The crackling sound you may hear is a discharge of static electricity. As the charge on an object increases, its **voltage** rises. A high voltage can make the charge jump to the ground or to the nearest conductor connected to the

ground – which could be you. An object can be discharged by connecting it to the ground. If the object is negatively charged then the extra electrons flow to the ground. A positively charged object is discharged as electrons flow onto it.

The most impressive static discharge that you are likely to see is lightning. A static charge builds up in thunderclouds. It is discharged to the ground, often through a large building or tall tree. A lightning flash usually consists of several static discharges one after another. The temperature inside the flash can be 30 000 °C.

In operating theatres, great care is taken to ensure that static charges cannot build up. This is because anaesthetics can release explosive vapours. What could happen, and why, if charges build up?

Fig. 50.4 Lightning is a violent discharge of static electricity. Some buildings have a pointed lightning conductor on the top, which is connected to the ground by a thick copper strap. The charged cloud attracts an opposite charge to the point. Charge is transferred at the point. This helps to reduce the chance of a strike. If lightning does strike, it passes down the strap and not through the building.

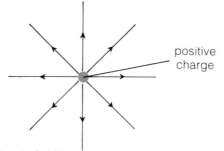

positive charge

Fig. 50.5 Electric field lines

Charged objects produce electric fields

Like magnets, charged objects can affect other objects without coming into contact with them. A charged rod can attract or repel other rods without coming into contact with them. The space around a charged object in which it can affect other objects is called its **electric field.**

As with magnetic fields, field lines around charged objects can be drawn to show the direction of the force. Arrows are drawn on the field lines to show the direction in which the force would act on a positively charged object.

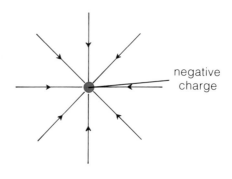

negative charge

A photocopier uses a static charge to transfer an image to the page

When a page to be copied is placed on the top of a copier, it is lit by a lamp. A lens focuses the image of the page onto a rotating **drum**. The drum is covered with a light-sensitive or **photoconductive** layer, which is charged with static electricity. Where light hits the surface, the photoconductive layer conducts electricity and the charge flows away. Charge is left behind in places corresponding to the black parts of the original page. As the drum moves round, a

copy of the black parts of the page is left as a charged pattern on the drum. Black **toner** dust is then applied to the drum. The toner is given a charge opposite to that of the charge on the drum, so the toner is attracted to those parts of the drum that are still charged. A black powder image of the page forms on the drum.

To print this image, a sheet of paper is rolled over the drum. The paper picks up a dusty copy of the original image.

The paper then passes between heated rollers that melt or **fuse** the black toner onto the surface of the paper.

The copier can be made to enlarge or reduce the size of the image by changing the position of the lens.

Laser and similar computer printers use the same principles. The computer forms an image of the page directly on the drum. A laser or other bright light is used to light up an image straight from the computer onto the drum.

Fig. 51.1 The basics of photocopying or xerography.
The lamp transfers an image of the page to the drum, which leaves a static charge. The drum collects toner dust and transfers it to the paper. The toner is melted onto the page.

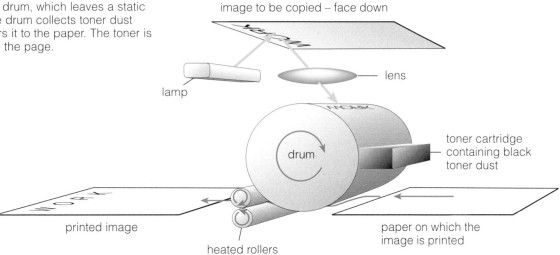

image to be copied – face down

lens

lamp

drum

toner cartridge containing black toner dust

printed image

heated rollers

paper on which the image is printed

Fig. 51.2 Changing a computer card. The strap prevents damage to the chips by static charges.

Electronic components can be damaged by static

Most computer chips are sensitive to static electricity. As you move around and brush against surfaces, a static charge can build up. Nylon carpets in particular are very good at giving people static charges. The charge that builds up can generate a voltage of thousands of volts. A computer chip may be destroyed if you touch it before removing any charge on you.

You can make sure that all the charge flows away first by wearing a conducting band connected to the ground. You have **earthed** yourself. You will then be at the same voltage as the chip and the computer, if that is earthed as well. If you have gained a positive charge, it is cancelled by electrons flowing up to you from the Earth.

Fig. 51.3 When an aircraft is refuelled, the plane must first be earthed so there is no chance of a static discharge igniting the fuel. The earthing wire here is attached to the nosewheel.

Powders and liquids flowing in pipes can cause static charge problems

The plastic pipe from a vacuum cleaner can collect charge as all the air and dust particles rush through it. You can see small hairs and fluff sticking to the outside of the pipe, attracted by this charge. The charge builds up because the pipe is an insulator and is not earthed. This can be a problem with industrial powders flowing in pipes, as a large static charge may build up on the pipes. Powders mixed with air can make explosive mixtures. For example, a cloud of custard powder would burn very quickly if ignited by a spark produced from a static discharge.

When pipes are carrying fuels, it is vital that a charge cannot jump between a nozzle and a fuel tank. If everything is conducting and earthed, sparks can be prevented. Aircraft and racing car filling systems are usually electrically connected to the vehicle before any pumping of fuel starts.

Ordinary car filler pipes and petrol pump hoses have to be carefully designed to make sure that any charge can safely leak away and not build up. It is not possible to rely on people attaching a connector to their cars before filling up. A series of petrol station fires in Germany, involving a particular make of car, are thought to have been caused by static buildup on the pipework near to the filler cap,

Fig. 51.4 A racing car being refuelled.

which was isolated from the rest of the car by plastic.

With the increased use of plastics in cars, it is more likely that charge could build up somewhere. It is recommended that hoses, tyres and garage forecourts should be conducting. This ensures that the car and fuel pipe are at the same voltage.

Question

1 Formula One racing regulations state that:
- All fuel system fittings should be metal.
- The car must be earthed before refuelling begins.
- All metal parts of the refuelling system as far back as the supply tank must be connected to earth as well.

Why is this?

The Van de Graaff generator

The Van de Graaff generator is a machine for producing large charges. A motor or handle drives a rubber belt around two rollers. The friction between the rollers and the belt charges the belt. The action of the points collects charge. If electrons are *repelled* from the points to the outside of the dome it becomes negatively charged. If electrons are *attracted* to the points from the dome, the dome becomes positively charged. Van de Graaff generators can be designed to produce either negatively or positively charged domes.

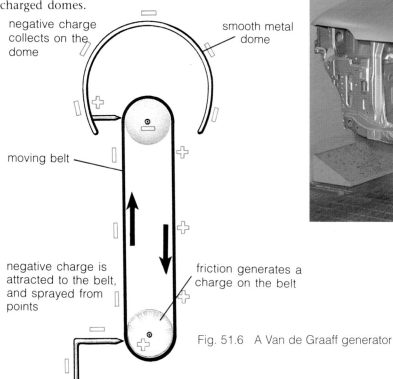

negative charge collects on the dome

smooth metal dome

moving belt

negative charge is attracted to the belt, and sprayed from points

friction generates a charge on the belt

Fig. 51.6   A Van de Graaff generator

Fig. 51.7   The car body is charged to attract paint from this sprayer. This gives a much more even coating than other methods of painting, and the paint reaches all parts of the car's surface. Great care has to be taken to keep dust particles out of the air around the car, or they too will be attracted on to its charged surface.

Fig. 51.8   A power station dust extractor. In a coal-burning power station, the waste gases contain a lot of dust and ash. Around 30 tonnes of this flue dust may be produced each hour. The flue gases are passed through a **precipitator** to remove the dust. As the gas passes through the negatively charged wires, it picks up a negative charge itself. The charged dust particles are then attracted to the positively charged plate. The dust collects on the plate, and can be collected and taken away. 99% of the dust in the flue gases can be removed like this, before the gases are released into the atmosphere.

unit providing high voltage between wires and plates

vibrator shakes the wires to dislodge any dust from them

vibrator shakes the plates to dislodge the dust

dust falls from the plates into this hopper

negatively charged wires

positively charged plate

Questions

1 a How would you charge a piece of polythene with static electricity?

b Explain how the polythene becomes charged, in terms of electrons.

c How does a charged perspex rod attract a small piece of foil?

2 If you slide out of a car seat on a dry day, you can get an electric shock when you touch the car bodywork. If you hold on to the bodywork as you slide out, this does not happen. Why is this?

Questions

1 Which block has the greatest unbalanced force?

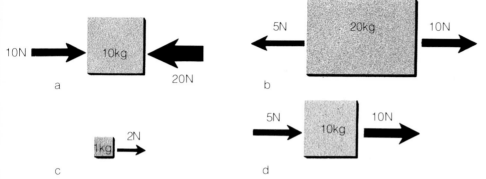

a b c d

2 A sky diver with a mass of 65 kg falls through the air. As the diver goes faster, air resistance increases. Eventually a steady velocity is reached.

 a What force must the air resistance exert?

 b Draw a diagram showing all the forces acting on the sky diver.

3 Draw all the force-reaction pairs of force and reaction for someone standing on a pair of stepladders. There are at least four.

---

4 a On the surface of the Earth, the gravitational field strength is 10 N/kg. Explain what this means.

 b On the Moon, a 10 kg mass has a weight of 16.7 N. What is the gravitational field strength on the Moon?

 c Which has the greater weight – a 1.8 kg mass on Earth, or a 10 kg mass on the Moon?

 d Which of the masses in part **c** would do more damage if it were thrown at you?

EXTENSION

5 The diagram shows an 'executive toy'.

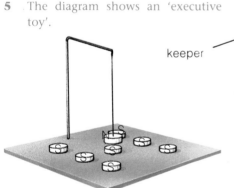

If the pendulum is swung, how will it move? Why?

6 A magnet left on its own slowly loses its magnetism. At each pole, the atomic magnets repel one another. Eventually, some change direction.

To prevent this happening, keepers are used. Keepers are made of iron, and are not permanently magnetised. Two keepers are used for two bar magnets. The diagram shows the arrangement used.

keeper

Complete the diagram, by labelling two North poles and two South poles. Some of them may be induced poles.

7 The diagram shows the apparatus used in an experiment. The surface was tilted, and the angle measured at which the block on the surface began to slip. Different surfaces were tested.

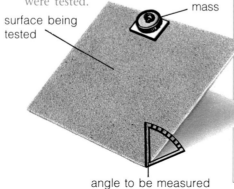

surface being tested

mass

angle to be measured

---

7 cont'd   The results are shown below:

Surface material	Angle at which block began to slip
glass	9°
ceramic tile	10°
polished metal	11°
cushioned vinyl	16°
polished wood	16°
unpolished wood	20°
cork	25°
carpet	30°

 a Which material gave the most grip?

 b Which material would you choose for making the surface of a children's slide? Give your reasons.

 c Which material would be best to cover the floor in a hospital corridor? Give your reasons.

 d Ceramic tiles are often used on steps, or around swimming pools. What precautions must be taken in the design of these tiles?

 e What is the force which prevents the block from slipping down the slope?

 f What is the force which pulls the block down the slope?

EXTENSION

8 The diagram shows a section through a ship. The total volume of the ship is 1000 m³. The total mass of the ship and its load is 400 000 kg.

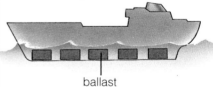

ballast

 a What is the weight of the ship and its load?

 b What upthrust must be provided by the water to keep the ship afloat?

 c What is the average density of the ship?

 d Why is the ballast carried in the bottom of the ship?

 e If the ship sails in salt water, would it need more or less ballast than if it was in fresh water? Explain your answer.

 f If the ship sails in warm water, would it need more or less ballast than if it was in cold water? Explain your answer.

ENERGY

52 ENERGY

We cannot do anything without energy. When something happens, energy is transferred. Energy is not used up, but is passed on to something else, or changed into a different form.

Fig. 52.1 Nuclear energy is released in the explosion of an atomic bomb. Into what forms of energy is it transferred?

Things happen when energy is transferred

Energy is needed in order to do things. Nothing happens without energy. Nothing happens, either, when energy is just stored. Things only happen when energy is **transferred**. The energy may be transferred from one object to another. Or it may be transferred from one form into another. When energy is transferred, things happen.

For example, petrol contains chemical energy. If the petrol remains in a car's fuel tank, nothing happens. But when the chemical energy is converted to thermal (heat) energy in the car's engine, movement occurs.

Most things that happen involve many different energy transfers. Each transfer causes something to change. Usually, the energy ends up in the surroundings. This usually means that there is a small temperature rise in the surroundings.

Internal energy is energy in a substance

In everyday speech, we often talk about 'heat energy' in an object. A hot object has 'heat energy'. However, a better term for the energy in an object is **internal energy**.

Energy transfers in firing a cannonball

Kinetic (movement) energy of a match is converted to internal energy in chemicals in the match head. This ignites them. The chemical energy is released as heat.

The heat raises the internal energy of the wick. The wick ignites. Chemical energy in the wick is released as heat.

The heat raises the internal energy of the gunpowder. The gunpowder burns very rapidly. Its chemical energy is released as heat. This raises the internal energy of gas molecules in the gun barrel. Their kinetic energy increases. The gas expands rapidly.

As the gas expands, the internal energy of the gas is transferred to kinetic energy in the cannon ball.

If you lift something and let it go it falls. The higher you lift it the harder it hits the ground when it drops. Lifting something gives it gravitational energy.

As the cannon ball rises, it gains gravitational energy, but loses kinetic energy.

At the highest point of its path, the cannon ball has maximum gravitational energy. For a split second it is vertically motionless. It has lost kinetic energy.

As the cannon ball falls, it loses gravitational energy and gains kinetic energy.

When the cannon ball lands, it has lost all its gravitational energy. On impact, its kinetic energy is converted to heat energy. The heat warms the ball, the air and ground.

Fig. 52.2

Mechanical energy has two main forms

Potential energy is stored energy. Something which has a store of energy has potential energy. A raised pendulum has potential energy. **Gravitational energy** is potential energy.

As the pendulum falls, it loses its potential energy. The potential energy changes to kinetic energy. At the bottom of the swing, it has lost all the extra potential energy it had at the top of its swing. As it rises up the other side, it gains potential energy again.

Eventually, the pendulum slows down and stops. All the potential energy has been lost to the air. The energy has been spread out into the surroundings.

Kinetic energy is energy of motion. A swinging pendulum has kinetic energy. It has most kinetic energy at the bottom of its swing. This is when it is going at its fastest. As it rises up the other side, the kinetic energy is converted back to potential energy. At the top point of its swing the kinetic energy is zero.

As the pendulum swings, air molecules bounce off it. The air molecules gain energy. The pendulum loses energy. After each swing, the pendulum rises a little less.

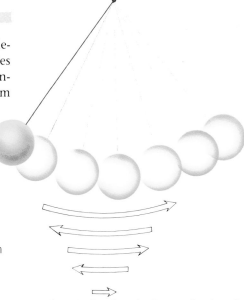

Fig. 52.3 At the top of its swing, a pendulum bob has zero kinetic energy and high potential energy. At the bottom of its swing it has high kinetic energy and low potential energy.

Questions

1 List all the energy transfers involved in firing a cannon ball.

2 A pendulum is set swinging in a vacuum. Will it continue to swing forever? Explain your answer.

3 Here is a list of some forms of energy:

chemical, electrical, gravitational, heat, internal, kinetic, movement, nuclear, sound, thermal.

a Which forms are similar, or the same?

b Describe six possible energy transfers that can be useful, giving the form that the energy starts with and how it ends up.

Energy enables something to do work. Both energy and work are measured in joules.

Work is done when a force moves an object

It takes energy to drag a block up a slope. If people pull a block up a slope, the energy comes from food in their bodies. This energy is chemical potential energy. Some of the chemical energy is changed to gravitational potential energy as the block is raised higher.

As the block is moved, a force is applied to it. We say that **work** has been done. This is the scientific use of the word 'work'. Work is done only if a **force** is **moving** an object.

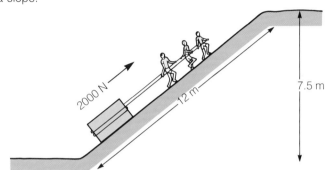

Fig. 53.1 Work is done in pulling a block up a slope.

Work is a transfer of energy, and is measured in joules

It is easy to calculate how much work is done.

work done = force applied x distance moved in the direction of the force

In Figure 53.1, the force being used is 2000 N. The block has been pulled a distance of 12 m. So the work done is 2000 N x 12 m, which is 24 000 newton-metres.

'Newton-metres' are usually called **joules**. So the work done in pulling the block up the slope is 24 000 joules. The symbol for joules is **J**.

Note that the equation

work done = force applied x distance moved in the direction of the force

is often written as

$$W = F \times d \qquad \text{or} \qquad W = Fd$$

Unfortunately, this can sometimes cause confusion, because W is also the unit of power, the watt. You must be careful not to confuse these two uses of the symbol W. (There is more about power in Topic 54.)

DID YOU KNOW?

Mass can be thought of as a form of energy. Einstein's famous equation, $E = mc^2$ means energy = mass x constant². The constant is a very large number. According to this equation, a 1 kg mass is actually 90 000 000 000 000 000 J.

Gravitational energy = mass x *g* x height

When the block reaches the top of the slope, it has gained gravitational energy. If it is pushed off, it falls back down and loses this energy.

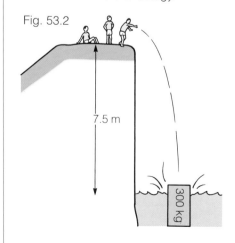

Fig. 53.2

If the block weighs 300 kg, the force pulling down on it – its weight – is 300 kg x 10 N/kg. This is 3000 N.

The block falls 7.5 m. So the energy transferred is:

force	x	distance	
3000 N	x	7.5 m	= 22 500 J

Notice that the amount of gravitational energy the block has is nothing to do with how it got to the top of the cliff. It might have been pulled up the slope by people, or it might have been lifted straight up by a crane. It makes no difference. The gravitational energy of the block depends only on its weight and how high up it is. Gravitational energy = mass x *g* x height.

Not all energy produces useful work

If you look back, you will see that the work done by the people who dragged the block up the slope was 24 000 J. But the gravitational energy that the block gained was only 22 500 J. 1500 J seems to have gone missing somewhere.

The block is pulled up the slope against a **gravitational** force and against a **frictional** force. The work done against gravity is not wasted. It becomes gravitational energy in the block. It can be released again when the block falls off the cliff.

The work done against friction is wasted. It is lost to the surroundings. The surroundings will be a little warmer after the people have pulled the block up the slope. This energy is very difficult to get back again.

EXTENSION

Energy transfer always involves some wastage

Whenever energy is transferred, some is wasted. When a block is pulled up a slope, friction causes energy to be lost as heat. The energy does not disappear. It is transferred to the surroundings.

Eventually, all the energy from most energy transfers ends up in the surroundings. It usually warms up the surroundings. This might be useful, but often this is just wasted energy.

Fig. 53.3   Energy transfers in pulling a block

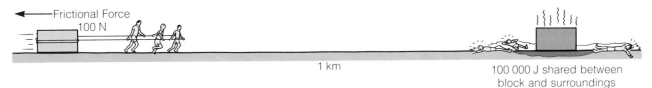

← Frictional Force 100 N

1 km

100 000 J shared between block and surroundings

Energy never disappears

Most energy transfers 'lose' energy. But the energy does not disappear. The energy goes into the surroundings.

> **Energy is never created or destroyed.** It is just transferred from one form to another. This is the **principle of the conservation of energy**.

As energy transfers continue, the energy ends up as heat in the surroundings. The 'surroundings' include the whole universe. This is enormous, so we hardly notice the temperature rise. The energy is effectively 'lost' to us.

Combustion transfers chemical energy into heat

Sometimes, we actually *want* a lot of energy to be released to the surroundings. A fire transfers chemical energy to internal energy (heat) in the fire and surroundings. A tiny fraction is used to raise the potential energy of the smoke particles as they go up the chimney. Since heat is what we want from the fire, we do not think of this as a wasteful energy transfer.

Muscles transfer chemical energy to work and heat

Muscles transfer chemical energy into work. Muscle cells contain a substance called glycogen. Glycogen contains chemical energy. When the muscle does work the glycogen is broken down to glucose. The chemical energy in the glucose is transferred to work in the muscle fibres. This actually involves many different energy transfers between different molecules in the muscle.

At each of these different energy transfers, some energy goes into the surroundings as heat. This is why you get hot when you exercise vigorously. Your body has special regulatory mechanisms to make sure you do not get too hot.

Fig. 53.4   Energy transfers in a coal fire

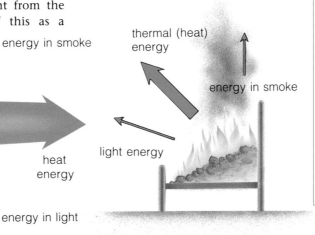

energy in smoke

chemical energy in coal

heat energy

energy in light

thermal (heat) energy

energy in smoke

light energy

Questions

1   A block is dragged up a slope for 12 m. 1500 J is wasted in dragging this block against a frictional force. What is the size of this force? (Remember work = force x distance).

2   Draw a diagram to show all the energy transfers in lighting a gas fire using matches.

3 a If you weigh 60 kg, what force is required to lift you?

  b If you run upstairs, and arrive 2.5 m higher than you started, how much gravitational energy have you gained?

  c Estimate how much energy you think you would have used to get there.

  d Where has a lot of the energy gone?

--- EXTENSION ---

4   A steam engine transfers 2 % of its heat into useful work. How much energy would be wasted if a steam-powered crane lifted a 200 kg block by 3 m?

5   A builder drops his sandwich from the top of a 300 m tower.

  a The sandwich weighs 100 g. How much energy does it lose when it falls?

  b Where does this energy go?

  c The sandwich contains 200 000 J of energy. If the builder weighs 70 kg is it worth the builder climbing down to get the sandwich?

54 POWER

Power is the rate of doing work. Power is measured in watts. A powerful system is one which is producing a lot of energy, or which does so in a short time.

Power tells you how quickly energy is transferred

Imagine a weight-lifter lifting a 400 kg mass. With enough levers or pulleys, anyone could lift this. But it would be a slow process. The weight-lifter can do it quickly.

The energy required to lift 4 000 N by 2.5 m is 10 000 J. This is less than half the chemical energy in 1 g of peanuts! So the amount of energy used by the weight-lifter is not large. But he does transfer the energy from himself to the 400 kg mass very quickly. He might take 1 s or 2 s to transfer the energy.

If you lifted the 400 kg mass with pulleys, it would take about 10 s to lift it. The same energy would be transferred, but much more slowly. So we say that the weight-lifter is more powerful. Power tells us how quickly energy is transferred.

$$\text{Power} = \frac{\text{work done in joules}}{\text{time taken in seconds}}$$

$$\text{or} \quad \frac{\text{energy transferred in joules}}{\text{time taken in seconds}}$$

The units of power are joules per second, or **watts**. The symbol for watts is **W**. So 1 W is the same as 1 J/s.

The weight-lifter transfers 10 000 J in 2 s. So his power output is:

$$\frac{10\ 000\ \text{J}}{2\text{s}} = 5000\ \text{W}$$

This is a very high power output. No-one could keep up this power output for very long!

Power is still sometimes measured in **horsepower**. A horse transfers about 750 J of energy per second. This is a power output of 750 W. So one horsepower is 750 W.

Fig. 54.1

Power must not be confused with energy

A 500 000 000 W or 500 MW (megawatt) power station produces 500 million J of energy every second. This is electrical energy transferred through heat from chemical energy in the fuel. This is a lot of energy per second! This energy output would heat 1400 kg of water to boiling point every second. So a power station is suitably named. It is very powerful.

A 100 000 000 000 W or 100 GW (gigawatt) laser system is 200 times more powerful than a power station. The laser flash lasts only 0.001 millionths of a second. The total energy transferred, however, is very small. It is about 100 J. This would raise the temperature of a teaspoon of water to boiling point! Yet, because the energy transfer is concentrated into such a short time, we say that the laser system is very powerful. Very powerful systems, like this laser system, often rely on short bursts of energy. They cannot run continuously.

Racing cars are much more powerful than normal cars. They can transfer energy very quickly. The racing car transfers chemical energy from its fuel at a much faster rate. So it uses up fuel more quickly than a normal car. For example, 40 cm³ of petrol might take a small car 1 km. The same amount of fuel would probably take a racing car about 100 m!

Fig. 54.2 Didcot Power Station has four 500 MW coal-burning generators. It also has four 25 MW gas turbines, which are sometimes used to meet a sudden extra demand on its electricity production.

Sprinters need to transfer energy quickly for a short time

Sprinting and long distance running make very different demands on an athlete. Sprinters need to be more powerful than long distance runners.

A sprinter trains his or her muscles to deliver a very quick burst of energy. In a top class 100 m race, the runners may transfer chemical energy to kinetic energy at a rate of almost 1500 W. They cannot keep this up for long. Some of the very best sprinters can only produce this sort of power output for the first 60 m or so.

A long distance runner needs to train muscles to transfer energy over a very long period of time. The power output is much lower, but takes place over a longer time. The total amount of energy transferred by a marathon runner is much greater than that transferred by a sprinter.

INVESTIGATION 54.1

Calculating power output

$$\text{Power} = \frac{\text{energy transferred}}{\text{time taken}} = \frac{\text{force x distance}}{\text{time taken}}$$

If you move a force through a distance and find the time that it takes you, you can calculate your power output.

Pedalling an exercise bicycle
Use a newton meter to measure the force that you exert. (Hook a newton meter to the pedal and pull at right angles to the crank until it is moving.) Measure the length of the pedal crank in metres. Pedal hard for one minute, and count the number of turns that you complete.

Force used = F newtons

Distance moved = circumference of circle x number of turns = $2\pi r$ x number of turns

Energy transferred = force used x distance moved = F newtons x $2\pi r$ x number of turns

So power output = F newtons x $2\pi r$ x number of turns/time taken in seconds

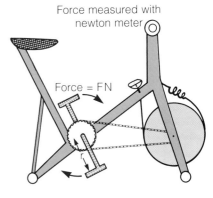
Force measured with newton meter

Force = FN

Running upstairs
Find your weight in newtons. Measure the vertical height of the stairs. Run up the stairs as fast as you can. Time yourself in seconds.

Energy transferred = force used x distance moved = your weight in newtons x height of stairs in metres

Power output = your weight x height of stairs/time taken in seconds

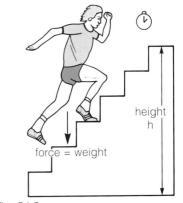
force = weight

height h

Fig. 54.3

Questions

1 Student A pedalled an exercise bicycle against a force of 100 N for 2 min. During this time the pedals turned 100 times. The pedals were 20 cm long.
Student B ran upstairs. She weighed 60 kg. She climbed 5 m in 10 s.
a Who was the more powerful?
b Who would be able to keep up this rate of energy transfer the longest?

2 A builder carries 24 bricks to the top of a 3 m wall. Each brick weighs 25 N.
a How much energy is transferred to the bricks?
b In what form is this energy after transfer?
c If the builder weighs 800 N, and takes 15 s, calculate the power that he delivers.
A motor driven conveyor is an alternative way of carrying the bricks to the top of the wall. The conveyor produces 1000 W of useful output.

d How long would it take the conveyor to lift the bricks?
e Which is the better way of lifting the bricks? Why?

3 A litre of petrol contains 35 000 000 J. If a car travels at a speed of 25 m/s, 1 dm^3 of petrol lasts 13 min.
a What power output does this represent?
b In 13 min, the car travels 20 km. If all the chemical energy is transferred, what force pushes against the car?

— EXTENSION —

55 HEAT

Heat is thermal energy as it is being transferred. If objects are heated, they gain internal energy. This can change their temperature, or the arrangement of their particles.

Heat flows from hot bodies to cold ones

A pan of cold water on a hot cooker hob gets hotter. Because the hob is hotter than the pan, energy is transferred from the hob to the pan. The energy being transferred from the hob to the pan is **heat energy**. Heat energy flows from hot bodies to cold ones.

Strictly speaking, we can only use the term 'heat' for energy *as it is being transferred*. The energy *in* the hot cooker hob is not heat energy. A better name for it is **thermal energy** or **internal energy**. The hot cooker hob has higher internal energy than the cold water. Heat flows from bodies with high internal energy to bodies with lower internal energy.

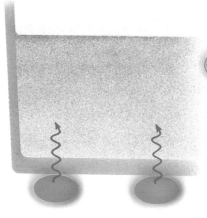

Fig. 55.1 Heating water on an electric hob. Heat energy flows from the rings, through the pan, and into the water.

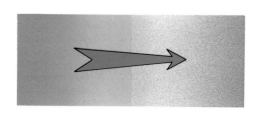

Heat energy flows from hot areas to colder ones.

Heating does not always increase temperature

Temperature is a measure of the speed of movement of the particles in a substance. The higher the kinetic energy of these particles, the higher the temperature. When heat flows into a substance, it may **increase the kinetic energy** of the particles. The temperature goes up. This is what we expect to happen. When we heat something, we expect it to get hotter.

But this is not always the case. Heat flowing into a substance may just **change the arrangement of the particles**. Instead of increasing their kinetic energy, it increases their potential energy. The particles get further away from each other. This is what happens to water at 100 °C. The particles fly away from each other and the water boils. It changes state from a liquid to a gas. While this is happening, the temperature of the water does not change. The heat energy flowing into it does not raise its temperature. It changes its state. This heat energy is called **latent heat**.

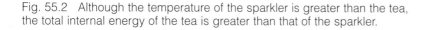

So when heat energy flows into a substance, it may increase its temperature, or change the arrangement of its particles, or both. Both of these changes involve a change in the internal energy of the substance. Heating a substance always raises its internal energy.

Temperature and internal energy are not the same

Temperature is related to internal energy. If the temperature of a substance increases, its internal energy increases. But temperature and internal energy are not the same. A hot spark has a higher temperature than a cup of tea. The individual particles in a hot spark have a higher internal energy than the individual particles in a cup of hot tea. But there are *more* particles in the cup of tea. Although each particle in the tea has a lower energy than the particles in the spark, the *total* energy in the cup of tea is greater. So although the temperature of the spark is greater than the cup of tea, the total internal energy of the cup of tea is greater than the spark.

Fig. 55.2 Although the temperature of the sparkler is greater than the tea, the total internal energy of the tea is greater than that of the sparkler.

Measuring changes in internal energy

Heating can cause a rise in temperature. The more particles there are to heat, the more energy is needed to produce the same change in temperature.

For example, it takes 4.2 J of energy to raise the temperature of 1 g of pure water by 1 °C. To raise twice as much water by this amount takes twice as much energy. So it takes 8.4 J of energy to raise the temperature of 2 g of pure water by 1 °C.

The energy needed to raise 1 g of a substance by 1 °C is called the **specific heat capacity** of that substance. The specific heat capacity of water is 4.2 J/g °C.

Specific heat capacities can be measured using the apparatus shown in Figure 55.3. Energy is transferred to the substance in the form of an electric current.

For example, it is found that 4550 J are needed to raise the temperature of 1 kg of aluminium by 5 °C.

The specific heat capacity of aluminium is the number of joules needed to raise 1 g of it by 1 °C.

So the specific heat capacity of aluminium = $\dfrac{4500\ \text{J}}{5\ °\text{C} \times 1000\ \text{g}}$ **= 0.91 J/g °C**

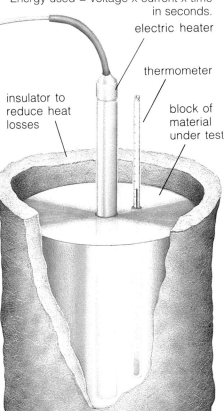

Fig. 55.3 Measuring specific heat capacity. To measure the electrical energy being transferred to the block, you can use a joule meter. If you do not have one, you need to measure the voltage (usually 6 V or 12 V) and the current (in amperes).
Energy used = voltage x current x time in seconds.

electric heater

thermometer

insulator to reduce heat losses

block of material under test

Water has a high specific heat capacity

The specific heat capacity of water is 4.2 J/g °C. This is a high value. It means that a lot of heat energy is needed to produce even a small temperature rise. This makes water a very useful substance for animals and plants. Your body is mostly water (65-70 %). So you have a high specific heat capacity. Large amounts of heat must flow into or out of your body before your temperature changes. This helps you to keep your body temperature constant.

Large bodies of water, such as lakes and the sea, do not change temperature rapidly. You have probably noticed how cold the sea is around Britain, even in summer! Large amounts of heat energy must flow into the sea before its temperature changes much. This makes life easier for the animals and plants which live in water. They do not have to cope with rapid temperature changes.

Questions

1 a An electric storage radiator contains 100 kg of blocks. The specific heat capacity of the blocks is 0.8 J/g °C. How much energy is needed to heat the blocks from 10 C to 70 °C?

b It takes 3 h for the radiator to cool down to 10 °C again. Does it give out more or less heat than a 1 kW electric fire over this time?

2 In a heating experiment, energy was transferred to 10 kg of water at a power of 42 W. The initial temperature of the water was 10 °C. The results were as follows.

Time (s)					
100	200	300	400	500	600
Temp. (°C)					
10.1	10.2	10.3	10.4	10.5	10.6

a What is the total temperature rise during the experiment?

b What is the total energy supplied during the experiment?

c What is the specific heat capacity of the water?

3 The specific heat capacity of water is 4.2 J/g °C. The specific heat capacity of concrete is 0.8 J/g °C.

a What mass of water will store 2.52 kJ for a 60 °C rise in temperature?
(Use the equation:
energy = mass x specific heat capacity x temp. rise)

b What mass of concrete will store the same amount of energy under the same conditions?

c The density of water is 1 g/cm³. The density of concrete is 2.5 g/cm³.
What volume of concrete would store the same energy as 100 cm³ of water?

d Is this the reason that concrete and not water is used in storage radiators? Explain your answer.

Stored energy is potential energy. Most of our stored energy comes from the Sun.

The Sun provides most of our energy

Sunlight falls on the Earth with a power of nearly 1 kW per square metre. Very little of this energy is used directly by humans. After a series of energy transfers, and perhaps millions of years, it appears in a form that we can store and conveniently transport. After all these transfers, much of this energy is wasted and appears as heat in the surroundings.

Fig. 56.1   Almost all energy on Earth originates from the Sun. Some of this energy becomes stored in forms which we can use.

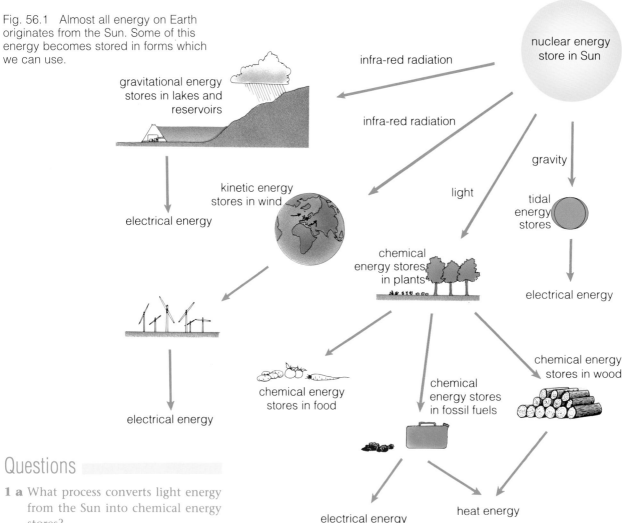

Questions

1 a What process converts light energy from the Sun into chemical energy stores?

b Explain how infra-red radiation from the Sun provides gravitational energy stores in lakes and reservoirs.

c Name one important source of electrical energy which is not included in this diagram.

d What type of chemical reaction converts chemical energy in fossil fuels into heat energy?

Human-made energy stores

Springs and weights can store energy. The weights in a grandfather clock or the spring in a clockwork motor store energy. The work they do is running the clock. A pile driver uses the stored gravitational energy in the weight to drive the piles into the ground. Piles are metal rods used to support building foundations and prevent earth collapsing.

Batteries store chemical energy. This is transferred into electrical energy when a current flows.

An electric central heating radiator has a core of special bricks. These are heated at night when electricity is cheap. Electrical energy is transferred to internal (heat) energy in the bricks. The energy is released throughout the day. In gas or oil central heating, the heat energy is stored in water and pumped around the house.

The Earth also stores energy like this. Rocks contain heat energy. This is called geothermal energy. Geothermal energy can be transferred to water pumped to the rocks. The hot water or steam produced is then pumped away to where it is needed.

Fig. 56.2 Part of the Joint European Torus (JET) research torus, shown here under construction. The large flywheels store kinetic energy for later release.

Fig. 56.3a Batteries store chemical energy, which is transferred into electrical energy when a current flows.

Fig. 56.3b The large mass of a pile driver stores gravitational potential energy at the top of its tower.

Questions

1 When you drink a glass of milk it gives you energy.
   a How did the energy get into the milk? Show all the energy transfers.
   b One argument for vegetarianism is that vegetarians make better use of the Sun's energy falling on to the Earth. Is this true? Explain your answer.

EXTENSION

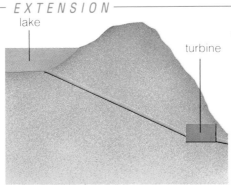

lake

turbine

2 A turbine is rather like a propeller. Water pushing past it turns the turbine. This movement can be used to generate electricity.

   a A turbine is 200 m below a lake. How much gravitational energy does each kilogram of water lose as it falls down to the turbine?
   b Where does this energy go? Show all the energy transfers.
   c If 12 500 kg of water fall through the pipe every second, what is the total energy transfer per second?
   d What power (rate of work) does the system produce?
   e Is this the same as the amount of electrical power that could be produced?
   f Where else might some of the energy go?

57 EFFICIENCY

The efficiency of a system describes how good it is at carrying out an energy transfer without wasting energy.

Efficient systems transfer energy without waste

Nearly all energy transfers produce some heat. Unless heat is the required form of energy, this is wasted. The efficiency of a system tells us how much energy is wasted in a system. The efficiency of a system is defined as:

$$\text{efficiency} = \frac{\text{useful output energy}}{\text{total input energy}}$$

This is a ratio, so it has no units.

If a system wastes no energy at all then it has an efficiency of one. This is often multiplied by 100 to give a percentage.

$$\frac{\text{percentage}}{\text{efficiency}} = \frac{\text{useful output energy}}{\text{total input energy}} \times 100$$

Calculating efficiency

A block is dragged up a slope with a force of 80 N. The block weighs 100 N. The slope is 20 m long and 10 m high.
Energy used = 80 N x 20 m = 1600 J.
Gravitational energy gained = 100 N x 10 m = 1000 J.
Energy 'wasted' = 1600 -- 1000 = 600 J.

Notice that the *useful* output energy is 1000 J. But the total energy output is 1600 J.

$$\text{Efficiency} = \frac{1000\,\text{J}}{1600\,\text{J}} \times 100 = \textbf{62.5 \%}$$

So 37.5 % of the energy used in dragging the block up the slope is wasted.

Power stations are about 30 % efficient

An electric fire is very efficient at converting electric energy to heat energy. 1 kJ of electrical energy is converted to 1 kJ of heat energy every second. So an electric fire is 100 % efficient.

But the power station which produced the electricity is much less efficient. For every 100 J stored in the fuel used by the power station, only 30 J of electrical energy is produced. The power station is only 30 % efficient. So perhaps an electric fire is not so efficient after all! 70 % of the energy in the fuel used to produce the electricity to heat the fire is wasted at the power station.

Plants and animals are inefficient energy converters

Plants convert sunlight energy to chemical energy in carbohydrate molecules. The process by which they do this is called **photosynthesis**. Photosynthesis is only 1 % efficient. Only 1 % of the sunlight energy is converted to chemical energy in the plant.

If the plants are fed to sheep, only 10 % of the energy in the plant is converted to energy in the sheep.

Sheep, like most animals, are about 10 % efficient in converting the energy in plants to energy in themselves. A lot of energy is lost as heat from the sheep's body.

If a human eats the sheep then only 10 % of the energy in the sheep is transferred to useful energy in the human. Humans are about 10 % efficient in converting the energy in their food into energy in themselves.

So, as a system for converting sunlight energy into useful work, food chains are not very efficient. Only a tiny fraction of the original energy in the sunlight ever reaches the animal at the end of the food chain. This is why food chains are usually quite short.

Fig. 57.1 Energy flows from the Sun, through plants, to animals.

Questions

1 A light bulb transfers electrical energy to light energy. Electrical energy is transferred to internal energy in the light bulb filament. The tungsten filament reaches 2500 °C. In a light bulb using 100 W of power, only 20 J of light energy is produced per second. How efficient is the light bulb?

2 The newer type of light bulb, which contains a coiled up fluorescent tube, is more energy efficient. A 25 W new bulb produces the same output as an old-style 100 W bulb. How efficient is this type of light bulb?

— E X T E N S I O N —

DID YOU KNOW?

If everyone in Britain replaced one light bulb in the house with an energy efficient light bulb, the energy saved would be enough to light the whole of Scotland.

Engines

A heat engine converts internal (heat) energy to useful work. Usually, the heat comes from burning a fuel. The earliest heat engine was the steam engine. The first steam engines were only 1 % efficient. A modern car petrol engine is about 20 % efficient. A diesel engine is 40 % efficient.

Heat engines usually have a lot of moving parts. So a lot of energy is lost because of friction. But even if friction could be completely eliminated, there is a limit on the engine's efficiency. The efficiency depends on the running temperature of the engine. The hotter the engine runs, the more efficient it will be. But the components of most engines cannot safely be heated to more than about 1100 °C. The maximum efficiency possible at this temperature is 75 %. Friction and other losses actually reduces this to between 25 % and 40 %.

A new type of engine, made of ceramics, has been built in Japan. This engine can be safely run at much higher temperatures than metal engines. It is much more efficient. It produces as much power from a 1600 cc engine as a normal 6000 cc engine.

Questions

1 An electric kettle uses 2.4 kW and runs for 100 s.
  a If the water it heats gains 228 kJ in this time, how efficient is the kettle?
  b Where has the wasted energy gone?
  c The kettle is made from polished stainless steel. Would it be more or less efficient if it was made from dark blue plastic?

2 A gas fire has a power input of 6.6 kW. It provides 4 kW of room heating.
  a How efficient is the gas fire?
  b What energy transfers take place?
  c In what form is the wasted energy?
  d Where does the waste energy go?
  e How does the efficiency of the gas fire compare with an electric fire?
  f In answering question **e**, what other energy transfers need to be considered?

— E X T E N S I O N —

3 A builder has two choices when he needs to lift bricks. He can pull them up himself using a pulley, or he can use an electric motor.
The diagrams show these two alternatives.

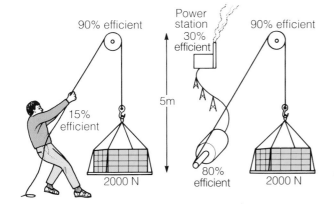

a What amount of chemical energy must the builder supply if he uses the pulley system to lift the bricks?

b What amount of electrical energy does the motor require from the power station to lift the bricks?

4 A food processor has a 400 W motor. A belt drives the blades around at 300 revolutions per minute. Each blade is 10 cm long.

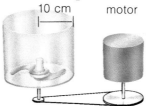

A food processor, with its outer casing removed.

a How many revolutions do the blades make in 1 s?
b How far does a blade travel in one second? (Circumference of a circle = 2πr).
c If the motor and blade system is 75 % efficient, what is the useful work output?
d How much energy does the system transfer as useful work in 1 s?
e What force can the blade tips exert? (Energy = force × distance.)
f Why do food processors need safety switches?

5 A diesel electric train has a diesel engine that drives a generator that drives the electric motors which drive the wheels.

a If the diesel engine is 40 % efficient, the generator is 75 % efficient and the electric motor is 80 % efficient, estimate the maximum efficiency of the train.
b Draw a diagram to show the energy transfers in the train.

Heat is transferred by conduction, convection and radiation

Heat energy can travel in three ways. **Conduction** transfers heat through any kind of material, although some materials are much better conductors than others. **Convection** transfers heat in liquids or gases. **Radiation** transfers heat even when no material is present at all. Heat from the Sun reaches us as radiation, which travels through space.

Metals are very good conductors of heat

Metals are made up of atoms which hold their electrons very loosely. Some of the electrons are free to move around the lattice structure. These electrons can carry the vibrational energy through the lattice more easily than phonons.

Mercury is a liquid so it ought to be a poor conductor of heat. But it is also a metal. The free electrons are able to carry the heat energy through the liquid. So mercury is actually a good heat conductor.

Another liquid metal – liquid sodium – is used as a heat conductor in nuclear power stations and some car engines. The valves in a car engine open and close to let gases in and out of the cylinders. The explosions in the cylinders make the valves very hot. Temperatures of around 700 °C are reached. It is important to stop the valves from overheating. In some cars the valve stems are hollow and contain sodium. When the valves get hot, the sodium melts. The molten sodium carries the heat away from the valve heads.

Fig. 58.2 Metals contain freely-moving electrons, which transfer heat energy easily through the metal. At higher temperatures these electrons have more kinetic energy and move faster.

Conduction transfers heat through materials

Conduction happens in all materials. Conduction is a direct transfer of the vibrational energy of atoms.

In solids the atoms are rigidly held together. If some atoms start vibrating more than others, the vibration will be passed through the structure. Vibrational waves or 'phonons' pass the increased vibration through the material. If you heat one end of a rod, the energy is passed down the rod through the bonds between the particles.

In liquids the particles are further apart. So they do not conduct heat so easily. Most liquids are poor conductors of heat.

In gases the particles are very far apart. So gases are very poor conductors of heat.

Fig. 58.1 Vibrating atoms in one part of a material pass on their vibrations to atoms close to them. This is how heat is conducted.

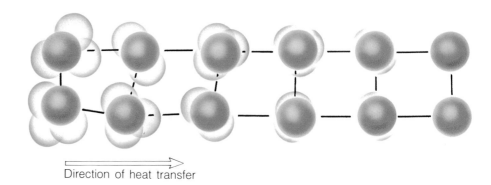

Direction of heat transfer

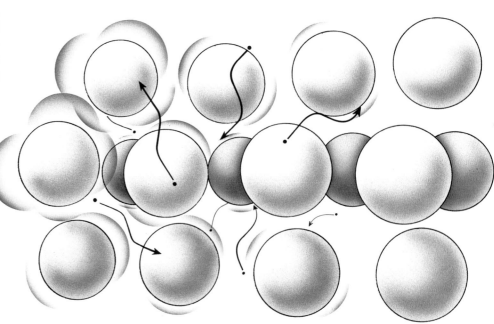

Substances which only conduct heat slowly are called insulators

Substances which are bad conductors of heat are called **insulators**. Materials which trap air inside themselves are good insulators. Air is a gas so it is a poor conductor of heat. Wood contains a lot of air, and is a good insulator. Fur is also a good insulator. Each hair contains trapped air. Fur can be made an even better insulator by raising the hairs on end, so that more air is trapped *between* the hairs.

The American space shuttle gets very hot as it travels quickly through the Earth's atmosphere. It is covered with special tiles which are good insulators. They stop the shuttle from overheating. You could heat one of these tiles with a blow lamp and pick it up straight away without burning yourself! Although the tile is very hot, it does not allow heat to travel from itself into your hand. It is a good insulator.

Convection transfers heat in fluids

If you drop a potassium permanganate crystal into a beaker of water, colour begins to spread into the water as the particles dissolve. If you heat the water, you can see the colour rise through the water.

Why does this happen? The heat energy transferred to the water increases the internal energy of the water and potassium permanganate particles. They get further away from each other. The water expands. This warmer water is now less dense than the colder water at the top of the beaker. So it rises upwards. People often say that 'heat rises'. This is not really true. The heat energy does not rise on its own. The *hot water* rises, and takes the energy with it.

So when one part of a fluid is hotter than another part, the hot part tends to move upwards. This movement is called **convection**. The currents pro-

Fig. 58.3   Convection transfers heat in fluids.

duced are called **convection currents**. Convection currents circulate around the fluid. They spread the energy through the fluid. Fluids include liquids and gases.

Winds are caused by convection currents

During the day, the land warms up more than the sea. This is because water has a high specific heat capacity. It takes a lot of heat to raise its temperature by only a small amount.

The warm air over the land rises.

warm air rises as the land quickly warms up

This happens on a larger scale, too. The air near the equator heats up more than air near the poles. The hot air over the equator rises. Cooler air from the north and south moves in to replace it. This sets up air movements which cause winds.

Cold air over the sea moves in to replace it. So during the day, breezes tend to blow from the sea on to land.

At night the land cools down faster than the sea. The warmer air over the sea rises. Cold air over the land moves in to replace it. So during the night, breezes tend to blow from the land on to the sea.

cool air moves in to replace the rising air

Fig. 58.4   During the day, sea breezes blow on to land.

Convection currents spread heat through a room

Central heating 'radiators' are badly named. They do not actually radiate much heat at all. They warm the air around them. This hot air rises. Cold air in the room moves towards the radiator to replace it. The radiator heats this cold air, so it rises. Convection currents are set up in the room, drawing cold air towards the 'radiator' and carrying warm air away from it.

Convection currents were used to ventilate mines. Air at the bottom of one shaft was heated. The hot air rose up the shaft. Cold air moved down another shaft and along the tunnels to replace the hot air.

Heat can be transferred as infra-red radiation

Changes in the internal energy of particles cause changes in the way in which atoms are arranged. These changes cause the atoms to emit energy in the form of **electromagnetic radiation**.

When an electric current flows through a lamp filament, it causes it to emit electromagnetic radiation. Some of this is high-energy radiation, which we see as light. But some of this radiation has a lower energy. It is called **infra-red radiation**. You cannot see infra-red radiation. But if it falls on to your skin it raises the temperature of your skin. Heat can be transferred by infra-red radiation. A filament lamp only produces 20 % of its energy as light. The remaining 80 % is infra-red radiation or heat. An electric fire produces even more of its energy as heat. Electric fires usually have a reflector behind the bars, to bounce the heat into the room.

Fig. 58.5 The element of an electric fire gets hot as an electric current flows through it, and produces infra-red radiation.

Black surfaces are the best radiators of heat

If two cans, one silver and one black, are filled with hot water, the black one will cool down faster than the silver one. This is because the black surface radiates heat away from the can faster than the silver surface. The best radiating surfaces are black.

A black surface looks black because it absorbs most of the light which falls on to it.

A lot of the light energy is radiated from the surface as heat, which you cannot see.

A good radiating surface also absorbs energy well. If you paint part of your hand black, and hold your hand in front of a fire, you will find that the black part feels hotter than the rest. The black part absorbs more heat.

Car radiators are used to keep the engine cool. They are painted black, so that they lose heat quickly. Kettles are usually made of shiny metal. This is so that they do not lose heat quickly.

Fig. 58.6 Black surfaces radiate and absorb heat better than white or silver ones. If you stood these two cans of water in front of an electric fire, which would get hotter faster? If you switched the fire off, which would cool down faster?

Radiation travels in straight lines

Radiated heat travels in straight lines. If you sit facing a fire, the radiated heat from the fire warms your face. But your back will feel cold.

So radiation is not a good way of heating a whole room. It only heats those parts of the room which the radiation can reach in a straight line. Convection is a better way of heating a space.

Radiation does not need to travel through a material

Electromagnetic radiation does not need any material to travel through. It can pass through a vacuum. The energy we receive from the Sun is radiated to us.

Fig. 58.7 The woman's body is in thermal equilibrium with the surroundings. Her back is absorbing extra energy from the fire so it will have settled at a higher

Everything radiates and absorbs at the same time

Objects radiate more energy as they get hotter. This energy is called **thermal radiation**. A paper clip heated in a Bunsen flame eventually radiates some energy that can be seen. As it gets hotter the colour changes and the metal starts to glow. It is also giving off a lot of energy that you can't see. Out of the flame it quickly cools. It is giving off more radiation than it is receiving from the room. When the temperature stops falling, the clip is absorbing as much energy from the room as it is radiating back. It has reached **thermal equilibrium** with its surroundings.

temperature than her front. The hot drink is losing more energy than it is absorbing from the room and is cooling down.

Questions

1 A central heating radiator is a thin aluminium or steel container full of hot water.
   a Why does its surface feel hot to the touch?
   b How does the radiator heat the air in contact with its surface?
   c What happens to the hot air?
   d Explain how the radiator heats the whole room.
   e If you hold your hands in front of the radiator, without touching it, how are your hands heated?
   f Why are radiators not painted matt black?
   g Why are the inlet and outlet pipes of a radiator at the bottom, not the top?
   h Why do radiators have a small outlet valve at the top?
2 An electric oven has heating elements in its walls, which release heat energy as electricity flows through them.
   a Give the name of the process by which heat is transferred through the walls of the oven.
   b Explain this process, in terms of the molecules in the oven walls. A diagram may help you to do this.
   c How is the heat transferred from the walls of the oven into the air inside it?
   d The inside surfaces of the oven walls are usually coloured black.

Suggest why this is done.
   e If you are cooking a dish which needs an especially high temperature, your recipe may tell you to put it on the top shelf of the oven. Why should you do this?

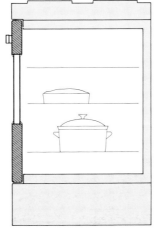

   f The diagram shows a section through the door of the oven. It is double glazed – which means it is made of two sheets of glass with an air gap between them. Explain how the double glazing cuts down the amount of heat which is lost from the oven.

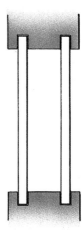

3 A crystal of potassium permanganate is dropped into a can of water.
   a If the can was left alone, the colour from the crystal would gradually spread through the water. Name, and then briefly describe, the two processes which would cause this to happen.
   b If the can was heated from below, convection currents would be produced in the water. Make a diagram to show the pattern of these currents. Explain why the currents make this pattern.
   c Two of these cans were painted, one red and one black. They were then heated until the water in them was at 80 °C. Would you expect the water in the two cans to cool at the same rate? Explain your answer.
4 Explain why:
   a Air is a poor conductor of heat.
   b On a hot day, you may feel cooler if you wear white clothes than if you wear dark ones.
   c It is more comfortable to stir a hot stew with a wooden spoon than with a metal one.
   d Hot air balloons designed to travel very long distances are often coloured silver.

CONTROLLING HEAT FLOW

Insulators slow down the transfer of energy from hot to cold bodies.

Fins are used for cooling by radiation

Most animals are poikilothermic. This means that their body temperature is the same as their surroundings. If it is hot, they are hot. If it is cold, they are cold. Reptiles are poikilothermic. If the air temperature is high, their body temperature is also high. This makes their metabolic reactions take place at a faster rate. They are more active. It is thought that the large fins on the backs of dinosaurs like *Dimetrodon* might have helped them to warm up quickly in the morning. If they stood sideways on to the sun, the large area of the fin would absorb radiated heat. This would raise their body temperature, and make them more active. Butterflies do this, too. On a cool day, a butterfly will rest with its wings outstretched, at right angles to the sun's rays. This warms its body and makes it more active. Large fins can also be used to *lose* heat by convection, conduction and radiation. *Dimetrodon's* fin might have been used for this on hot days. It would have stood end on to the sun's rays. Air-cooled engines also have fins. Heat is lost from the large surface area of the fins into the air.

Fig. 59.1 *Dimetrodon*. The large fin is believed to have been useful in controlling body temperature. Energy would be absorbed as radiation, or lost by radiation and convection in the air.

Insulation saves energy

Mammals and birds are **homeothermic** animals. They can keep their body temperature high, even when outside temperatures are low. They use food to generate heat inside their bodies. This uses a lot of food, so homeothermic animals need to make sure that they do not lose too much of this heat. As it is, they have to eat a lot more food than poikilothermic animals. It is important to save as much as internal energy as possible so as not to waste food which might be difficult to obtain.

Layers of fat or blubber around the animal's body act as insulators. They slow down loss of heat energy. Fur and

Fig. 59.2 The thick layer of fat, and the fur which traps air, insulates the polar bear.

feathers are also good insulators. Thick layers of fur trap air, which is a very poor conductor of heat. In cold weather, animals and birds may fluff up their fur or feathers. This traps even more air, and increases the insulation. Hu-

INVESTIGATION 59.1

Penguins

Penguins living in very cold regions around the Antarctic often huddle together to keep warm. Using test tubes full of water to represent penguins, design an experiment to find out if huddling together helps penguins to retain their body heat.

mans do not have much fur! We make up for it by wearing clothing. Wool is a good insulator because wool fibres trap air. Woollen clothing stops the heat generated inside your body from escaping. Cotton is not such a good insulator because the fibres don't trap much air. Cotton is good to wear in summer when you actually want to *lose* heat from your body. If you really want to keep warm, many thin layers of clothes are better than one or two thick ones. Air is trapped between each layer. The more layers, the better the insulation.

Small objects lose their internal energy faster than large ones

Fig. 59.3   How surface area changes with volume.

This block has sides of 1 cm. What is its surface area? What is its volume?

This block has sides of 5 cm. What is its surface area? What is its volume?

Which block has the greatest ratio of surface area to volume? What can you say about the way in which this ratio changes as things get bigger?

Energy is lost through surfaces. The larger the surface area, the faster the rate of energy loss.

A small animal has a larger surface area for its size than a big animal. A polar bear, for example has a very large volume. It also has a large surface area. But most of a polar bear is 'inside'. Its volume is large compared to its surface area. A shrew has a very small volume. It also has a small surface area. But a lot of a shrew is 'outside'. A shrew's surface area is large compared to its volume. Both polar bears and shrews are homeothermic animals. They generate heat inside their bodies. A polar bear is much bigger than a shrew, so it generates a lot more heat. And not very much of this heat escapes from the polar bear, because its surface area is small compared to its volume. But the small shrew generates much less heat. And a lot of this heat escapes from its relatively large surface area.

So small animals have problems keeping warm. They lose heat quickly through their relatively large surface area. A shrew has to eat its own weight in food every day, just to generate enough heat to keep warm. A polar bear can manage by just eating one seal every few days.

Human babies have large surface areas compared to their volumes. They lose heat easily. Small babies must be wrapped up well in cold temperatures.

Vacuum flasks reduce heat flow between the contents and the air

A vacuum flask is made of two containers inside one another, and separated by a vacuum. (There is usually another covering on the outside, to make it stronger and to look attractive.) The surfaces of both containers are shiny. They are made from glass or stainless steel.

These shiny surfaces reduce energy transfer by radiation. The vacuum between them prevents heat loss by conduction. (Remember – conduction only happens through materials.) A small amount of energy is conducted up the sides of the inner wall. This is kept as small as possible by making the sides of the walls very thin.

So if hot coffee is put into a vacuum flask, it is difficult for heat to be transferred from the coffee to the air. The coffee stays hot. If liquid nitrogen at –196 °C is put into a vacuum flask, it is difficult for heat to be transferred from the air into the liquid nitrogen. The liquid nitrogen stays cold. Vacuum flasks are just as good at keeping things cold as keeping things hot. A flask of ice-cold orange juice will stay cold for a long time, even on a hot day.

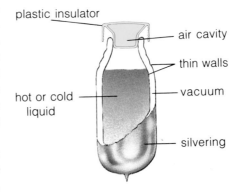

Fig. 59.4   A vacuum flask, with its outer covering removed.

Questions

1   Explain why:
 a Thin people tend to feel the cold more than fat people.
 b It is important for old people to eat well in winter.
 c Homeothermic animals need to eat far more than poikilothermic animals of similar size.
 d Supermarkets usually display frozen goods in chest freezers rather than upright freezers. (Clue – think what happens when the door is opened.)
 e Several layers of thin clothes will keep you warmer than two layers of thick ones.
 f Large penguins are found at the South Pole and smaller ones are found further away.

60 FRICTION

Whenever there is motion there is some friction, except in space. This means that all energy transfers involving movement produce some heat as well.

Friction is a force which results from surfaces in contact

The particles in materials often attract one another when the materials are in contact. So when one surface is dragged over another, work has to be done against this force of attraction. The energy transferred often causes heating. Rough materials have a larger frictional force than smooth ones. This is because the ridges and grooves on the surfaces catch on one another. There is also attraction between the particles.

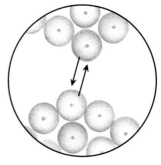

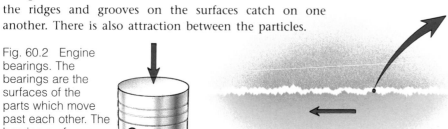

Fig. 60.1 Friction acts when surfaces move over one another.

Fig. 60.2 Engine bearings. The bearings are the surfaces of the parts which move past each other. The bearing surfaces and the crankshaft are kept apart by oil under pressure.

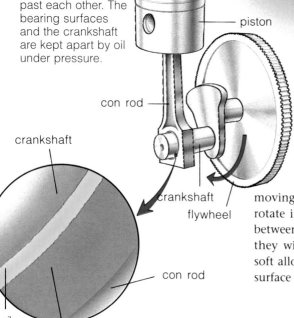

piston

con rod

crankshaft

crankshaft flywheel

con rod

oil

bearing

Friction produces heat

When a match is dragged across the side of the matchbox, the friction between the match head and the rough surface raises the internal energy of the chemicals in the match head. This ignites them. The chemicals in the match head burn.

This is a useful heating effect. But often friction is a nuisance. It wastes energy and causes wear. In a car engine the moving parts are kept apart from each other by a layer of oil. The oil is a **lubricant**. A lubricant is a substance which reduces friction between two moving surfaces. As the piston moves up and down it causes the crankshaft to rotate inside the connecting rod (con rod). The con rod and crankshaft have oil between them to keep the surfaces separate and reduce friction. But, all the same, they will sometimes touch. So the con rod is lined with a **bearing** made of a soft alloy. If the moving surfaces make contact, the soft alloy will not damage the surface of the crankshaft. The soft bearing can be replaced if it wears out.

Braking systems make use of friction

A car braking system uses friction to slow down the rotation of the wheels. Asbestos-based pads or shoes push against metal discs. Friction between the pads and the discs slows down movement between them. It transfers the kinetic energy of the car to internal energy in the metal. If the car is travelling fast, the kinetic energy is very high. It could be as much as 1.5 MJ. So the amount of energy transferred is also very high. The discs can get very hot. Powerful cars and motor-bikes have ventilated brake discs. This allows some of this energy to be lost to the air. This means that the kinetic energy of the moving car or motorbike can be transferred to thermal energy more quickly, without overheating the brakes. It is important that brake pads or shoes should not conduct heat to the brake mechanism. Heat can damage the brake mechanism and melt rubber or boil the fluid. The pads are made of asbestos, which is a good insulator. Some brakes are designed to be able to operate safely even when the discs are so hot that they are glowing!

DID YOU KNOW?

Manufacturers often quote the 'Cd' of a car. The Cd, or drag coefficient, shows how 'slippy' the *shape* is. A pea and a football are the same shape, so they have the same Cd. But the frictional force on a car also depends on its *size*. To know how much frictional force there is on a car, you need to know its size as well as its Cd. Some lorries have a Cd that is smaller than some cars!

148

Fluids can also cause friction

We often use liquids as lubricants. Liquids such as oil can reduce friction between solid surfaces. But liquids can also *cause* friction. 'Fluids' include gases and liquids. Anything moving through a fluid experiences fluid friction. The particles of the fluid have to be pushed aside by the moving object.

A fan can be used to blow air over a car to simulate the car moving through air. The layer of air next to the car is still. This is because the air molecules are attracted to the surface of the car.

Further away from the car the air molecules move faster. So layers of air at different distances from the car are moving at different speeds. These layers have to slide over one another. The molecules in each layer attract one another. This produces friction between the layers.

You can see the same effect as water flows in a river. At the edge, there is friction between the water and the bank. In the centre, the water flows faster.

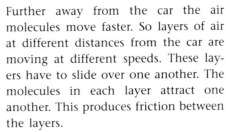

Fig. 60.3 To reduce air friction, it is important to have a smooth flow of air over the surface of a car. The unbroken stream of smoke shows how successfully this has been achieved.

Fig. 60.4 Friction in a river. Where the water flows near the bank it is slowed down. The water flows faster in the middle. The water in the middle flows past water near the edges, and friction acts between the flows of different speeds. This slows all of the water.

Air can be used as a lubricant

Instead of oil, air can be pumped between moving surfaces to reduce friction. This produces a very smooth bearing. A hovercraft moves on a cushion of air which keeps it above the surface over which it is moving. This reduces friction. But a lot of energy is required to 'levitate' the hovercraft. So not much energy is saved. The real benefit of this system is that the hovercraft can move over most surfaces.

Viscosity is internal friction in a fluid

In any moving fluid, there is friction between the layers which are moving at different speeds. This friction inside the fluid is called **viscosity**. Thick liquids, such as syrup, have high viscosity. We say that they are **viscous**. Thin liquids are less viscous. Low viscosity engine oils are more runny than some other oils. They still reduce the surface friction in the engine, but they cause less fluid friction than a thicker oil. This makes the car easier to get going in cold weather.

EXTENSION

At constant speed, force produced by a car's engine balances air resistance

When a car is travelling at a constant speed, the forces on the car balance. The force pushing the car forwards is the force provided by the engine. The force pushing the car backwards is air resistance. These forces are moving, so work is being done.

Imagine a car travelling at 160 km/h, with its engine supplying 44 kW of power. At this speed, the car travels about 44 m in a second. The engine supplies 44 kJ every second. So we can calculate the force provided by the engine as follows:

```
energy in joules = force in newtons x distance in metres
so force in newtons = energy in joules
                      ────────────────
                      distance in metres
```

$$\text{force} = \frac{44\ 000}{44} = 1000\,\text{N}$$

This is the forward force produced by the engine. It balances the air resistance on the car. So the air resistance is also 1000 N.

This frictional force on the car produces a lot of heating. 44 kJ of energy is transferred every second – a power of 44 kW. This is enough to raise 100 g of water to boiling point every second! The air and the car's body are warmed by this energy transfer.

Questions

1 a List as many examples as you can where friction is a nuisance and as many examples as you can where friction is useful.

  b How do we try to increase or reduce friction in each case?

2 A match is struck with a force of 4 N. The distance that the match head moves across the box is 5 cm.

  a How much energy is used in striking the match?

  b If it takes 0.5 s to pull the match across the box, what power does it take to strike the match?

3 A manufacturer claims that his thinner oil will save petrol if used in your car.

  a How can using oil save petrol?

  b How would you test the manufacturer's claim?

61 PRESSURE

Pressure increases with applied force. If a force is concentrated on a small area, this produces a large pressure.

Pressure is force divided by area

If you push hard on a drawing pin, you can push the point into a wooden table top. If you pushed the pin when it was upside down, the point would go into you. Why?

Look at Figure 61.1. The *force* between your finger and the pin is nearly the *same* as the force between the pin and the table top. But the *area* over which this force acts is *different*. The force between your finger and the pin is spread all over the top of the drawing pin. This is quite a large area. But the force between the pin and the table is concentrated in the point of

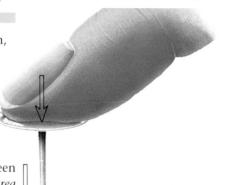

Fig. 61.1   Pushing on a drawing pin

Fig. 61.2   The surface area of the end of a brick is 60 cm². The surface area of the base of a brick is 200 cm². Each brick weighs 5 kg. What pressure is each of these bricks exerting on the ground?

The pressure depends on the force applied, and the area over which this force is spread. **Pressure = force**

area

Pressure is measured in **pascals (Pa)**. 1 pascal is a force of 1 N spread over an area of 1 m² (using the force acting at right angles to the surface).

Pressure in pascals = force in newtons
area in m²

For example, if a force of 10 N is applied to an area of 0.1m², then the pressure is 10 N = 100 Pa
0.1 m²

the pin. This is a much smaller area. When a force is spread over a large area, it produces a small **pressure**. If the same force is concentrated over a small area, it produces a larger pressure.

If you squeeze a gas, it becomes squashed. Try squeezing a bicycle pump with your finger over the end. The plunger will go in a long way, as the gas particles squeeze closer together. If you try again with water in the pump instead of air, the plunger will not move much. If you squeeze a liquid, it hardly changes volume at all.

In a gas, the particles are far apart and can easily be pushed closer. In a liquid, the particles are already very close. Large forces are needed to push them even a little closer. Because the liquid is not squashed, most of the force at one end is passed on to the other. You can feel this with the bicycle pump. A large force on the bicycle pump full of water requires a strong thumb over the end! The water seems to push very hard on

Fig. 61.3   Liquids such as water transmit forces from one place to another.

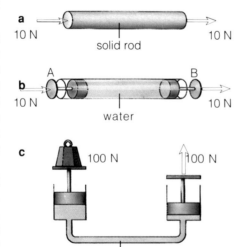

your thumb.

Pressure changes with the depth of a liquid. At the bottom of a swimming pool you can feel the extra weight of the water pushing on your eardrums. The extra weight of water above you increases the sideways pressure into your ears as well as onto your head. Pressure is the same in all directions at the same depth. This makes liquids and pipes a useful way of transmitting forces from one place to another.

Figure 61.3 shows water in a pipe. Piston A is pushing on the water, producing pressure. The water is not squashed and the pressure at the other end is the same. The pressure on piston B is the same as the pressure on A.

When you push on a liquid contained in a pipe, it is like pushing on a solid. The difference is that the liquid can transmit the force around corners. This is a really useful system of transmitting forces from one place to another. It is called a hydraulic system. 'Hydraulic' means 'to do with water'.

Hydraulic systems can produce a greater force from the same pressure

You will remember that pressure depends on force and area.

Pressure = force / area

So Force = pressure x area

So if you exert a particular pressure on a *large* area, you will produce a *larger* force than if the same pressure was exerted on a small area. By changing the area that a liquid acts against, large forces can be produced. An example is in the braking system of a car.

In a car it is important to be able to stop quickly. Friction pads are pushed against moving discs or drums to slow the car down. The earliest type of car linked the pads to the brake pedal with cables and levers. But the best way to get forces to go round corners is to use hydraulics. A hydraulic system does not have complicated joints to wear out. The pistons in the calipers are larger than the piston in the master cylinder. The pressure in the pipes is the same. So the force on the pads is larger than the force on the pedal, because the pressure pushes on a larger area.

In use, friction can make the disc brake very hot. The fluid in the brake system is a special oil, not water. Can you suggest why this is?

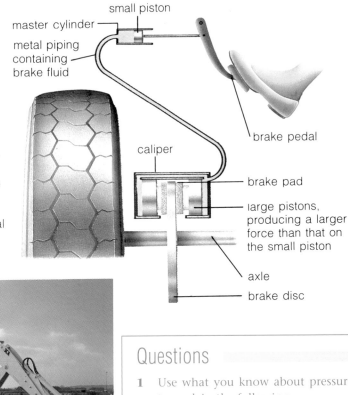

Fig. 61.4 The hydraulic brake system of a car. A relatively small force exerted on the brake pedal produces a larger force on the brakes. The brake pads are pushed against the disc, causing a frictional force which stops the axle turning.

Fig. 61.5 Forces are transferred around this digger through hydraulic pipes. The wide tyres help spread the load when driving on soft ground.

Questions

1 Use what you know about pressure to explain the following.
   a Snow shoes and skis stop you sinking into snow.
   b A dam is always built with the base thicker than the top.
   c Reindeer have broad feet.
   d A brick wall should be built on a concrete foundation at least twice the width of the wall.
   e Ice skates can do a great deal of damage to floor surfaces and to fingers.

2 a A person weighs 600 N. They are wearing shoes with a total area of 0.02 m^2. What pressure do they exert on the floor?
   b If the same person wears stiletto heels, with an area of 0.00003 m^2, what pressure do they exert on the floor?
   c What effect would this have on a wooden floor?

3 A brake master cylinder piston has an area of 1 cm^2. The piston in the brakes has an area of 4 cm^2. The master cylinder is pushed with a force of 600 N (see Figure 61.4).
   a What is the pressure on the master cylinder piston?
   b What is the pressure on the brake piston?
   c What is the force applied to the brake pads?

Air exerts pressure

The Earth is surrounded by a layer of air about 120 km deep. The air pushes in on objects at ground level with a pressure of $100\,000 \text{ N/m}^2$. This is 100 000 pascals, or 100 kilopascals. A kilopascal, kPa for short, is 1000 pascals.

We are not crushed by this air pressure (atmospheric pressure) because the air inside us pushes outwards as hard as the air outside us pushes inwards. So we are not aware of the pressure at all.

As with a liquid, the pressure varies with depth. We tend to notice more that pressure decreases as you go up. You are already at the bottom of the atmosphere! Pressure changes can make your ears 'pop' when you fly.

Vacuum cleaners and lungs fill by lowering the air pressure inside them

Vacuum cleaners and lungs use air pressure to fill themselves with air. A vacuum cleaner works by removing some of the air inside itself. This reduces the air pressure inside it. So the air pressure outside is greater than the air pressure inside it. The air pressure outside the vacuum cleaner pushes air into it, taking bits of dirt with it.

When you breathe in, your rib and diaphragm muscles make the volume inside your chest larger. So the air inside you is spread over a larger space. This makes its pressure smaller. The air pressure outside you is now greater than the air pressure inside you. Just as with a vacuum cleaner, the air pressure outside you pushes air into your lungs.

PRESSURE, VOLUME AND TEMPERATURE

Decreasing the volume of a gas increases its pressure and temperature.

Temperature changes with the speed of molecules

Atoms and molecules are always moving. As particles get hotter, they move faster. The temperature of a particular particle can tell us how quickly it is moving.

Look at Figure 62.1a. It shows a container full of gas. A piston seals the top. The piston does not drop down because the gas molecules keep hitting its surface and bouncing off. This produces a force on the piston. We say that the gas exerts a pressure on the piston.

You are surrounded by a mixture of

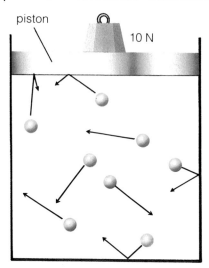

gases called air. The molecules in the air produce **atmospheric pressure**. As you sit reading this book, millions of molecules are bouncing off your face with a velocity of about 3000 km/h. This produces the same force as a 100 kg mass resting on your face!

So why do you not feel it? The force is spread evenly all over your face. And there is an equal and opposite force pushing outwards from inside you to balance the atmospheric pressure. So you are not aware of it at all.

Fig. 62.1a   Gas molecules bouncing against a surface produce a pressure.

EXTENSION

If the volume decreases, the pressure increases

If the piston in the diagram is pushed downwards, the gas particles have less room to move around in. The volume of the gas is decreased. But there are still the same number of particles. And they are still flying around and bumping into the piston. Because they have less space, they will hit the piston and the sides of the container more often than before. So they produce a greater force on the piston. The pressure is greater. The greater pressure produced by the gas is balanced by the greater force on the piston. This is why you have to push hard on the piston to push it downwards.

So an increase in pressure reduces the volume of a gas. A decrease in pressure increases the volume of a gas. For a fixed volume of gas, if pressure, P, is doubled then volume, V, will halve. This means that the volume multiplied by the pressure will always give the same number, as long as the temperature does not change.

initial pressure x initial volume
= final pressure x final volume
or   P x V = constant

This is called Boyle's Law.

Fig. 62.1b   If you squash the gas, the molecules hit the surfaces more often, producing a greater pressure.

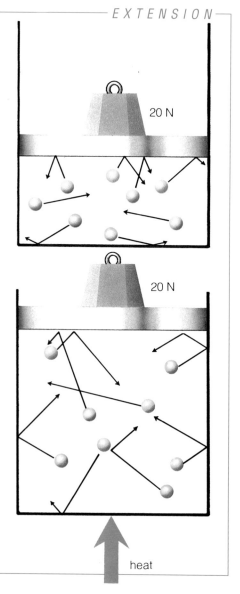

Temperature changes volume or pressure

Imagine that the same container is now heated. The temperature rises. The gas particles move faster and faster. They hit the piston and sides of the container more often and harder. So the force on the piston increases. If you do not push down harder on it, the piston will move upwards. So now the gas has more room. Its volume has increased. Heating a gas can increase its volume.

But you could stop the piston moving upwards. You could push down harder on it. If you did this, the volume of the gas would not increase.

But the pressure of the gas would be greater. So heating a gas can increase the pressure if the volume is restricted. Boyle's Law only applies if the temperature is fixed. Doubling the temperature (in kelvin) doubles the pressure if the volume is not allowed to change. At 0 K, a gas would have no pressure as its molecules are not moving.

$$\frac{P \times V}{T \text{ (in kelvin)}} = \text{constant}$$

Fig. 62.1c   If you heat the gas, the molecules move faster, producing a greater pressure.

heat

A manometer can measure gas pressure

A manometer is a tube with a U-bend. It contains a liquid. The container of gas with the pressure to be measured is attached to one end. The pressure of the gas pushes against the liquid in the tube. The height of liquid it can support is a measure of the gas pressure.

Fig. 62.2 A U-tube manometer measures the difference in pressure on the two ends of the liquid in the tube. A person blowing down one side of the tube causes the pressure on this side to increase, so the liquid moves round. The distance between the top of the liquid on the two sides gives a measure of pressure.

Constant volume gas thermometers are very accurate

A manometer can also be used to measure temperature. It works because when a gas increases in temperature, its pressure rises. It can support a higher column of liquid.

The constant volume gas thermometer has a thin-walled glass container and a manometer. The glass container is full of gas. The manometer contains mercury. If the temperature rises, the pressure of the gas in the container increases. The mercury is kept at position B, so the volume of the gas cannot change. The height h is a measure of the pressure, and therefore the temperature of the gas. Constant volume gas thermometers are used when a very accurate temperature measurement is required.

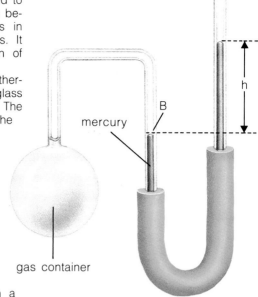

Fig. 62.3 A constant-volume gas thermometer

Calculating changes

You can use the pressure and volume equation to find what happens when pressure or volume changes. If you use the **suffix** 1 to represent initial or starting values and 2 to represent final or finishing values

then $P_1 \times V_1 = P_2 \times V_2$

By putting in the values you know, you can work out the answer. It is best to write everything out carefully so that you don't make any mistakes in substituting your numbers. In this example you have to work out which volumes to use as initial and final values.

A football contains $0.2\,m^3$ of air at one atmosphere pressure. An air pump adds another $0.3\,m^3$ of air. Assuming that the volume of the ball does not change, what is the new pressure in the ball?

For these problems you can use whatever units the question gives you.

This is because P and V are on both sides of the equation. You don't have to convert atmospheres to pascals. (The same would apply if volume was in another unit such as litres.) Notice that $0.3\,m^3$ of air was **added**, so a total of $0.5\,m^3$ at 1 atm pressure in the pump and ball became $0.2\,m^3$ at the new pressure in the ball.

Write down what you know:

$P_1 = 1\,atm$
$V_1 = 0.5\,m^3$
$V_2 = 0.2\,m^3$
$P_2 = ?$

Now put the numbers in:

$1\,atm \times 0.5\,m^3 = P_2 \times 0.2\,m^3$

and so

$P_2 = \dfrac{1\,atm \times 0.5\,m^3}{0.2\,m^3} = 2.5\,atm$

So the new pressure is 2.5 atmospheres.

Fig. 62.4 Apparatus to investigate the pressure and volume of a gas. As the pump varies the pressure, the volume of the enclosed gas can be measured.

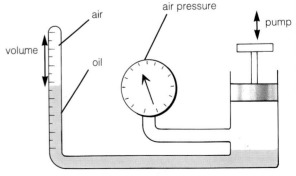

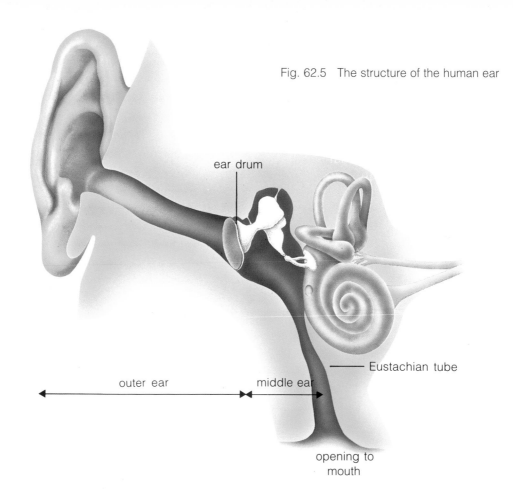

Fig. 62.5 The structure of the human ear

ear drum

Eustachian tube

outer ear

middle ear

opening to mouth

Questions

1 In an experiment the glass container of a gas thermometer was heated (see Figure 62.3). The pressure was measured at different temperatures. The manometer reading was converted to a pressure. The table shows the results.

Temperature (°C)	Pressure (kPa)
−30	89
−10	96
+10	103
+30	111
+50	118
+70	126
+90	133
+110	140

a Plot a graph of these results.

b What pressure would be read if the vessel was immersed in boiling water?

c Atmospheric pressure is 100 kPa. What temperature would the vessel have to be in order to produce this pressure?

d What could you dip the glass container into to make the levels on both sides of the manometer the same?

2 A bicycle pump has a volume of 150 cm³. If you keep your thumb on the end and can squeeze the air down to 1 cm³, what would the pressure be? Assume the pressure of the air starts at 100 kPa. If the hole at the end has an area of 0.0001 m², would you be able to keep your thumb over it? (Hint: calculate the force for this pressure.)

3 Figure 62.5 shows the structure of the human ear. Both the outer and middle ear are full of air. They are separated from each other by the eardrum. The Eustachian tube leads from the middle ear into the back of the mouth. Normally, the lower end of this tube is kept closed. It opens when you swallow or yawn. Under normal circumstances, the air pressure on both sides of the eardrum is the same.

a Imagine that you are taking off in an aeroplane. As you go higher, the air pressure gets less. What happens to the air pressure in your outer ear?

b If you do not swallow, what happens to the air pressure in your middle ear?

c What effect will this have on your eardrum?

d If you swallow hard, your ears may 'pop' and feel more comfortable again. What do you think happens when your ears 'pop'?

4 An oil can is connected to a vacuum pump and the air inside it removed. The surface area of the oil can is 0.4 m². Atmospheric pressure is 100 000 Pa.

a What is the total inwards force on the can?

b What mass is this equivalent to?

c What will happen to the can?

d Why does this not happen when the can is open to the air?

AIR AND WATER

The Earth is surrounded by a layer of gas which we call air. Air is a mixture of several different gases.

Air is a mixture of gases

The Earth's atmosphere contains several different gases. Most of it – 78 % – is **nitrogen**. Almost all of the rest – 21 % – is **oxygen**. The rest is made up of very small amounts of **carbon dioxide**, **argon** and other **noble gases**, and **water vapour**.

Table 63.1 Gases in the air

gas	percentage
nitrogen	78
oxygen	21
argon	1 (just under)
carbon dioxide	0.04
neon, krypton, xenon, water vapour	very small amounts

Carbon dioxide is needed for photosynthesis

Only a tiny proportion of the air – about 0.04 % – is carbon dioxide. Yet without this gas, there would be no plants or animals on Earth! Plants use carbon dioxide to make food, in the process of photosynthesis. All the food which animals eat is made from carbon dioxide from the air.

Carbon dioxide is used to make fizzy drinks. It is also used in some types of fire extinguishers. Because it is a heavy gas, it forms a 'blanket' over the flames, stopping oxygen getting in.

You can test a gas to see if it is carbon dioxide by using limewater. Limewater is a solution of calcium hydroxide in water. When carbon dioxide is mixed with it, calcium carbonate (chalk) is formed. This is not soluble in water, so it makes the limewater look cloudy.

Fig. 63.1 The Earth's atmosphere. The atmosphere gets thinner and thinner as you go upwards. 80% of the atmosphere is in the lowest layer, the troposphere.

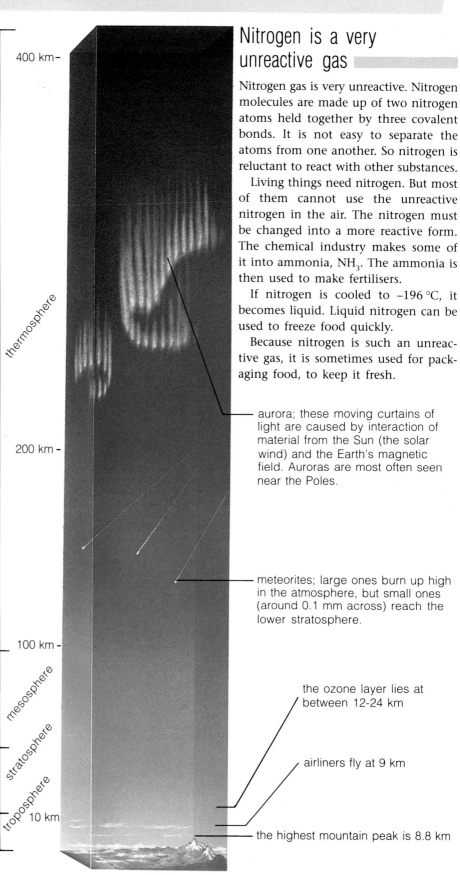

400 km –

200 km –

100 km –

10 km

thermosphere

mesosphere

stratosphere

troposphere

Nitrogen is a very unreactive gas

Nitrogen gas is very unreactive. Nitrogen molecules are made up of two nitrogen atoms held together by three covalent bonds. It is not easy to separate the atoms from one another. So nitrogen is reluctant to react with other substances.

Living things need nitrogen. But most of them cannot use the unreactive nitrogen in the air. The nitrogen must be changed into a more reactive form. The chemical industry makes some of it into ammonia, NH_3. The ammonia is then used to make fertilisers.

If nitrogen is cooled to –196 °C, it becomes liquid. Liquid nitrogen can be used to freeze food quickly.

Because nitrogen is such an unreactive gas, it is sometimes used for packaging food, to keep it fresh.

aurora; these moving curtains of light are caused by interaction of material from the Sun (the solar wind) and the Earth's magnetic field. Auroras are most often seen near the Poles.

meteorites; large ones burn up high in the atmosphere, but small ones (around 0.1 mm across) reach the lower stratosphere.

the ozone layer lies at between 12-24 km

airliners fly at 9 km

the highest mountain peak is 8.8 km

Oxygen is needed for burning and respiration

Like nitrogen, oxygen gas is made up of molecules in which two atoms are joined together with covalent bonds. But oxygen is a reactive gas. Oxygen reacts easily with many substances to produce oxides. This can be a nuisance. For example, when it reacts with iron it forms iron oxide (rust).

Oxygen is needed for burning. When something burns, it combines with oxygen in the air and releases heat energy. Respiration inside living organisms is a similar process. Oxygen reacts with food substances and releases energy. This is why most organisms need to take in oxygen when breathing.

Substances burn more readily in pure oxygen than they do in air. This fact is used in the test for oxygen. Light a wooden splint, and blow it out. Then put it into the gas you think might be oxygen. If it is, the splint will burst into flame again.

INVESTIGATION 63.1

Measuring the percentage of oxygen in air

When copper is heated in air, it reacts with the oxygen in it, forming copper oxide. If you use plenty of copper, then *all* the oxygen in a sample of air will react. By comparing how much air you had to start with, and how much you have left when all the oxygen has been used up, you can find out how much oxygen there was in your sample.

1 Set up the apparatus as in the diagram. It is really important that everything fits tightly, with no air gaps. Notice that one syringe has its plunger pushed right to the end, so that it contains no air at all. The other one contains $100 \, cm^3$ of air.

2 Heat the copper wire strongly. As you heat it, push the plungers of the two syringes back and forth several times. The copper will gradually turn black, as it combines with the oxygen in the air to form copper oxide.

3 When you think that all the oxygen has had a chance to react with the copper, stop heating and let the apparatus cool down to room temperature again.

Then push all the air into one syringe, and measure its volume. It should be less than you started with. Write down the volume.

4 To make sure that all the oxygen really has been used up, heat the copper again, and push the syringes back and forth as before. Cool, and measure the volume of air. If it is the same as last time, then all the oxygen has gone. If not, you must repeat instructions 2 and 3.

5 Work out how much oxygen there was in your sample of air.

6 You had $100 \, cm^3$ of air to start with. Work out the percentage of air which was oxygen.

Questions

1 Why is it best to use copper wire or turnings in this experiment, rather than lumps of copper?

2 Why is it important for the apparatus to be airtight?

3 Why should you push the syringes back and forth while heating the copper?

4 Why must you wait for the apparatus to cool down before measuring the new volume of air?

5 What gases will be present in the air inside the syringe at the end of the experiment?

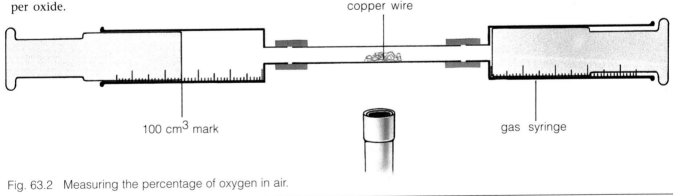

copper wire

$100 \, cm^3$ mark

gas syringe

Fig. 63.2 Measuring the percentage of oxygen in air.

Questions

1 List the gases present in the air, giving the percentages of each.

2 **a** Give two uses of nitrogen.
  b What is the boiling point of nitrogen?
  c Why can living organisms not use nitrogen gas from the air?

3 **a** Give two uses of oxygen.
  b What is the molecular formula for oxygen gas?
  c Give two ways in which the chemical reactions involved in burning and respiration are similar.
  d How can you test for oxygen?

4 **a** Why is carbon dioxide a very important gas for living organisms?
  b How can you test for carbon dioxide?

64 OZONE

The layer of ozone in the atmosphere, high above the Earth's surface, protects us from ultraviolet light.

Ozone is found in the atmosphere

Ozone is a gas. Whereas oxygen gas has two oxygen atoms in its molecules, ozone has three. The molecular formula for ozone is O_3.

Ozone is formed from oxygen. Quite a lot of ozone is formed near the ground. For example, ozone is produced when sunlight interacts with nitrogen oxides from vehicle exhaust fumes. This ozone may contribute to the formation of smog over cities where there is a lot of traffic.

So high concentrations of ozone in the air near ground level are not good for us. But high above us, between 20 km and 35 km above the ground, there is an area where large concentrations of ozone occur naturally. This high-level ozone, up in the stratosphere, is called the **ozone layer**. It is very important to all living things on Earth.

Ultraviolet light damages cells

Various forms of electromagnetic radiation reach the Earth from the Sun. Some of this is **ultraviolet radiation**. Ultraviolet radiation can damage DNA inside living cells. If you sunbathe, your skin reacts to the ultraviolet radiation by producing extra melanin. Melanin is a dark brown pigment that absorbs the ultraviolet light before it can penetrate cells deep in your skin. The production of melanin is a defence mechanism used by your skin to stop damage to your DNA.

Recently, the number of cases of a disease called **skin cancer** have been increasing. It is likely that this is happening because more and more people can afford to go on holidays to sunny places. If a person with a naturally light-coloured skin works indoors and doesn't get much sunlight on their skin, then they have no natural protection from the ultraviolet rays when they first go into strong sunshine. The DNA in some of the cells in their skin may be damaged. Many years later, these cells may begin to divide uncontrollably, forming a lump or **tumour**. Most cases of skin cancer can be cured, but sometimes the illness is fatal. You can read more about skin cancer in Topic 225.

Ultraviolet light can also damage your eyes. It can make the lens go cloudy, so that light cannot pass through it. This is called a **cataract**.

Ultraviolet can damage plant cells, too. It can harm the pigments that are needed to absorb light for photosynthesis. Some crop plants do not grow so well in very strong ultraviolet light.

Ozone absorbs ultraviolet light

The layer of ozone in the stratosphere protects us from ultraviolet light. Much of the ultraviolet light that hits the upper layers of the atmosphere is absorbed by ozone. The ozone stops too much ultraviolet light getting down to the ground. If it wasn't for the ozone layer, there would be many more cases of skin cancer and cataracts, and many plants would not grow well.

The ozone layer has been damaged

Human activities have damaged the ozone layer. Figure 64.1 shows two satellite images, taken in 1979 and 1990. The concentration of ozone is measured in Dobson units, and if you look at the key at the side of the photographs you can see that there is much less ozone present in the second picture than in the first.

These measurements were made over the South Pole. This is where most of the damage is occurring. However, in recent years, the amount of ozone over the North Pole has been decreasing as well. Sometimes, these ozone 'holes' spread as far south as Britain. The average amount of ozone over Britain in March is about 365 Dobson units. But in March 1996 the amount was only 195 Dobson units.

Why is this happening? The culprit is a group of chemicals called **chlorofluorocarbons**, or **CFCs**. CFCs have been widely used as coolants in refrigerators, and in aerosols. When they get into the air, they remain unchanged for over 100 years. This gives them plenty of time to travel right up to the higher layers of the atmosphere containing ozone. Then they react with the ozone, and break it down. CFCs are destroying the ozone layer.

Can we protect the ozone layer?

Many countries do not use CFCs any more. Most developed countries have stopped using them completely. But some developing countries still use them, because it is expensive for them to find substitutes. And even if we stopped using any CFCs at all, it would still take at least 100 years for them to disappear from the atmosphere, because they are very stable substances.

But at least we have reduced the use of CFCs, and this will give the ozone layer a chance to recover. Ozone is constantly formed in the stratosphere, so – given time – the amount of ozone in the ozone layer should gradually increase, so long as we do not keep on releasing CFCs.

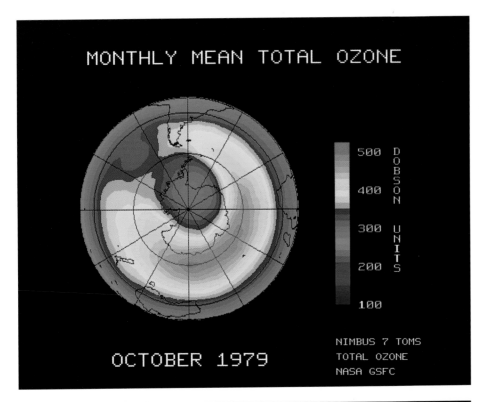

MONTHLY MEAN TOTAL OZONE

OCTOBER 1979

500
400
300
200
100

DOBSON UNITS

NIMBUS 7 TOMS
TOTAL OZONE
NASA GSFC

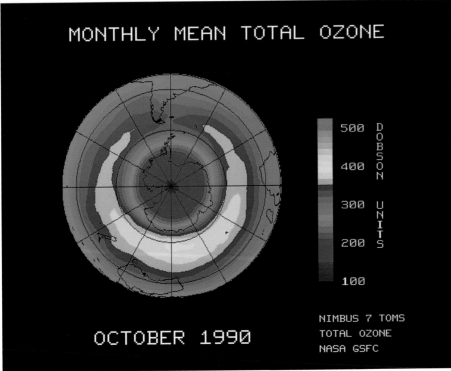

MONTHLY MEAN TOTAL OZONE

OCTOBER 1990

500
400
300
200
100

DOBSON UNITS

NIMBUS 7 TOMS
TOTAL OZONE
NASA GSFC

Fig. 64.1 Maps of the ozone concentrations over Antarctica, made using data collected by the Nimbus-7 weather satellite.

Questions

1 a What is ozone?

b Describe one harmful effect of ozone near ground level.

c Explain why ozone high in the atmosphere is important to living things.

2 Suggest explanations for each of the following facts:

a Even if everyone stopped using CFCs immediately, the amount of ozone in the ozone layer would continue to decrease for many years.

b People with naturally light-coloured skin are more likely to get skin cancer than people with naturally dark-coloured skin.

c Although there has been a general increase in the incidence of skin cancer in recent years, this increase has been especially large in Australia.

3 A survey carried put in England in 1997 showed that most people believed that global warming and the hole in the ozone layer are the same thing.

Write an explanation, which could be understood by a person who has had little or no science education, to help them to understand that global warming and the hole in the ozone layer are really very different, with different causes.

You may like to use diagrams to make your explanation clearer.

4 Use Figure 64.1 to answer these questions.

a In which month were both of these images taken? Suggest why it is important to compare images taken at similar times of year, in order to determine whether the amount of ozone in the Earth's atmosphere is changing.

b Describe two differences between the two images.

c Why is the term 'hole' in the ozone layer not strictly correct?

65 THE CARBON CYCLE

Carbon dioxide is taken from the air by green plants, and put back into the air by all living organisms and by burning.

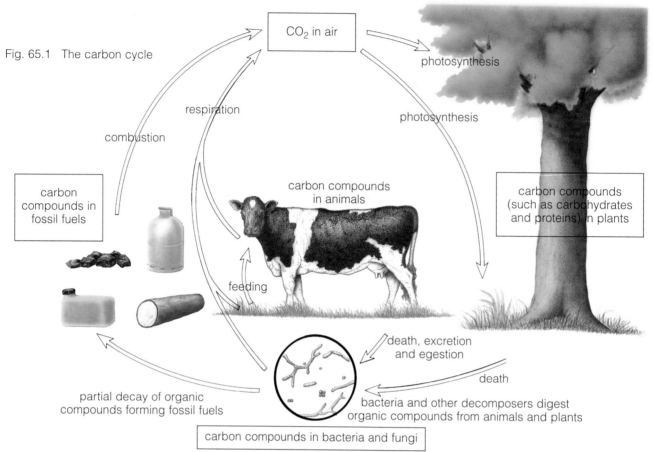

Fig. 65.1 The carbon cycle

Carbon atoms are cycled

Many of the molecules from which living organisms are made contain carbon atoms. Carbohydrates, fats and proteins all contain carbon atoms. These carbon atoms are passed from one organism to another, into the atmosphere and into the soil. The possible pathways a carbon atom could take are shown in Figure 65.1. This is called the **carbon cycle**.

Photosynthesis uses up carbon dioxide

Only about 0.04 % of the air is carbon dioxide. Plants use carbon dioxide from the air when they photosynthesise. They use it to make food. The carbon from the carbon dioxide becomes part of the food molecules. Plants can only photosynthesise during daylight, because photosynthesis needs energy from sunlight.

Respiration produces carbon dioxide

All livings things need energy. They get their energy from food. When the energy is released from food, carbon dioxide is produced. This is where the carbon dioxide which you breathe out comes from.

All living things respire. So all living things produce carbon dioxide. Even plants produce carbon dioxide. They do it all the time. But during the day, they use it up in photosynthesis faster than they make it in respiration. So in the daytime plants take in carbon dioxide. At night they give it out.

Decomposers release carbon dioxide

Dead plants and animals contain a lot of carbon. So do their waste materials, such as urine, faeces and fallen leaves. All of these substances can be used as food by **decomposers**, such as bacteria and fungi. The decomposers break down the molecules in the dead bodies and waste materials, and use them to build their own bodies. They use some of the molecules for respiration. When they do this, they release carbon dioxide.

Burning produces carbon dioxide

When things burn, they react with oxygen in the air. The fuels which we burn all contain carbon. The carbon reacts with oxygen in the air to form carbon dioxide.

How did the carbon get into the fuels? Fossil fuels, such as coal, oil and gas, were formed from plants and bacteria. The plants took carbon dioxide from the air. So the fuels contain carbon.

160

The enhanced greenhouse effect and global warming

Carbon dioxide in the air behaves rather like the glass of a greenhouse. It lets the Sun's rays through on to the surface of the Earth. Some of these rays heat the Earth's surface. Some of this heat escapes from the Earth and goes back into space. But the carbon dioxide in the atmosphere traps some of the heat and stops it escaping. This is called the **greenhouse effect**. Without the carbon dioxide, the Earth would lose much more heat and would be much colder than it is. Earth would be a frozen, lifeless planet.

Photosynthesis takes carbon dioxide out of the air. Respiration and burning put carbon dioxide into the air. These processes balance each other, so the amount of carbon dioxide in the air stays approximately the same.

But humans may now be upsetting the balance. We burn more and more fossil fuels. We also cut down and burn growing trees. This releases extra carbon dioxide into the air, which increases the greenhouse effect. As a result, the Earth may be getting warmer. This is called **global warming**.

Carbon dioxide is not the only gas which acts like a blanket around the Earth. **Methane** has a similar effect. Although there is much less methane in the atmosphere than carbon dioxide, the amounts are increasing and its effect on global warming is significant. Methane is produced in especially large amounts by cattle, by termites, and sometimes from the mud in paddy fields where rice is grown.

Does global warming matter? We do not know to what extent global warming will happen, or exactly what its effects will be. If it does happen, then it will change weather patterns on the Earth. Some places will become drier, while others will become wetter. Global warming may cause a lot of ice at the poles to become liquid water, which would increase sea levels. This could flood many major cities.

some energy is reradiated from the Earth's surface

some radiation escapes back into space

radiation from Sun reaches Earth

some radiation is reflected or absorbed by greenhouse gases

Fig. 65.2 The greenhouse effect. Radiation from the Sun passes through the atmosphere on to the Earth's surface. Here, some is reflected, while some is absorbed and reradiated as heat. Some of this reradiated energy escapes into space, but some is retained in the atmosphere. Gases such as carbon dioxide, ozone and methane increase the amount of energy retained. This warms the Earth.

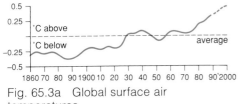

Fig. 65.3a Global surface air temperatures.
The horizontal dotted line shows the average air temperature on the Earth's surface. It is calculated from all the measurements taken, all over the world, between 1950 and 1979.

Eight of the nine warmest years this century have occurred in the 1980s. In 1988, the global average temperature was 0.34 °C above the long-term average. We do not know whether this has been caused by increased CO_2 emissions, or whether there is some other natural cause.

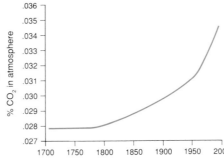

b Atmosphere carbon dioxide levels. In recent years the CO_2 level has begun to rise much more steeply than before. The levels of carbon dioxide in the 18th and 19th centuries have been measured from bubbles of air trapped in ice at the Poles.

Questions

1 a What is the percentage of carbon dioxide in the atmosphere?

b Which process removes carbon dioxide from the atmosphere?

c Which processes return carbon dioxide to the atmosphere?

2 Discuss the causes, and possible results, of the enhanced greenhouse effect. What do you think that humans can do to keep the damage to a minimum?

— EXTENSION —

161

66 THE WATER MOLECULE

Water is a very strange substance. It has unusual properties because of the structure of its molecule. Water molecules are attracted to each other.

The most common substance on the surface of the Earth

363 000 000 km² of the Earth's surface is covered by water. Without water, there would be no life on Earth. Your body is about 70 % water. Life probably began in water. Many animals and plants can live only in water. On Earth, there is plenty of liquid water – although there are some places where it is in short supply. But none of the other planets in the Solar System have liquid water on their surfaces. It is either too hot, or too cold. Water is liquid only over a very narrow temperature range – between 0 °C and 100 °C. If the Earth was just a little closer to the Sun, most of the water would turn to water vapour. If it was just a little further away, most of it would turn to ice.

Fig. 66.1 This picture of the Earth, taken from Apollo 17, gives an idea of the huge area of the Earth's surface covered by water. The white areas are clouds, which are also made up of water.

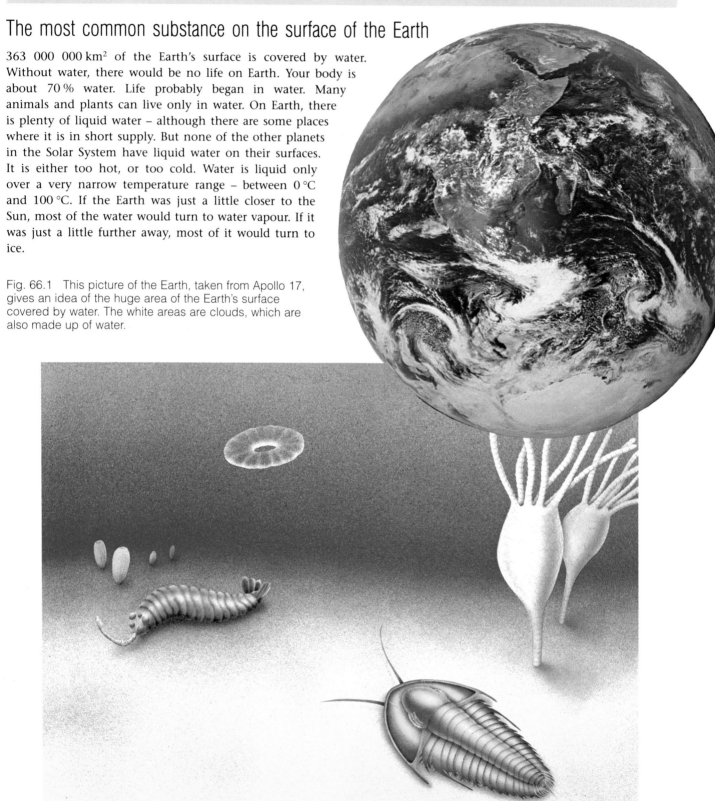

Fig. 66.2 Life in the sea about 530 million years ago. At this time, all living things were aquatic. Nothing at all lived on land. Cells are mostly water, and die if they dry out. The first living organisms to have bodies which could retain water in their cells when on dry land, evolved about 400 million years ago.

Water molecules have dipoles

Water is a compound of hydrogen and oxygen. Each water molecule is made up of two hydrogen atoms, and one oxygen atom. The atoms are held strongly together by covalent bonds. The molecular formula for water is H_2O.

A covalent bond is an electron-sharing bond. Each hydrogen atom in a water molecule shares its electron with the oxygen atom. But the electrons are not spread evenly around the molecule. They spend more time close to the oxygen atom than the hydrogen atoms. Electrons have a negative charge. The oxygen atom in a water molecule gets more than its fair share of the electrons. So the oxygen atom has a slight negative charge. The hydrogen atoms get less than their fair share of the electrons. So they have a slight positive charge.

So one part of a water molecule has a slight negative charge, and two parts have a slight positive charge. This is called a **dipole**. Water molecules have dipoles.

It is important to remember that the *overall* charge on a water molecule is nil. The positive and negative charges are equal, so they cancel each other out.

Water molecules are attracted to one another

Figure 66.5 shows what happens in a group of water molecules. The positive parts of one water molecule are attracted to the negative parts of other water molecules. In an ice crystal, the water molecules arrange themselves in a regular pattern, or **lattice**. The attractions between the water molecules hold them in position.

If the ice is warmed, the water molecules begin vibrating faster. At 0 °C, they are moving vigorously enough to break out of the lattice pattern. They move freely around each other. The ice has melted. The water is now liquid. But the water molecules are still attracted to each other. As they move around, they are briefly attracted to each other. Liquid water is a mass of moving water molecules, briefly attracted to each other as they pass by.

slight negative charge

slight positive charge

slight positive charge

Fig. 66.3 A water molecule has no overall charge. But there is a slight negative charge in the region of the oxygen atom, and a slight positive charge near the hydrogen atoms.

Fig. 66.4 This strip has been charged with static electricity by rubbing it with a cloth. The charge on the strip attracts the stream of water from the tap, because the charges on the water molecules are not evenly distributed.

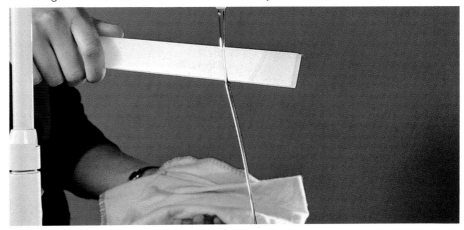

Fig. 66.5

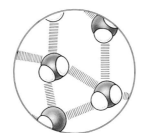

Steam: the water molecules are so far apart from each other that the forces between them are very small.

Ice: the water molecules are held firmly together in a lattice. The positive parts of each water molecule are attracted to the negative parts of its neighbours.

Water: the molecules move about, briefly attracted to each other as they pass by.

If the water is heated, the molecules move faster and faster. At 100 °C, they are moving so fast and so energetically that they can break away from each other completely. The attractive forces are no longer strong enough to hold them together. The water molecules fly away from one another, into the air. The water boils. It turns into gaseous water, or **steam.**

It is the attractive forces between water molecules which make it a liquid at normal temperatures. Most compounds with molecules similar to water, such as carbon dioxide and methane, are gases. Without the attractive forces between its molecules, water would be a gas at normal temperatures, and there would be no life on Earth.

67 LATENT HEAT

When water changes from a liquid into a gas, a lot of energy is used. This overcomes the attractive forces between the molecules, and rearranges them. This energy is called latent heat.

A lot of energy is used to change water from a liquid to a gas

Imagine that you have a beaker of pure water with a thermometer in it. You heat the water with a Bunsen burner. The temperature increases.

What is actually happening to the water molecules in your beaker? They are moving faster and faster as you heat them. This is why the temperature increases.

But when the temperature gets to 100 °C, something different starts to happen. Now the heat from the Bunsen burner, instead of making the molecules move faster, is used to break the attractive forces between them. The water molecules become free. They fly off into the air. The water boils.

All the time that the water is boiling, your thermometer goes on reading 100 °C. All the heat energy from the Bunsen burner is being used to break the attractive forces between the water molecules. The temperature stays constant. It is as though you are pouring heat energy into the water, and it just disappears! When you heat something, you expect its temperature to increase. But when liquid water is turning into gas, its temperature does not increase. The heat energy becomes 'hidden' in the water. Another word for 'hidden' is 'latent'. We call this 'disappearing' heat energy **latent heat**. A fuller name for it is **latent heat of vaporisation**.

Water evaporates well below boiling point

Imagine that it has been raining. Puddles have formed. It stops raining, and the sun begins to shine. After a few hours, the puddles have disappeared. The water in them has evaporated.

But the water in the puddles has certainly not boiled! Water can evaporate at temperatures well below boiling point.

In a puddle of water, the water molecules are moving around randomly. Some have more energy than others and move faster. At the surface of the water, the most energetic particles will have enough energy to be able to break away from the other water molecules. They will escape into the air. The temperature of the puddle of water is related to the average speed of movement of the water molecules in it. If the faster moving molecules escape, then the average speed of movement of the particles goes down. So the temperature of the water decreases.

So liquid water can become a gas without boiling. This is called **evaporation**. Evaporation is not the same as boiling. Boiling takes place *throughout* a liquid. All the particles are so hot that they have enough energy to break away from one another. Evaporation takes place at the *surface* of the liquid, at temperatures well below boiling point. Only the most energetic particles can break away from the others.

— E X T E N S I O N —

When water evaporates, it cools things down

When you get out of a swimming pool or your bath, you may feel cold. But you would not feel cold if you were dry. It is the water evaporating from your body which cools you down. The water evaporates as the most energetic water molecules escape from the others. This lowers the average temperature of the water on your skin. The water cools, and so does your skin. Moreover, the change from a liquid to a gas takes energy. If the water happens to be sitting on your skin when it does this, it will take heat energy from your skin. So your skin feels cooler.

We, and some other mammals, use this fact to cool our bodies. Human cells work best at a temperature of 37 °C. In hot weather, or when we exercise, the body temperature may go above this. Sweat glands in the skin produce a watery liquid which lies on the skin surface. The water in the sweat evaporates. As it does so, it takes heat energy from the skin. This cools the skin down.

Plants may use a similar method to cool their leaves. In hot climates, it would be all too easy for a plant's leaves to get so hot that the cells would be damaged. But if it has plenty of water, a plant can keep itself cool. Water is allowed to evaporate from its leaves. It evaporates through small holes on the underside of the leaves, called **stomata**. As the water evaporates, it takes heat energy from the leaf cells, and cools them down.

Fig. 67.1 Huge quantities of water evaporate from rain forests. The evaporation helps to keep the plants cool.

The effect of evaporation on temperature

You need to get fully organised with this experiment before you begin, because once the tubes have cotton wool on them you need to start taking temperatures straight away. Ideally, each tube should have its cotton wool put on at exactly the same time. Draw a results chart before you start.

1 Fill five boiling tubes with tap water, leaving room for a bung to go in.

Support all five tubes in clamps on a retort stand.

2 Surround one of the tubes with dry cotton wool. Surround another with cotton wool soaked in warm water. Surround a third with cotton wool soaked in cold water. Surround the fourth one with cotton wool soaked in ethanol. Leave the fifth one with no covering.

3 Put the bungs with thermometers into each tube, and immediately take the temperature of each. This is Time 0. Take the temperatures every 2 min for at least 20 min. Carry on for longer if you have time.

4 Record your results in the way you think best. A line graph is a good idea.

Questions

1 Why were all five tubes supported in clamps, rather than lying on the bench or being held in your hand?

2 Which tube cooled most slowly? Try to explain why.

3 Which tube cooled fastest? Explain why.

4 Was there much difference between the rate of cooling of the tube with cold wet cotton wool and the tube with warm wet cotton wool? If so, can you explain this?

5 Do you think that your experiment gave a fair comparison between the five tubes? How could you have improved it?

Refrigerators use evaporation to produce cooling

When a liquid evaporates it takes in energy and cools its surroundings. When the gas condenses back to a liquid, the latent heat is released. This is used to take heat from inside a fridge, and release it outside.

A liquid which evaporates easily is used. Liquids which evaporate easily are called **volatile** liquids. The liquid used in most fridges is a type of CFC called 'Freon'. The liquid evaporates in the coils around the ice box or cold plate inside the fridge. This causes cooling. The gas formed is pumped away. It is pressurised in the condenser on the back of the fridge. Here the gas condenses back into a liquid. As it condenses it releases the heat energy it has taken in. So heat energy has been taken from inside the fridge, and released outside it. Because the pump is working hard to push the liquid around, more energy is released from the back of the fridge than is taken from inside it. If you leave the fridge door open, the pump will be working very hard. So your kitchen will eventually become hotter!

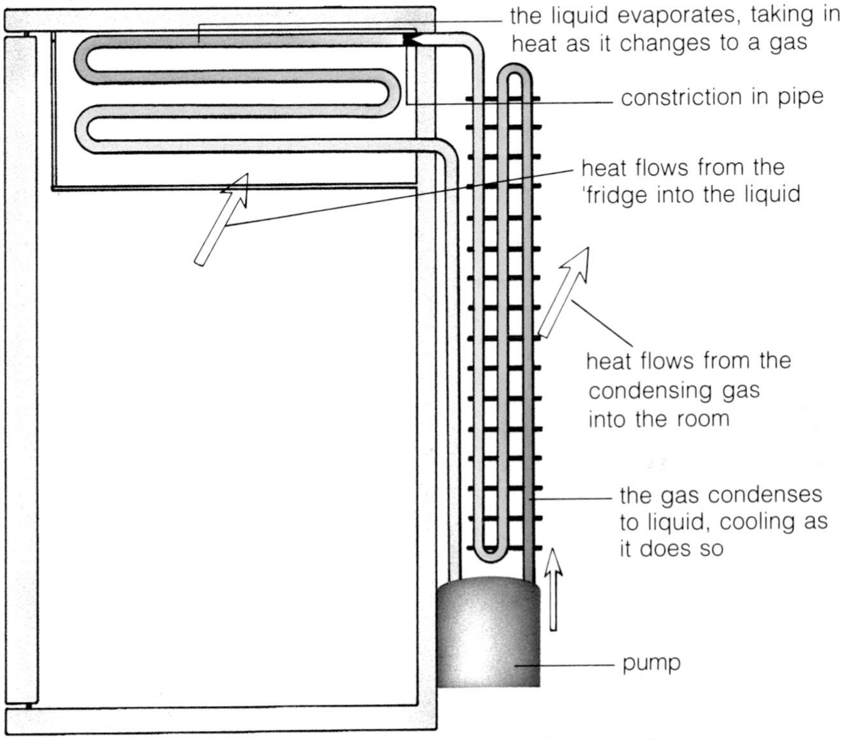

- the liquid evaporates, taking in heat as it changes to a gas
- constriction in pipe
- heat flows from the 'fridge into the liquid
- heat flows from the condensing gas into the room
- the gas condenses to liquid, cooling as it does so
- pump

Fig. 67.2 How a refrigerator works. The pump and the pipes leading from it (the condenser) are at the back of the 'fridge. Can you suggest why there is a constriction in the pipe carrying the liquid from the condenser into the coils around the ice box?

— EXTENSION —

Questions

1 Why does the temperature of boiling water not change, even if you continue heating it?

2 a Explain what is meant by **latent heat of vaporisation**.

b When you are going to have an injection, you will probably have ethanol put on to your skin to kill any germs. The ethanol quickly disappears and your skin feels cold. Why is this?

c Explain how mammals and plants use the latent heat of vaporisation of water to cool themselves.

3 Someone has fallen into a river on a cold day. You manage to get them on to the bank. They are still conscious and breathing, but exhausted. What should you do next, and why?

Liquids often behave as if they have a weak 'skin'. This is caused by the attraction of the molecules in the liquid for each other.

Surface tension pulls water molecules together

A pond skater can walk on water. A dry needle can be made to float on a water surface. Why is this?

Water molecules are attracted to one another. In the middle of a container of liquid, a molecule has other molecules all around it. It is attracted equally from all directions. The forces balance out. At the surface, a molecule has no water molecules above it. So all the attractive forces are sideways and downwards. There is an overall force pulling the surface molecules inwards. This force keeps the surface together. The surface acts like a skin, with the molecules sticking together. This 'skin' is under tension, and the effect is known as **surface tension**.

It is surface tension which causes water to form droplets. In a raindrop, the molecules in the surface are pulled together. The tension in the surface pulls the surface into as small an area as possible. The shape with the smallest surface area for a given volume is a sphere. So water has a tendency to form spherical droplets. In the same way, liquid mercury forms small balls.

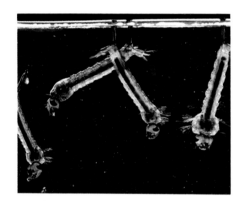

Fig. 68.2 Mosquito larvae breathing. Their water-repellent hairs break through the surface film.

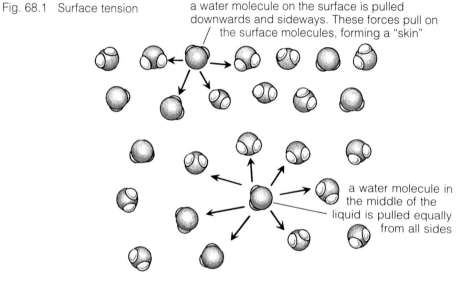

Fig. 68.1 Surface tension

a water molecule on the surface is pulled downwards and sideways. These forces pull on the surface molecules, forming a "skin"

a water molecule in the middle of the liquid is pulled equally from all sides

Many water animals rely on surface tension

Mosquito larvae live in water. But, like all insects, they have to get their oxygen from air. When they need air, they come to the surface 'tail first'. They float just under the water surface. Water-repellent hairs break through the surface film. The hairs are pulled apart by surface tension forces. Oxygen from the air can then diffuse into the larva.

If detergent is added to the water, it reduces the surface tension. The mosquito larvae's breathing method no longer works. Detergent is sprayed on to ponds and lakes in places where malaria is common. The detergent prevents the larvae from breathing. Mosquitos carry the organism which causes malaria, so killing mosquito larvae can help to keep malaria under control.

Other insects carry air stores under water with them. The water beetle *Dytiscus* does this. It holds a bubble of air underneath its wing cases. The surface tension of the water stops it from getting in between the water-repellent surfaces of the wing cases.

The camphor beetle lives on the surface film of small ponds. It walks on water. It produces a chemical from the tip of its abdomen which lowers the surface tension of the water behind it. The high surface tension at its head end now pulls it forward so it skims over the water like a tiny speedboat.

Surface tension prevents water from penetrating the tightly woven fibres of a tent canvas. Once the fragile film is broken by touching the canvas, the water drips through.

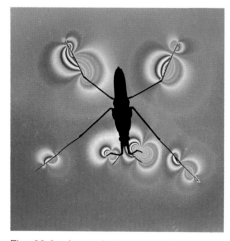

Fig. 68.3 A pond skater, standing on the surface film of pond water.

Attraction is not always between the same kind of molecules

Surface tension exists because molecules in a liquid are attracted to one another. This attraction is called **cohesion**. Cohesion makes water form droplets.

But water molecules are not only attracted to other water molecules. They may be attracted to other molecules. Molecules to which water molecules are attracted are called **hydrophilic** molecules. Glass is hydrophilic. Drops of rain cling to a glass window. The water molecules are attracted to the glass. This attraction is called **adhesion**. Water molecules are attracted to

Fig. 68.4 A drop of water spreads out on a clean sheet of glass, whereas mercury forms a rounded globule.

A meniscus is caused by adhesion and cohesion

Water in a glass container forms a **meniscus**. The shape of the meniscus is caused by adhesion between the water and the glass. The water is more strongly attracted to the glass than to itself. So where the water touches the

glass more than they are attracted to other water molecules. The adhesion is stronger than the cohesion. So the water wets the glass. It does not stay in round droplets, but spreads out over the surface of the glass.

Mercury behaves rather differently. Mercury particles have stronger cohesion than adhesion to glass. Mercury particles are attracted more strongly to other mercury particles than they are attracted to glass particles. So mercury forms small droplets, and does not spread out on glass.

Fig. 68.5 Mercury forms a convex meniscus in a glass container, because its particles are more attracted to each other than to the glass. Water forms a concave meniscus, because the molecules are more attracted to glass than to each other.

glass, it 'climbs up' the glass.

Mercury does the opposite. Cohesion between mercury particles is stronger than adhesion between mercury and the glass. So a mercury meniscus is 'upside down'.

Water creeps upwards by capillarity

If water is put into a narrow glass tube, adhesion between the water and the glass causes the water to rise up the edge of the glass. If the tube is narrow enough, the water level actually rises up the tube. The narrower the tube the higher the level it rises to. This is called **capillarity**.

Capillary action makes water creep between the fibres of a kitchen towel. Without capillary action, you could not mop up spills.

Soil particles are separated from each other by small air spaces. Soil particles are hydrophilic. They attract water molecules. So adhesion between water and soil particles is strong. Water moves up between the soil particles by capillarity. This is very important in providing plant roots with water. In soils where the spaces between the soil particles are too big, the water cannot rise very high. Plants may find it difficult to get enough water.

Capillarity can be a problem. Bricks contain tiny air spaces. Water can creep up through these spaces from the ground. This causes dampness in the house. A barrier called a **dampcourse** is put in just above ground level to stop this happening. It is made of a substance through which water cannot rise, such as tarred felt, slate or plastic.

Questions

1 Explain why:
  a Water forms a concave meniscus, while mercury forms a convex meniscus when in a glass container.
  b When water is poured out of a glass container the glass stays wet.
  c If a fly falls into water, it has great difficulty in getting out.
  d Water on a dirty car spreads out, but if the car is polished the water collects in beads.
  e Clay soils, made up of small particles, hold water better than sandy soils, which are made of large particles.

2 Could you mop up a spill of mercury with a cloth? Explain your answer.

3 The diagram shows a toy boat. It floats on water. A small pellet of soap is fixed at the back of the boat, so that it dips into the water. Soap reduces the surface tension of the water.

  a Where is the surface tension greatest?
  b Are the forces on the boat balanced?

  c In which direction will the boat move?

4 A wire frame is dipped into soap solution. The lower film is then burst. What shape will the remaining film take up? Explain your answer in terms of surface tension.

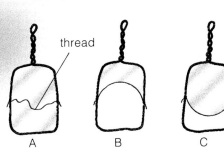

A suspension is cloudy, and can be separated by filtration

If you stir some chalk into a beaker of water, you get a murky liquid. The tiny chalk particles float in the water. They may stay there for several hours, but most of them will eventually fall to the bottom of the beaker. Some, though, will stay floating in the water. They make it slightly cloudy. The mixture of chalk particles and water is called a **suspension**. Although the chalk particles may be too small for you to see, they are big enough to be trapped in filter paper. If you **filter** the suspension, you can separate the water and the chalk particles. The water is called the **filtrate**. The chalk particles on the filter paper are called the **residue**.

Fig. 69.1 A suspension of powdered chalk in water. Suspensions look cloudy.

Fig. 69.2 Separating powdered chalk from water by filtration. The clear liquid which runs through is the filtrate. The material which is trapped on the filter paper is the residue.

Fig. 69.3 Salt (sodium chloride) solution and copper sulphate solution. Solutions look clear.

Fig. 69.4 Separating salt from water by evaporation. As the water boils away, the salt is left behind in the evaporating dish.

A solution is clear, and cannot be separated by filtration

If you add salt to a beaker of water, and stir, the salt seems to disappear. It forms a **solution**. The solution is clear. Some solutions, such as copper sulphate solution, are coloured. But they are always clear.

You cannot separate a salt solution by filtering. If you pour it through filter paper, the salt and the water both go through the paper. (How can you separate the salt and the water?)

In a solution, the substance which dissolves is called the **solute**. In salt water, salt is the solute. The liquid that the solute dissolves in is called the **solvent**. In salt water, water is the solvent.

Substances such as salt and copper sulphate, which will dissolve in water, are said to be **soluble** in water. Substances such as chalk particles are **insoluble** in water.

Water is an excellent solvent

A very large number of different substances will dissolve in water. Water is a very good solvent. Why is this?

To be able to answer this question, we need to understand what happens when something dissolves in water. Sodium chloride is a good example.

Sodium chloride, NaCl, is an ionic compound. It is made up of sodium ions, Na^+, and chloride ions, Cl^-. The ions are held strongly together, because the positive and negative charges attract one another.

Water molecules have dipoles. Part of each water molecule has a slight negative charge, and part has a slight positive charge. When sodium chloride is stirred into water, the water molecules are attracted to the sodium ions and chloride ions. Figure 69.5 shows what happens. Each ion becomes surrounded by water molecules, attracted towards the ion. This is why the sodium chloride seems to disappear when it dissolves in water. Each sodium and chloride ion is separated from the others. They are in between the water molecules.

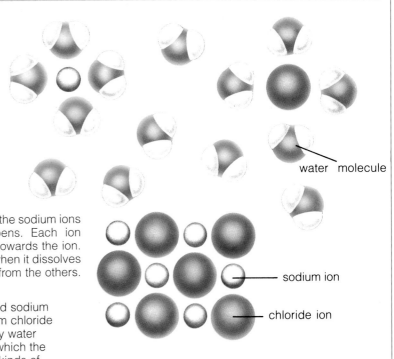

water molecule

sodium ion

chloride ion

Fig. 69.5 Sodium chloride dissolving in water. In solid sodium chloride, each ion is held firmly in the lattice. If sodium chloride is mixed with water, each ion becomes surrounded by water molecules. Can you see the difference in the way in which the water molecules arrange themselves around the two kinds of ion? Can you explain this?

Immiscible liquids can form emulsions

When you add another liquid to water, it may mix with it completely. Ethanol will do this. Ethanol and water are said to be **miscible**. The ethanol dissolves in the water. A completely clear ethanol solution is formed.

Fig. 69.7 Separating petrol and water using a separating funnel. The water has been coloured so that it shows up more clearly.

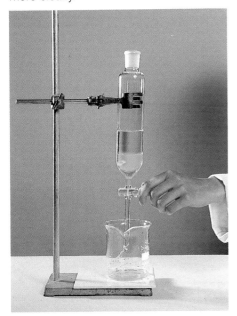

Fig. 69.6 Some liquids will mix together, while others will not. On the left, oil is floating on water – they do not mix. The centre tube contains a solution of methanol and water. Methanol is soluble in water. The tube on the right contains an emulsion of oil and water. The two liquids have been shaken strongly together, so that the oil has broken into tiny droplets which float in the water.

But if you add petrol to water, the two liquids will not mix. They are said to be **immiscible**. The density of petrol is lower than that of water, so the petrol floats on the water. Figure 69.7 shows how the petrol and water can be separated.

If you add cooking oil to water, the oil floats on the water. Oil and water are immiscible. But if you shake the oil and water mixture vigorously, you will end up with a milky-looking liquid. This is called an **emulsion**. As you shake the mixture of oil and water, you break up the oil into tiny droplets. The droplets are small enough to float in the water. They make the water look cloudy. Your emulsion of oil and water will probably separate if you leave it to stand. But many emulsions will stay for a long time. Milk is an emulsion. It is made up of tiny droplets of fat floating in water. Emulsion paint is another example. In this case, the tiny droplets in the water are coloured oils.

70 MORE ABOUT SOLUTIONS

Substances are more likely to react together when they are in solution than when they are in solid form.

Substances are more ready to react in solution than in solid form

If you mix dry copper sulphate crystals with dry sodium hydroxide, nothing happens. But when you mix a copper sulphate *solution* with a sodium hydroxide *solution*, they react together to form copper hydroxide and sodium sulphate. Why is this?

In solid copper sulphate, each Cu^{2+} ion is strongly attracted to the SO_4^{2-} ions around it. The bonds between the copper and sulphate ions are very strong. The ions are reluctant to move from their positions in the lattice.

The same is true of solid sodium hydroxide. The Na^+ ions are strongly attracted to the OH^- ions. So solid copper sulphate and solid sodium hydroxide do not react together.

But when they are in solution it is a different story. Now the ions are all separated from each other. Each ion is surrounded by a group of water molecules. Everything is continually moving around. So copper ions, sulphate ions, sodium ions, hydroxide ions and water molecules will keep bumping into each other.

When a copper ion bumps into a hydroxide ion, they are strongly attracted together. They form an insoluble compound, copper hydroxide. Because it is not soluble, the copper hydroxide makes the water look cloudy. The appearance of this pale blue cloudiness tells you that a reaction has happened in the solution. The copper hydroxide eventually falls to the bottom of the vessel and so is said to form a **precipitate**.

EXTENSION

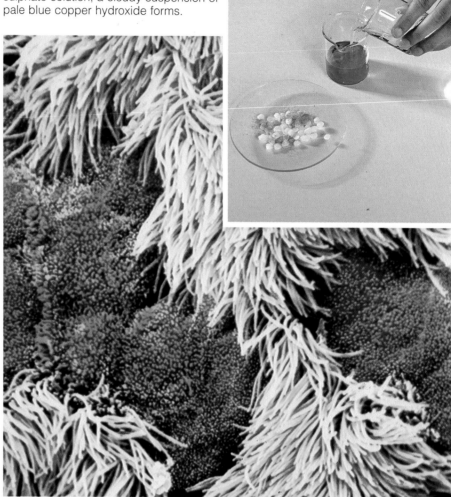

Fig. 70.1 Copper sulphate powder and sodium hydroxide pellets do not react together when dry. But if sodium hydroxide solution is added to copper sulphate solution, a cloudy suspension of pale blue copper hydroxide forms.

Fig. 70.2 Cells from the lining of the trachea. Like all cells, they are largely composed of water.

Reactions in living cells happen in solution

Your body is about 70 % water. If you lose much of this water, you suffer from dehydration. If a very large amount of water is lost, you may die.

One of the reasons you need so much water is to allow reactions to take place in your body. You are alive because chemical reactions are happening in your cells. There are hundreds of different reactions which are needed to keep you alive. One that you will meet in this book is **respiration**. Respiration is a chemical reaction which releases energy from food. It happens in every cell in your body.

These chemical reactions will happen only if the substances taking part in the reaction – the **reactants** – are in solution. Your cells are jelly-like solutions of many different substances in water. If the water was not there, the chemical reactions would not take place. You would quickly die.

DID YOU KNOW?

The longest time that anyone has ever lived without food and water is 18 days. An 18-year old man in Austria was put into a police cell on 1st April 1979, and forgotten. When he was discovered on 18th April, he was very close to death

Solubility is a measure of how much of a substance will dissolve

If you put 100 g of water into a beaker, add some sodium chloride and stir, it will dissolve. But if you go on and on adding sodium chloride, you will eventually get to a point where no more will dissolve. You have made a **saturated solution** of sodium chloride.

The amount of sodium chloride that will dissolve in 100 g of water is called the **solubility** of sodium chloride. Different substances have different solubilities.

Fig. 70.3 A SodaStream forces pressurised carbon dioxide into your drink. The colder the drink, the more carbon dioxide will dissolve, and the fizzier it will be.

INVESTIGATION 70.1

Finding the solubility of different substances

You are to find out the solubility of three different substances at room temperature. Plan your investigation carefully. Decide what apparatus you will need, how you will carry out your investigation, and how you will record your results. Check your plan with your teacher.

When you know exactly what you are going to do, carry out the experiment. Write it up in the usual way.

Solubility of solids increases with increasing temperature; solubility of gases decreases with increasing temperature

If you heat a saturated solution of sodium chloride, you will find that you can 'persuade' a little more sodium chloride to dissolve in it. The solubility of sodium chloride increases as the temperature increases. This is true for most solids. The amount of a substance which can dissolve in 100 g of water at different temperatures can be shown on a graph called a **solubility curve.**

Gases, however, become *less* soluble at higher temperatures. Oxygen is slightly soluble in water. Fish and other water animals use this dissolved oxygen. They take it in through their gills, and use it for respiration. In a pond on a cool day, there is usually plenty of oxygen dissolved in the water. But on a hot day, the hot, fast-moving oxygen molecules can escape from the water into the air. Less oxygen remains in solution in the water. Fish may get short of oxygen, and have to come to the surface to get air.

Fig. 70.4 Waste water from a coal- and oil-burning power station discharges into the River Medway. The water is warm, creating problems for living organisms in the river.

Questions

1 Explain what is meant by the terms:
   a precipitate **b** saturated solution
   c solubility **d** solubility curve

2 Describe, in your own words, why substances are more likely to react together when they are in solution, than when they are in solid form.

3 The chart shows the solubility of copper (II) sulphate at different temperatures.

a Draw a solubility curve for copper (II) sulphate. Temperature goes on the horizontal axis.

b What would happen if you cooled a saturated solution of copper (II) sulphate from 70 °C to 10 °C?

Temperature (°C)	0	10	20	30	40	50	60	70
Number of grams which will dissolve in 100 g of water	14	17	20	25	29	34	40	47

Soaps and other detergents are used to help water to remove dirt. They work by reducing surface tension.

Soap was the first detergent

When water is used for washing, it dissolves dirt. But some dirt will not dissolve in water. Oily dirt will not dissolve. Detergents help oil to dissolve in water. Soap was the first detergent. People have been using soap for a very long time. The first kinds of soap were made by boiling animal fat with wood ash. This is how the Romans made soap. Now most soap is made from oils which smell more pleasant, such as palm oil. The oil is boiled with potassium hydroxide solution.

Fig. 71.1 Soaps, made from plant and animal fats.

detergent molecule

water

oil drop

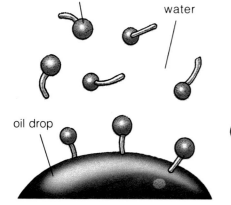

How does detergent affect surface tension?

1 Take a small piece of fabric. Using a dropper pipette, carefully place one drop of water on to its surface. Draw a diagram of the water droplet on the cloth.

2 Take a second piece of the same fabric. Mix a little detergent with water in a beaker. Carefully place one drop of the detergent solution on to the surface of the cloth. Draw a diagram of the water on the cloth.

Questions

1 Why does the water without detergent form a rounded droplet on the cloth?

2 What happens to the drop of water plus detergent on the cloth?

3 Use your answers to questions 1 and 2 to explain how detergents can help when washing clothes.

Detergent molecules are attracted to both water and oil

Oil will not dissolve in water. Oil and water molecules are not attracted to one another. Oil molecules are said to be **hydrophobic**. Hydrophobic means 'water-hating'.

hydrophobic tail

Detergent molecules are made up of two parts. They have a head and a tail. The tail is hydrophobic. It is not attracted to water molecules. But it *is* attracted to oil. The head of a detergent molecule is attracted to water molecules. The head is **hydrophilic**. Hydrophilic means 'water-loving'.

Figure 71.2 shows how detergent molecules can help oil to float away in water. The hydrophobic tails are attracted to the oil, and bury themselves in it. The hydrophilic heads are attracted to the water. The detergent molecules pull the oil into a rounded droplet which can float in the water. An emulsion is formed. An emulsion is a mixture of tiny droplets of one liquid in another. When you wash a greasy plate with a solution of washing-up liquid, this is what happens. The detergent helps the grease to form an emulsion in your washing-up water. The grease comes off the plate, and into the water.

hydrophilic head

Fig. 71.2a A detergent molecule

Fig. 71.2b Detergent removing grease (oil) from a surface. The hydrophobic tails of the detergent molecules are attracted to oil. The hydrophilic heads are attracted to water, so the molecules arrange themselves with their tails in the oil and their heads in the water. The heads repel each other, swinging apart and pulling the oil into a spherical droplet. The droplet of oil, surrounded by detergent molecules, floats off into the water.

Fig. 71.3 Many foods contain emulsifiers. These stop fat and water in the food from separating out into layers.

Bile contains detergents

A similar thing happens in your intestine. Fat or oil in your food will not dissolve in the watery liquids inside your digestive system. So a liquid called **bile** is poured into your intestine. Bile contains detergent-like substances called **bile salts**. Bile salts help the fat in your food to form tiny droplets in the liquids inside your intestine. They are said to **emulsify** the fats. This makes it much easier for the fat to be digested.

Fig. 71.4 Most products which we use for washing are soapless detergents.

Many modern detergents are soapless detergents

Early detergents were made from sodium or potassium hydroxide and animal or plant oils. But animal and plant oils are better used for human food. Now, most detergents are made using by-products of oil refining. They are called **soapless detergents**. Soapless detergents are better than soap in another way, too. Soap forms scum with hard water. This wastes a lot of soap. Soapless detergents do not form a scum.

Fig. 71.5 Foaming caused by detergent pollution.

Detergents and pollution

The first soapless detergents caused pollution problems. They could not be broken down by bacteria. They left houses and factories in waste water, and flowed into the sewage system. Sewage treatment did not break them down. So they eventually got into rivers and streams. Here, they caused foaming, which looks ugly and can harm animals and plants in the water.

Modern soapless detergents can be broken down by bacteria. They are called **biodegradable** detergents. The bacteria in sewage treatment plants break them down. So modern detergents are less likely to cause foaming in rivers.

But detergents do still cause pollution problems. Many detergents have phosphate salts added to them. This improves their dirt-removing power. These phosphates can get into streams and rivers. They act as fertilisers, so that water plants grow more than usual. When the plants die, bacteria in the rivers feed on them. The bacteria use up all the oxygen in the water, so that there is none for fish and other animals.

Questions

1 a What is a detergent?
  b Explain the difference between soap and soapless detergents.
  c What are the advantages of soapless detergents?

2 a What is an emulsion?
  b How do bile salts help you to digest fatty foods?
  c Emulsifiers are often added to foods which are made of mixtures of watery and fatty substances. Ice cream is an example. What do these emulsifiers do? What might happen if they were not added?

3 Explain how detergents can cause pollution. What can be done to reduce this problem?

72 THE WATER CYCLE

Water moves in a continuous cycle from the land and sea into the atmosphere and back again. Human activities can disrupt the balance of this cycle, and cause drought and flooding.

Water evaporates from land and sea

A very large part of the Earth's surface is covered with liquid water. Some of this water evaporates. It goes into the air as water vapour. Water also evaporates from plants' leaves. This process is called **transpiration**. Transpiration is an extremely important way of putting water vapour into the air.

Water vapour condenses as it goes into the atmosphere

When water evaporates into the air, it is in the form of a gas. You cannot see it. But as it rises, the gas cools. When it is cool enough, it changes from a gas into a liquid. This process is called **condensation**. The liquid water forms tiny droplets. They form clouds. Sometimes, if it is cold enough, the water will form tiny ice particles in the cloud.

Water falls from clouds as rain, hail or snow

Eventually, the droplets in the cloud become so large that they fall to the ground. This is called **precipitation**. Precipitation may fall as rain, hail or snow. Some of the water falls on to land. It may sink into the soil, where it will eventually find its way into streams and rivers. These flow into lakes or the sea. The water can evaporate from the surfaces of any of these bodies of water. Some of the water which falls on to land will be taken up by plant roots. It travels up through the plant, and is lost to the air by transpiration.

Deforestation can cause soil erosion and flooding

Trees are very important to the water cycle. But in many parts of the world people have cut trees down. If this is done on a large scale, it is called **deforestation**.

Deforestation happens because people want the land for growing crops, keeping animals such as cattle, and for building roads and houses. They may also want to use the wood for building, making tools and making furniture. In Europe, huge areas of woodland were cut down long ago. Today, deforestation is happening most rapidly in tropical rain forests, such as those in Brazil and some parts of south east Asia.

If there are trees where the rain falls, much of the rain hits the trees before it hits the ground. If there are no trees or other plants, the rain falls directly on to the ground. This can damage the soil. It presses the top layer of soil tightly together. So as more rain falls, it cannot sink into the soil. It runs over the surface. As it does this, it wears away the soil. The soil is carried along with the water, into streams and rivers. Plant roots also help to stop soil washing away when it rains. The roots bind the soil together. Tree roots are especially good at this, as they go very deep into

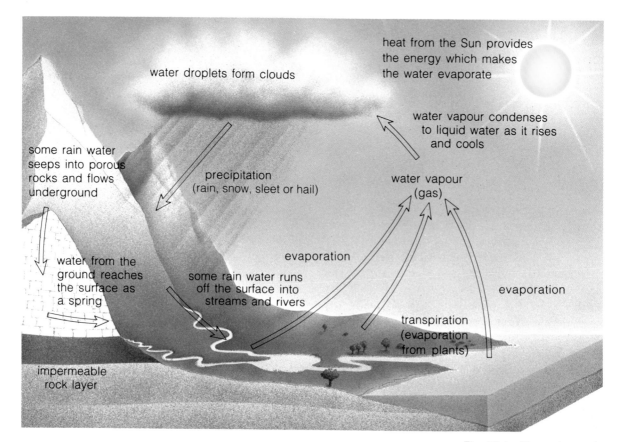

Fig. 72.1 The water cycle

the soil. Without trees, soil easily washes away.

This **soil erosion** is damaging in two ways. Firstly, it means that soil is lost from the land. Soil takes hundreds or thousands of years to form. If it is lost like this, it is very difficult to replace. Secondly, the soil fills up the streams and rivers. When it rains, the water cannot flow so easily in them. The water overflows from the riverbed. This can cause very serious flooding.

Fig. 72.3 Bangladesh has suffered very serious flooding in recent years. This is thought to have been caused by removal of trees from the hills on which rain falls. The water runs off the bare hillsides into rivers, carrying mud which fills the riverbeds and makes them overflow.

Fig. 72.2 Water running over bare ground washes away the soil. Tree roots help to stop this happening. This erosion has happened because prospectors for gold have cut down many trees in a rain forest, allowing rain to wash away the river banks.

Fig. 72.4 Overgrazing in dry scrubland in Sahel, Niger, is turning the area into a desert. There are too many animals and these destroy the vegetation by grazing and trampling, leaving the soil open to erosion, and reducing the amount of moisture in the already dry air. Humans also cut down trees to use for fuel. Little can now be grown on this land, but careful farming practices, which look after the soil, can provide food for large numbers of people from a small area of land.

Lack of trees reduces evaporation

Lack of trees does not only cause soil erosion and flooding. It can also cause droughts. If there are plenty of trees, they soak up a lot of the rain water through their roots. The water then evaporates from their leaves in the process of transpiration. This keeps the air moist. Clouds can form, and more rain can fall.

But if there are no trees, the rain which falls runs directly into rivers. Not very much will evaporate into the air. The air becomes dry. Clouds do not form and less rain falls.

Questions

1   Discuss the ways in which human activities can disrupt the water cycle. What damage might this cause to humans and other living organisms? What can be done to prevent this damage?

EXTENSION

175

Questions

1 Read the following passage, and then answer the questions at the end.

Ozone-friendly chemical plants

It has been known for some time that chlorofluorocarbon gases, often known as CFCs, are harmful to the Earth's atmosphere. CFCs are destroying the Earth's ozone layer. CFCs are also one of the main contributors to the greenhouse effect, which is causing a warming of the atmosphere.

CFCs are widely used as coolants in refrigerators, as aerosol propellants, and for blowing bubbles in polystyrene foam. But now ICI, which makes about 10 % of the world's CFCs, has announced plans to build plants to make harmless replacements for the damaging CFCs. The Dupont group, which is the world's largest producer of CFCs, is also building a similar plant. The cost of each new plant is around £30 million.

The new chemical to replace CFCs is called HFC 134a. It is non-toxic, chemically stable, and does not contain any chlorine. It is the chlorine in CFCs which is responsible for damaging the ozone layer. But HFC 134a will cost five times as much to produce as CFCs. HFC 134a can be used instead of CFCs in refrigerators, air conditioners and aerosol propellants. But a suitable replacement has not yet been found for the use of CFCs in expanded polystyrene.

---

a What is CFC?

b Give three uses of CFCs.

c Why is it important that the production and use of CFCs should be reduced?

d It has been known for some time that CFCs are harmful. Suggest why manufacturers have only recently begun to produce alternatives. You should be able to think of several reasons.

e Developing countries such as India and China are still using large amounts of CFCs. Why is it unlikely that they will quickly switch to the new alternatives?

2 Use the information in Table 63.1 on page 156 to draw a pie chart to show the relative proportions of the gases in the air.

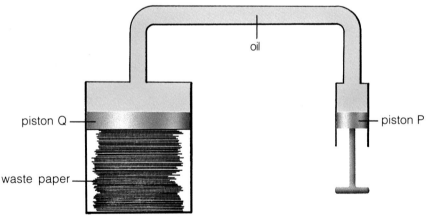

3 The diagram shows a hydraulic press used to crush waste paper.
The area of piston Q is ten times that of piston P. Pressure is applied to the oil by pressing up on piston P.

a How does the pressure on the oil above piston P compare with that above piston Q?

b How does the downward force on piston Q compare with the upward force on piston P? Explain your answer.

c Why is oil in this machine more suitable than air?

d When piston P moves up 20 cm, explain why piston Q moves down 2 cm.

4 River water contains dissolved gases. The gases were removed from the water and analysed. The results of the analysis were:

Gas	Volume of gas in 1 m³ of water
oxygen	28 cm³
nitrogen	70 cm³
other gases	2 cm³

a What was the total volume of the gases dissolved in 1 m³ of water?

b Name one of the 'other gases' that might have been in the water.

c Give one factor which might lower the amount of dissolved oxygen in a river.

d What effect would this loss of dissolved oxygen have on fish?

5 The graph shows how the density of 1 kg of water changes with temperature.

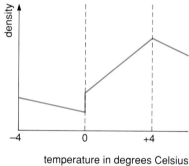

temperature in degrees Celsius

a What happens to water at 0 °C?

b At which temperature shown on the graph will 1 kg of water have the biggest volume?

c If a deep lake is covered with ice, what will be the temperature of the water at the bottom of the lake?

d Why do pipes burst when the water in them freezes?

CHEMICAL REACTIONS

500 ml
± 5%

600 ml
PYREX®

400

300

No. 1000

200

Acids are sour-tasting liquids. Alkalis can neutralise acids. The pH of a liquid is a measure of how acidic or alkaline it is.

Acids are sour-tasting liquids

Acids taste sour. Some common examples include lemon juice and vinegar. These are **weak** acids. But some of the acids that you use in the laboratory are **strong** acids. These include hydrochloric acid and sulphuric acid. Acids are corrosive. Strong acids can rapidly burn your skin.

Tasting a liquid is not a good way to find out if it is an acid! A much better way is to use an **indicator**. An indicator is a substance which changes colour depending on whether a substance is an acid or an alkali. Many natural substances are good indicators. Red cabbage juice and blackcurrant juice are two examples. **Litmus** is a purple dye which comes from lichens. It turns red in acids. Litmus paper is paper which has been soaked in litmus solution and then dried.

Fig. 73.5 Red cabbage juice is a natural indicator. The beaker contains a liquid obtained by liquidising red cabbage leaves. The juice has been pipetted into an alkali on the left, tap water (neutral) in the centre, and acid on the right.

Fig. 73.1 Some examples of weak acids

Fig. 73.2 Strong acids

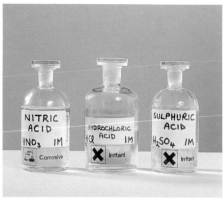

Fig. 73.3 Some examples of weak alkalis

Fig. 73.4 Strong alkalis

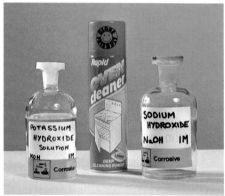

Alkalis turn litmus paper blue

Alkalis are the chemical 'opposite' to acids. Some examples of alkalis are potassium hydroxide, calcium hydroxide, sodium hydroxide and ammonium hydroxide. Like acids, strong alkalis can burn your skin. They feel 'soapy' if you get them on your skin. This is because the alkali reacts with oils in your skin to make soap. You can test for alkalis using litmus paper. It turns blue in an alkaline solution.

Alkalis are soluble in water. A more general name for a substance that is 'opposite' to an acid is **base**. An alkali is therefore a soluble base. Most alkalis are the hydroxide compounds of metals. Ammonium hydroxide is an exception to this.

The strength of an acid or alkali is measured on the pH scale

The **pH scale** runs from 0 to 14. A substance with a pH of 7 is **neutral**. This means that it is neither acidic nor alkaline. Water is the best-known example of a neutral substance. It is neither acidic nor alkaline and therefore has a pH of 7. (And by the way, 'pH' isn't a misprint, it really is meant to be a small p and a capital H.) A substance with a pH of below 7 is acidic. If it has a pH of above 7, it is alkaline. Strongly acidic substances have a very low pH. The lower the pH, the stronger the acid. Similarly, the higher the pH, the stronger the alkali.

pH can be measured using a pH meter. This is simply dipped into the solution you are testing, and gives a reading on a digital display. But pH meters are expensive and can be delicate. A more common method of measuring pH in a school laboratory is to use **universal indicator**. This is a mixture of indicators, which has different colours across the entire pH range. You can use it as a liquid or as paper.

Fig. 73.6 A pH meter. The electrode is placed into the solution whose pH you want to measure.

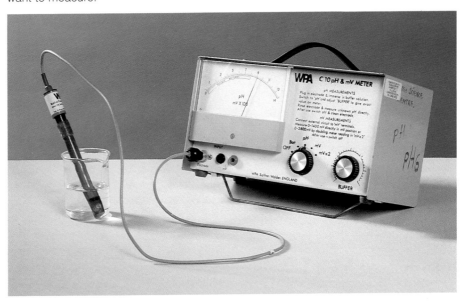

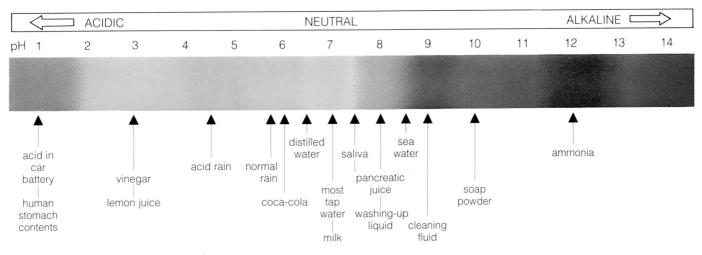

Fig. 73.7 The universal indicator colour range

Questions

1 Divide part of a page in your book into two columns. Head one column ACIDS and the other column ALKALIS. Then decide which of the following pairs of statements fits in each column, and write them in. Keep the two statements in a pair opposite each other in your chart.

  a A taste sour
   B feel soapy on your skin

  b A turn litmus paper blue
   B turn litmus paper red

  c A turn universal indicator green to purple
   B turn universal indicator orange to red

  d A have a pH of over 7
   B have a pH of below 7

  e A examples include lemon juice and vinegar
   B examples include sodium hydroxide and calcium hydroxide

2 In the expression 'pH' the p is a mathematical symbol and the H is a chemical symbol. Find out what each represents.

74

STRONG AND WEAK ACIDS AND ALKALIS

Acids are acidic because they contain H⁺ ions

All acids contain hydrogen ions (H^+). This is what makes them acids. The more hydrogen ions they contain, the more acidic they are. A good definition of an acid is a *substance which produces H⁺ ions when it dissolves in water*.

Strong acids produce a lot of H⁺ ions

Hydrogen chloride is a colourless gas. It has no effect on dry litmus paper. It is not acidic. But when it dissolves in water, it forms **hydrochloric acid**. Hydrochloric acid is a strong acid. It is very corrosive, and can burn your skin. It turns litmus paper and universal indicator solution red. It has a pH of 1–3.

What happens when hydrogen chloride dissolves in water? As the hydrogen chloride molecules dissolve, they split or **dissociate** into hydrogen ions and chloride ions. We can show this as follows:

$$HCl(aq) \rightarrow H^+(aq) + Cl^-(aq)$$

The (aq) means 'dissolved in water'. 'Aq' is short for 'aqueous', which means 'to do with water'. Hydrogen chloride dissociates very easily when it dissolves in water. It produces a lot of hydrogen ions. This means that it is a **strong** acid.

Weak acids produce only a few H⁺ ions

Lemon juice contains citric acid. Citric acid molecules also dissociate when they dissolve in water. They also produce hydrogen ions. But they do not dissociate very easily. So they do not produce very many hydrogen ions. Citric acid is a weak acid. It has a pH of 3–6, and turns universal indicator orange or yellow.

Both strong and weak acids can be concentrated or dilute

If a lot of hydrogen chloride dissolves in a small amount of water, a concentrated solution is made. It is called **concentrated hydrochloric acid**. If a small amount of hydrogen chloride dissolves in a lot of water, a dilute solution is made. It is called **dilute hydrochloric acid**. In the same way, you can have either concentrated or dilute solutions of a weak acid like citric acid.

The **strength** of an acid tells you how easily it dissociates to produce hydrogen ions. The **concentration** of an acid tells you how much water it contains. Both strong and weak acids can be dilute or concentrated. Concentrated acids can be made dilute by adding water. Dilute acids can be made concentrated by evaporating off most of the water in them.

Fig. 74.1 Strong, weak, concentrated and dilute acids. The water molecules are not shown.

STRONG ACID e.g. HCl(aq) WEAK ACID e.g. CH₃COOH(aq)

CONCENTRATED - a lot of hydrogen ions and negative ions in a certain volume of water

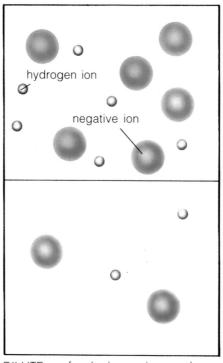 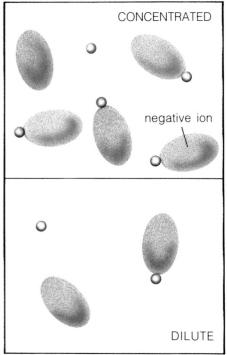

DILUTE - a few hydrogen ions and negative ions in a certain volume of water

Hydrogen chloride forms ions in solution

Hydrogen chloride gas is a covalent compound. The hydrogen atom and chlorine atom share electrons. They are held together by a covalent bond.

But when hydrogen chloride dissolves in water, the hydrogen and chlorine atoms separate. The hydrogen atoms lose one electron each, becoming hydrogen ions. The chlorine atoms gain one electron each, becoming chloride ions. So some covalent compounds can form ions when dissolved in water!

EXTENSION

Strong and weak acids

In this investigation you are going to compare two acids. Hydrochloric acid, HCl(aq), is a strong acid – it produces a high concentration of H^+ ions. Ethanoic acid, $CH_3COOH(aq)$, is a weak acid – it produces a lower concentration of H^+ ions.

1   Put a piece of magnesium ribbon into each of two test tubes. Cover one piece with hydrochloric acid, and the other with ethanoic acid. Compare the rates of reaction.

2   Repeat 1, but using marble instead of magnesium.

3   Pour HCl into a test tube to depth of 1 cm. Add one drop of universal indicator solution. Note the colour and pH.

4   Mix $1\,cm^3$ of HCl and $9\,cm^3$ of water in a test tube. This is Solution A. Test its pH as in 3.

5   Mix $1\,cm^3$ Solution A and $9\,cm^3$ water. This is Solution B. Test its pH.

6   Mix $1\,cm^3$ Solution B and $9\,cm^3$ water. This is Solution C. Test its pH.

7   Repeat steps 3–6 using ethanoic acid instead of hydrochloric acid.

8   Write up all your results in the way you think is best. You might like to use a comparison table.

9   Summarise the evidence you have for saying that ethanoic acid is a weaker acid than hydrochloric acid.

All alkalis contain OH⁻ ions

An alkali is a substance which produces hydroxide (OH⁻) ions when it dissolves in water.

Like acids, alkalis can be strong or weak. The more hydroxide ions it produces, the stronger the alkali.

Sodium hydroxide is an example of a strong alkali. When it dissolves in water, it readily dissociates to produce a lot of hydroxide ions.

$$NaOH\ (aq) \rightarrow Na^+\ (aq) + OH^-(aq)$$

Strong alkalis have a pH of 12–14. They turn universal indicator dark blue or purple.

Ammonia solution is an example of a weak alkali. When ammonia dissolves in water, the ammonia molecules react with the water molecules to form ammonium ions and hydroxide ions:

$$NH_3(aq) + H_2O(l) \rightarrow NH_4^+(aq) + OH^-(aq)$$

However, only a small proportion of the ammonia molecules do this, so only a low concentration of OH⁻ ions are produced. So ammonia solution is a weak alkali. Weak alkalis normally have a pH of 8–11. They turn universal indicator dark green or blue. The (l) after water is a **state symbol**. It tells you that the water is a liquid. You have already met the state symbol (aq) for aqueous or dissolved in water. Other state symbols commonly used are (s) for solid and (g) for gas.

Measuring the pH of soil

The pH of soil affects the plants which grow in it. Farmers and gardeners need to know how acidic or alkaline their soil is, so that they know which plants they can grow.

You cannot just add universal indicator to a mixture of soil and water, because the colour of the soil muddles your result. You need to filter the soil first.

1   Take your first soil sample. Put two spatulas of it into a test tube. Add $10\,cm^3$ of water. Cork the tube, and shake really well to mix.

2   Set up a filter funnel and paper as shown in Figure 74.2. Filter your soil and water mixture.

3   Test the filtrate with universal indicator, and record the result.

4   Repeat with your other soil samples.

Questions

1   Does it matter what sort of water you add to your soil sample? Why?

2   Most plants which we grow as vegetables, such as cabbages and carrots, like a neutral soil. Is your local soil suitable for growing these vegetables?

3   Do any of your soil samples have an acid pH? If so, what sort of soil is it? Where does it come from?

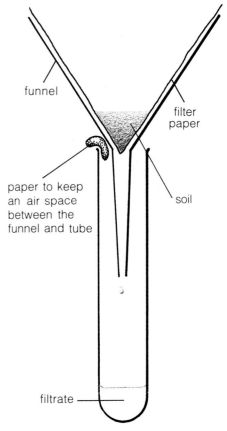

Fig. 74.2   Filtering a soil suspension

funnel

filter paper

paper to keep an air space between the funnel and tube

soil

filtrate

NEUTRALISATION OF ACIDS BY ALKALIS

Acids and alkalis readily react together. The alkali neutralises the acid, and a new substance called a salt is formed.

Alkalis neutralise acids

If you add hydrochloric acid to some sodium hydroxide solution in a beaker, the solution gets hot. If you use equal amounts of equal concentrations of acid and alkali, you will find that the resulting solution does not affect litmus paper. It has a pH of 7. It is a neutral solution.

On page 180, you saw that a solution of hydrochloric acid contains H^+ and Cl^- ions. Sodium hydroxide solution contains Na^+ and OH^- ions. When these four kinds of ions are mixed up together in the beaker, the H^+ and OH^- ions react together to make water molecules:-

$$H^+(aq) + OH^-(aq) \rightarrow H_2O(l)$$

If there were equal amounts of H^+ and OH^- ions in the acid and alkali you mixed up, then all the H^+ ions from the acid will react like this. The H^+ ions have been removed from the acid. The acid has been neutralised.

Equations show what happens during a chemical reaction

So far, we have only looked at what has happened to two of the four kinds of ions which were mixed in the beaker. What about the Na^+ and Cl^- ions?

They do not really do anything at all. They just remain as ions, dissolved in the water in the beaker. But, if you heated the solution in the beaker so that the water evaporated, the Na^+ and Cl^- ions would form an ionic lattice. You would have **sodium chloride** – common salt – in the bottom of your beaker. Any ionic compound formed in this way (by replacing hydrogen ions in an acid with other positive ions) is called a **salt**. Sodium chloride is one of many different salts.

We can write a full equation for this reaction, including the sodium and chloride ions as well as the hydrogen and hydroxide ions, like this:

$$H^+(aq) + Cl^-(aq) + Na^+(aq) + OH^-(aq) \rightarrow Na^+(aq) + Cl^-(aq) + H_2O(l)$$

This is called an **ionic equation**. It shows what is happening to each kind of ion in the reaction.

We can also show the same reaction in terms of the molecules which are taking part in it:

$$HCl(aq) + NaOH(aq) \rightarrow NaCl(aq) + H_2O(l)$$

This is called a **molecular equation**.

Yet another way of showing what happens in the reaction is in words, without using symbols or formulae at all, like this:

hydrochloric acid + sodium hydroxide $\rightarrow$ sodium chloride + water

This is called a word equation.

Word equations, molecular equations and ionic equations all tell you what happens in a reaction. But molecular equations or ionic equations give much more information than word equations.

INVESTIGATION 75.1

Neutralising a weak acid with a weak alkali

In this investigation you will mix a weak acid (ethanoic acid) with a weak alkali (ammonia solution), and follow the changes with universal indicator solution.

1   Read through the investigation, and decide how you are going to record your results. Draw up a results chart if you decide to use one.

2   Fill approximately one-quarter of a boiling tube with ethanoic acid.

3   Fill another boiling tube approximately two-thirds full with ammonia solution.

4   Add universal indicator solution to each, to get a distinct, but not deep, colour.

5   Use a teat pipette to add the ammonia solution to the ethanoic acid drop by drop. After every ten drops, stir the mixture with a glass rod, and note the colour of the indicator and the pH number of the mixture.

Questions

1   Ethanoic acid is a weak acid, and ammonia solution is a weak alkali. What is meant by the word 'weak'?

2   What was the pH of the ethanoic acid?

3   What was the pH of the ammonia solution?

4   What happened to the pH of the acid solution as you added alkali to it?

5   Did you, at any stage, produce a neutral solution? If so, how many drops of ammonia solution did it take to produce this solution?

Neutralising a strong acid with a strong alkali

In this investigation you will mix a strong acid (hydrochloric acid) with a strong alkali (sodium hydroxide solution), and follow one of the changes with a thermometer.

1 Fill the burette to zero with the acid.

2 Use a measuring cylinder or pipette to put 20 cm³ NaOH(aq) into a conical flask.

3 Measure and write down the temperature of the NaOH(aq).

4 From the burette, add 2 cm³ of acid to the conical flask. Swirl the mixture. Measure and write down the new temperature of the alkali in the flask.

5 Continue adding acid to the flask, 2 cm³ at a time. Swirl the flask, measure and record the temperature after every addition. Do this until you have added a total of 40 cm³ of acid.

6 Thoroughly wash out the apparatus.

7 Draw a line graph of your results. Put volume of acid added on the horizontal axis and temperature on the vertical axis.

You will have seen that as acid is added to the alkali heat is given out. As the acid and alkali react, they release heat energy. A reaction like this one, which gives out heat, is said to be an exothermic reaction.

Questions

1 What evidence have you that a chemical reaction took place as the acid and alkali were mixed? Your results should show that the temperature gradually rose, and then gradually fell.

2 What was the highest temperature reached?

3 How much acid had been added in total at this point?

4 Explain why the temperature fell after this point.

Questions

Read the following information about tooth decay, and then answer the questions.

1 Explain what is meant by:
a dental caries
b plaque
c abscess
  The graph shows how the pH of someone's mouth changed during one day.

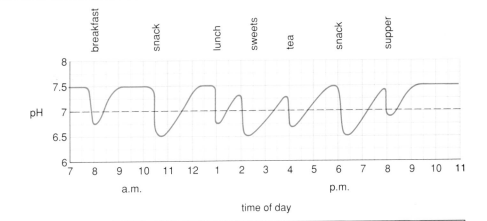

Tooth decay

Bad teeth can make you miserable. They look awful, can cause bad breath – and they hurt! Tooth decay, or dental caries, is caused by bacteria. Everyone has bacteria living in their mouth. If sugary foods are left on, or between, your teeth, some of these bacteria feed on it. They form a sticky film on your teeth, called plaque. The bacteria change the sugar into acid. The acid eats through the enamel covering of the teeth. If the amount of damage is slight, the tooth can repair itself. But if a tooth is exposed to acid for a long time, then a hole gradually develops in the enamel. Once it reaches the dentine, the hole enlarges rapidly. It may begin to be painful. If the decay reaches the pulp cavity, it is very painful. An abscess may result, which is an extremely painful, infected swelling beneath the tooth.

None of this need happen to you! If you never ate sweet foods, you would never get tooth decay. But most people like sweet things sometimes. And it is possible to eat them, and still enjoy perfect teeth. One of the golden rules is to eat your favourite sweet foods – and drink sugar-containing drinks – only two or three times a day. If your teeth are exposed to acid all day long, they do not have a hope of fighting off decay. But if the conditions in your mouth are acidic for only a short while, decay is less likely to happen. Another thing that you can do is to brush the plaque off your teeth twice a day. If you use a fluoride toothpaste, the fluoride will help your teeth to resist decay. Toothpastes also contain weak alkalis, which will neutralise the acid formed by bacteria in your mouth.

2 When a solution becomes more acidic, does the pH get higher or lower?

3 Explain why the pH in the person's mouth became lower after each meal.

4 What damage could be caused while the pH is low?

5 How could the person change their eating habits to lessen their chances of suffering from tooth decay?

6 Give three reasons why brushing regularly with fluoride toothpaste can lower the likelihood of suffering from tooth decay.

Molecular equations show the formulae of the substances involved in a chemical reaction. The numbers of each kind of atom must balance on each side of the equation.

Equations must be balanced

You have already seen that hydrochloric acid and sodium hydroxide solution react together to form sodium chloride and water. The word equation for this reaction is:

hydrochloric acid + sodium hydroxide → sodium chloride + water

The molecular equation for this reaction is:

$$HCl(aq) + NaOH(aq) \rightarrow NaCl(aq) + H_2O(l)$$

The molecular equation tells us more than the word equation. It tells us **how many** of each kind of atom react

together. Look first at the number of hydrogen atoms. On the left-hand side of the equation, there is one hydrogen atom in the HCl molecule, and one in the NaOH molecule. On the right-hand side of the equation, there are also two hydrogen atoms. They are both in the water molecule.

Check the numbers of each of the other kinds of atoms – chlorine, oxygen, and sodium – on each side of the equation. You should find that, for each of them, the numbers are the same on the left and the right. A molecular equation in which the numbers of each kind of atom are the same on the left and the right hand side is called a **balanced equation**. Equations must always be balanced.

Fig. 76.1 Sodium hydroxide and hydrochloric acid react to form sodium chloride and water.

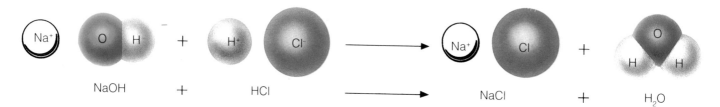

You sometimes need to put numbers in front of formulae to make equations balance

Another example of the neutralisation of an acid by an alkali is the reaction between **sodium hydroxide solution** and **sulphuric acid**.

The word equation for this reaction is:

sulphuric acid + sodium hydroxide → sodium sulphate + water

If we now begin to write a molecular equation, by writing chemical formulae instead of words, we get:

$$H_2SO_4 + NaOH \rightarrow Na_2SO_4 + H_2O$$

But there is something badly wrong with this equation. Look at the numbers of hydrogen atoms first of all. On the **left** side of the equation, there are two hydrogen atoms in the H_2SO_4 molecule, and one hydrogen atom in the NaOH molecule. That makes **three**

altogether. But on the right side of the equation, there are only **two** hydrogen atoms, both in the H_2O molecule. So this cannot be right! Hydrogen atoms do not just disappear like that. You cannot start off with three and end up with two! And if you look again at the equation, you will see that the numbers of sodium atoms do not balance out either. There are two on the right-hand side, but only one on the left-hand side. An extra sodium atom seems to have appeared from nowhere – which just is not possible. So something must be done to the equation to make it **balance** – to make the numbers of each kind of atom match up on either side.

Let us get the sodium atoms right first. There are two on the right-hand side of the equation, as sodium sulphate has the formula Na_2SO_4. So we need two sodium atoms on the left-hand side as well. We can do this by having **two** molecules of NaOH taking

part in the reaction:

$$H_2SO_4 + 2NaOH \rightarrow Na_2SO_4 + H_2O$$

But now the hydrogens and oxygens are wrong. There are four hydrogens on the left, and only two on the right. There are six oxygens on the left, and only five on the right. (Can you find them all?) We can sort this out in one go by having **two** water molecules on the right-hand side of the equation:

$$H_2SO_4 + 2NaOH \rightarrow Na_2SO_4 + 2H_2O$$

And that is the correct, balanced equation. If you add up the numbers of each kind of atom on each side of the equation, you should find that they match. Check it for yourself.

Equations must always balance. You must not lose or gain atoms between the beginning and end of a chemical reaction. The number of a particular sort of atom on one side of an equation must always equal the number of that sort of atom on the other side.

Another example of balancing an equation

When nitric acid reacts with calcium hydroxide solution, calcium nitrate and water are formed. How do you go about writing a balanced equation for this reaction?

First, write a word equation:

nitric acid + calcium hydroxide → calcium nitrate + water

Next, write down the correct formulae for each of the substances involved in the reaction. It is very important that you get these right. They are:

$$HNO_3 + Ca(OH)_2 \rightarrow Ca(NO_3)_2 + H_2O$$

Next, count up numbers of atoms on either side of the equation.

Calcium seems all right – there is one calcium atom on each side. But the hydrogen is *not* right. HNO_3 has one, and $Ca(OH)_2$ has two, so there are three hydrogen atoms on the left-hand side. On the right, there are only two hydrogen atoms. (Where are they?)

Count up the numbers of nitrogen and oxygen atoms. Do they balance?

The next thing to do is to *put a number in front of one of the formulae*, and see if it helps. Until you really get the hang of this, it is best just to guess and try it out. You can always rub it out if it does not work. In this case, it seems sensible to leave the $Ca(OH)_2$ and $Ca(NO_3)_2$ alone, as the Ca atoms balance already. So let us try:

$$2HNO_3 + Ca(OH)_2 \rightarrow Ca(NO_3)_2 + H_2O$$

Count atoms again. You should find that calcium and nitrogen balance, but that hydrogen and oxygen do not. On the left, there is a total of four hydrogen atoms, but only two on the right. The left has eight oxygen atoms, but only seven on the right. We can sort this out by putting a two in front of the water molecule:

$$2HNO_3 + Ca(OH)_2 \rightarrow Ca(NO_3)_2 + 2H_2O$$

Now the equation balances. Finish it off by putting in the state of each of the molecules involved:

$$2HNO_3(aq) + Ca(OH)_2(aq) \rightarrow Ca(NO_3)_2(aq) + 2H_2O(l)$$

Rules for balancing equations

1. Write a word equation.
   sulphuric acid + sodium hydroxide →
   sodium sulphate + water

2. Write correct molecular formulae.
   $$H_2SO_4 + NaOH \rightarrow Na_2SO_4 + H_2O$$

3. Balance with numbers in front of formulae if necessary.
   $$H_2SO_4 + 2NaOH \rightarrow Na_2SO_4 + 2H_2O$$

4. Add state symbols.
   $$H_2SO_4(aq) + 2NaOH(aq) \rightarrow Na_2SO_4(aq) + 2H_2O(l)$$

Questions

1. Hydrochloric acid and potassium hydroxide react and produce potassium chloride and water.
   a. Write a word equation for this reaction.
   b. Write an equation with the formula of each substance.
   c. Add large numbers in front of the formulae if necessary to balance the equation.
   d. Add correct state symbols after the formulae.

2. Ammonium hydroxide and nitric acid react and produce ammonium nitrate and water.
   a. Write a word equation for this reaction.
   b. Write an equation with the formula of each substance.
   c. Add large numbers in front of the formulae if necessary to balance the equation.
   d. Add correct state symbols after the formulae.

3. Sulphuric acid and magnesium hydroxide react and produce magnesium sulphate and water.
   a. Write a word equation for this reaction.
   b. Write an equation with the formula of each substance.
   c. Add large numbers in front of the formulae if necessary to balance the equation.
   d. Add correct state symbols after the formulae.

4. Write word equations, and then balanced equations, for:
   a. hydrochloric acid and ammonium hydroxide, reacting to form ammonium chloride and water
   b. hydrochloric acid and calcium hydroxide, reacting to form calcium chloride and water
   c. nitric acid reacting with sodium hydroxide
   d. a reaction between an acid and alkali which forms potassium nitrate and water

DID YOU KNOW?

Some balanced equations have confused scientists. The equation for photosynthesis:

$$6CO_2 + 6H_2O \rightarrow C_6H_{12}O_6 + 6O_2$$

would suggest that the oxygen might come from the carbon dioxide. But it actually comes from the water. Try to write an equation to show this.

BASES

A base is a substance which neutralises acids. Alkalis are one sort of base. An alkali is a base that is soluble in water.

A base is a substance which can neutralise an acid

There are many different substances which can neutralise acids. They include:

metal hydroxides e.g. sodium hydroxide
metal oxides e.g. copper oxide
metal carbonates e.g. zinc carbonate
metal hydrogencarbonates e.g. sodium hydrogencarbonate
ammonia solution
metals e.g. magnesium

All of these substances are **bases**. A base is a substance which can neutralise an acid.

Some bases are also **alkalis**. An alkali is a base which is soluble in water.

DID YOU KNOW?

Magnesium oxide is a base which is used to neutralise acids in the stomach. It is not absorbed before it can do this. It can easily raise the pH of the stomach by about one unit.

Bases neutralise acids by removing H⁺ ions

You have seen how metal hydroxides, such as sodium hydroxide, neutralise acids. The H$^+$ ions – which are what make a solution acidic – combine with the OH$^-$ ions from the hydroxide. Water is formed.

$$H^+(aq) \quad + \quad OH^-(aq) \quad \rightarrow \quad H_2O(l)$$

the acid part of the acid   the alkaline part of the alkali   a neutral compound

Bases neutralise acids by removing the hydrogen ions from solution. The following examples show how each of the six kinds of bases listed above neutralise acids.

Metal hydroxide + acid → metal salt + water

You have met this one before. An example is the neutralisation of hydrochloric acid by sodium hydroxide. Sodium hydroxide is a soluble base, or alkali.

hydrochloric acid + sodium hydroxide →
sodium chloride + water

$$HCl(l) + NaOH(aq) \rightarrow NaCl(aq) + H_2O(l)$$

INVESTIGATION 77.1

Metal hydroxide + acid → metal salt + water

An example of this is the reaction between hydrochloric acid and sodium hydroxide, forming sodium chloride and water.

1. Fill a burette to zero with hydrochloric acid.
2. Use a pipette and filler to place 25 cm³ sodium hydroxide solution in a conical flask. Add three drops of the indicator phenolphthalein, which will turn it pink.
3. Slowly add the acid to the alkali, swirling all the time, until the indicator turns completely colourless. This is called the **end-point**. The procedure you have just carried out is called a **titration.**
4. Now that you know roughly where the end-point is, repeat steps 2 and 3. This time, take more care as you near the end-point, so that you get a more accurate value of the amount of acid needed to neutralise the alkali.
5. You have now found out exactly how much acid neutralises 25 cm³ of alkali. Your flask contains salt solution, contaminated with a little indicator. If you want to get pure salt crystals, you now need to repeat the titration without adding indicator. Use 25 cm³ of alkali, and the amount of acid you know is needed to neutralise it.
6. You now have a solution of sodium chloride. To speed up the formation of the salt crystals, boil away approximately 90 % of the water and leave to stand.

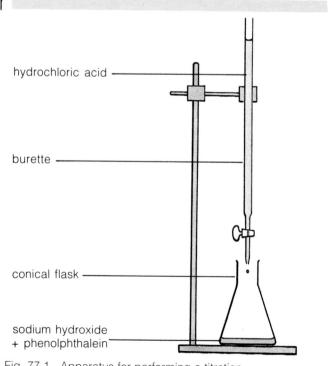

hydrochloric acid

burette

conical flask

sodium hydroxide + phenolphthalein

Fig. 77.1 Apparatus for performing a titration

Ammonia solution + acid → ammonium salt + water

When ammonia gas dissolves in water, some ammonium hydroxide forms.

$$NH_3(aq) + H_2O(l) \rightarrow NH_4OH(aq)$$

So the following reaction is really the same as the reaction between a metal hydroxide and an acid:

nitric acid + ammonium hydroxide → ammonium nitrate + water

$$HNO_3(aq) + NH_4OH(aq) \rightarrow NH_4NO_3(aq) + H_2O(l)$$

The salt formed in this reaction is ammonium nitrate. This is sold as fertiliser. As you can see from its formula, it contains a lot of nitrogen. Nitrogen is essential for plants to make proteins. Ammonium nitrate is the most widely used fertiliser in the world.

INVESTIGATION 77.2

You can carry out this experiment in just the same way as Investigation 77.1, but using ammonium hydroxide solution instead of sodium hydroxide solution. You will also need to use a different indicator. Use screened methyl orange instead of phenolphthalein. It will be green when you add it to the alkali, but turns grey when you reach the end point of the reaction.

Questions

1 Is ammonia an alkali?

2 Write word equations, and then balanced molecular equations, for the reactions which would occur between each of the following:
   a ammonia solution and hydrochloric acid
   b ammonia solution and sulphuric acid

— EXTENSION—

Fig. 77.2 Screened methyl orange in, from left, alkaline, neutral and acidic solutions.

Metal oxide + acid → metal salt + water

Metal oxides produce oxide ions, O^{2-}. The oxide ions react with the H^+ ions in acids, to form water.

$$2H^+ + O^{2-} \rightarrow H_2O$$

An example of such a reaction is the neutralisation of sulphuric acid by zinc oxide:

sulphuric acid + zinc oxide → zinc sulphate + water

$$H_2SO_4(aq) + ZnO(s) \rightarrow ZnSO_4(aq) + H_2O(l)$$

Notice that the zinc oxide has the symbol (s) after it. This means that it is in solid form, not dissolved in water. Zinc oxide is not soluble in water. It is a base, but not an alkali.

Questions

1 Is zinc sulphate soluble in water?

2 Write word equations, and then balanced molecular equations, for the reactions which would occur between each of the following:
   a hydrochloric acid and zinc oxide
   b sulphuric acid and copper (II) oxide
   c sulphuric acid and magnesium oxide

— EXTENSION—

INVESTIGATION 77.3

Metal oxide + acid → metal salt + water

You are going to add copper (II) oxide to dilute sulphuric acid, to form copper sulphate and water.

1 Put about 25 cm^3 of dilute sulphuric acid into a beaker. Warm the acid *gently*. This will speed up the reaction when you add copper oxide to it. When the acid is warm, put the beaker on the bench.

2 Add one spatula full of copper (II) oxide to the warm acid. Stir thoroughly. When all the copper (II) oxide has dissolved, add another spatula full.

3 Every now and then, test with universal indicator paper. When the copper (II) oxide seems *very* slow to dissolve, you are reaching the end of the reaction. Stop when no more copper (II) oxide will dissolve and the pH is nearly 7.

4 If you want a sample of the copper sulphate you have made, first filter the solution, to get rid of any copper (II) oxide that did not react. You now have a copper sulphate solution. Concentrate it by boiling off excess water and leave it to crystallise.

Metal carbonate or hydrogencarbonate + acid → metal salt + water + carbon dioxide

This reaction is often used to find out if an unknown substance is a carbonate or hydrogencarbonate. The carbon dioxide being given off in the reaction is usually very obvious. The solution fizzes (**effervesces**) as the gas is given off. The gas can be tested by passing it into lime water, which goes milky if it really is carbon dioxide.

An example of this reaction is the neutralisation of hydrochloric acid by calcium carbonate:

hydrochloric acid + calcium carbonate → calcium chloride + water + carbon dioxide

$$2HCl(aq) + CaCO_3(s) \rightarrow CaCl_2(aq) + H_2O(l) + CO_2(g)$$

Another example is the neutralisation of hydrochloric acid by sodium hydrogencarbonate. This reaction can happen in your stomach! If you have indigestion, it sometimes helps to swallow a spoonful of sodium hydrogencarbonate (baking soda). This neutralises the hydrochloric acid in your stomach. What do you think happens to the carbon dioxide produced?

hydrochloric acid + sodium hydrogencarbonate → sodium chloride + water + carbon dioxide

$$HCl(l) + NaHCO_3(aq) \rightarrow NaCl(aq) + H_2O(l) + CO_2(g)$$

Questions

1 Is calcium carbonate a base?
2 Is calcium carbonate an alkali?

---

3 Write word equations and then balanced molecular equations, for the reactions which would occur between each of the following:
   a sulphuric acid and zinc carbonate
   b nitric acid and potassium carbonate
   c hydrochloric acid and potassium hydrogencarbonate
4 You are given a white powder, which you think might be a carbonate or hydrogencarbonate. How could you find out? What happens if you are right?
5 Many indigestion tablets contain sodium hydrogencarbonate.
   a Why is this?
   b Given three different types of indigestion tablets, what experiments could you do to see which acts fastest (without swallowing them!)?

EXTENSION

Fig. 77.3 Spreading lime on to agricultural land. The lime could be calcium hydroxide, or calcium carbonate. These are bases, and will reduce acidity in soil. Lime also improves the texture of heavy soils.

INVESTIGATION 77.4

Metal carbonate + acid → metal salt + water + carbon dioxide

In this reaction, you will use magnesium carbonate and dilute sulphuric acid. The reaction is:

magnesium carbonate + sulphuric acid → magnesium sulphate + water + carbon dioxide

$$MgCO_3(s) + H_2SO_4(aq) \rightarrow MgSO_4(aq) + H_2O(l) + CO_2(g)$$

1 Put about 25 cm^3 of dilute sulphuric acid into a beaker.
2 Add one spatula measure of magnesium carbonate. It will seem to disappear as it reacts. The solution will fizz, as carbon dioxide gas is produced. (What do you think happens to the water that is produced? What happens to the magnesium sulphate which is produced?)

3 Continue adding magnesium carbonate until there is no more fizzing, and the magnesium carbonate does not disappear. The acid is now neutralised.
4 If you want a sample of the magnesium sulphate that you have made, do the following. First, take your beaker of neutralised acid – which now contains your magnesium sulphate solution. Filter it, to remove any magnesium carbonate which did not react. Then concentrate the solution, and leave to crystallise.

Metal + acid → metal salt + hydrogen

Many metals react with acids, neutralising them. In this case, the hydrogen ions do not combine with oxygen to form water. The hydrogen ions from the acid gain electrons from the metal atoms. Then the hydrogen atoms join in pairs to form hydrogen gas:

$$2H^+(aq) + 2e^- \rightarrow H_2(g)$$

The acid fizzes as the hydrogen bubbles out. You can show that the gas is hydrogen by putting a lighted splint into the test tube. If the gas is hydrogen, it gives a squeaky pop.

Some metals, like silver, do not react with acids. Others, such as sodium, react dangerously fast. Zinc, iron and magnesium all react at a steady speed.

zinc + sulphuric acid → zinc sulphate + hydrogen

$$Zn(s) + H_2SO_4(aq) \rightarrow ZnSO_4(aq) + H_2(g)$$

Questions

1 Write word equations and then balanced molecular equations, for the reactions which occur between:
   a magnesium and hydrochloric acid
   b iron and sulphuric acid (the iron compound formed is pale blue-green)
2 When metals react with nitric acid, different reactions may take place. Find out about these.

EXTENSION

INVESTIGATION 77.5

Metal + acid → metal salt + hydrogen

In this experiment you will use magnesium ribbon and dilute hydrochloric acid. The reaction is:

magnesium + hydrochloric → magnesium chloride
              acid   + hydrogen

1 Put about 25 cm³ of dilute hydrochloric acid into a beaker.
2 Add a strip of magnesium ribbon. The solution will fizz, as hydrogen gas is given off. You can test the gas with a burning splint if you like.
3 Continue adding magnesium ribbon, bit by bit, until the fizzing stops. The acid has now been neutralised.

Questions

1 What can you now do if you want to obtain a sample of the magnesium chloride you have made?
2 Do you think that it would make any difference to the speed of this reaction if you used magnesium powder instead of ribbon? Design an experiment to find out the answer to this question.

Summary – Learn these!

acid + metal → metal salt + hydrogen

acid + metal oxide → metal salt + water

acid + metal hydroxide → metal salt + water

acid + metal carbonate → metal salt + water + carbon dioxide

acid + metal hydrogencarbonate → metal salt + water + carbon dioxide

acid + ammonia solution → ammonium salt + water

More summary – Remember these!

When **hydrochloric** acid is neutralised, the salt formed is a metal **chloride**.

When **nitric** acid is neutralised, the salt formed is a metal **nitrate**.

When **sulphuric** acid is neutralised, the salt formed is a metal **sulphate**.

78 ACID RAIN

Acid rain is caused when sulphur dioxide and nitrogen oxide gases are released into the air. They dissolve in water droplets, and fall to earth as rain or snow.

Sulphur dioxide is emitted when fossil fuels are burnt

Fossil fuels, such as coal, natural gas and oil, contain sulphur. When they are burnt, the sulphur combines with oxygen. It forms sulphur dioxide. The sulphur dioxide is given off as a gas.

Sulphur dioxide is an unpleasant gas. It damages living things. Humans who breathe in a lot of sulphur dioxide over a long period of time have an increased risk of suffering from colds, bronchitis and asthma. Sulphur dioxide can kill plant leaves. It may completely kill the plant.

Fig. 78.1 The upper branches of this tree may have been killed by acid rain.

INVESTIGATION 78.1

The effect of sulphur dioxide on plants

Sodium metabisulphite solution gives off sulphur dioxide. You are going to test its effect on two kinds of important crop plants.

1 Fill four containers with damp compost. Press the compost down firmly, but not too hard. Make the tops level.
2 Put barley seeds on to the compost in two containers. Spread them out evenly. Put maize seeds on to the compost in the other two containers.
3 Cover all the seeds with enough compost to hide them completely. Label all four containers.
4 Leave the containers in a warm place until most of the seeds have germinated. Keep them well watered.
5 When the shoots of barley and maize are about 3–5 cm tall, make labelled diagrams of each of the four sets of seedlings.
6 Now take four small containers, such as watch glasses. Into two of them place a piece of cotton wool soaked in sodium metabisulphite solution. (TAKE CARE. Do not get it on your hands.) In the other two, place a piece of cotton wool soaked in water.
7 Place each container of seedlings in a large plastic bag, with one of the watch glasses. You should have one of your groups of barley seedlings with a container of sodium metabisulphite solution and one with a container of water. Do the same with the maize seedlings. Tie the plastic bags tightly, so that no air can get in or out.
8 Make labelled diagrams of each of your four sets of seedlings about 30 min after putting them into their bags, and again after one or two days.

Questions

1 Why were some seedlings enclosed with sodium metabisulphite solution, and some with water?
2 Were both types of plant affected in the same way by sulphur dioxide? Explain what these effects were, and any differences between the maize and barley seedlings.

Sulphur dioxide and nitrogen oxides form acid rain

Sulphur dioxide dissolves in water to form an acid solution. Nitrogen oxides also form an acid solution when they dissolve. Nitrogen oxides form when the nitrogen and oxygen in the air react together under the high temperatures and pressures inside a petrol engine. Sulphur dioxide and nitrogen oxides are carried high into the air. Here, they dissolve in water droplets in clouds. They make the water droplets acid. The clouds may be carried many miles before they drop their water. It falls as acid rain or snow.

Acid rain damages trees and aquatic animals

Acid rain can damage trees. Acid rain washes important minerals, such as calcium and magnesium, out of the soil. Acid rain falling on thin soils, such as those in the mountains of Scandinavia, can kill huge areas of forest.

As the acid water runs through the soil, it washes out aluminium ions. The aluminium runs into rivers and lakes. Aluminium is very toxic to fish. Some acid rivers and lakes now contain hardly any fish.

Acid rain damages buildings

Acid rain can also damage buildings. The acid reacts with carbonates in limestone. The limestone dissolves, so that the stone gradually crumbles away.

Fig. 78.2 Stonework in Lincoln Cathedral

What is the major cause of acid rain?

Normal rain is slightly acid anyway. This is because carbon dioxide in the air dissolves in rain drops to form carbonic acid. Carbonic acid is a very weak acid. So even ordinary rain has a pH a little below 7.

But the large amount of sulphur dioxide and nitrogen oxides which we are now releasing into the air are making rain much more acid than this. We are damaging plants, animals and buildings. Something must be done to stop this getting any worse.

Scientists are still not sure about the most important cause of acid rain. Is it power stations burning coal? Certainly these are very important. Large amounts of sulphur dioxide are released from these power stations. Many of them are now beginning to remove the sulphur dioxide from the smoke they produce, so that it does not go into the air. But this is expensive and means that we will have to pay more for the electricity they make. Or are cars the most important producers of acid rain? Car exhaust fumes contain nitrogen oxides. Since 1956 the amounts of sulphur dioxide emitted by coal burning have gone down. But the acid rain problem has got worse! Over this time the numbers of cars on the roads has increased enormously. So it looks as though cars are as much to blame as coal-burning factories and power stations.

Fig. 78.3 Acid rain

Combustion of fossil fuels, for example oil or coal in power stations and petrol in cars, releases oxides of sulphur and nitrogen into the air.

The sulphur and nitrogen oxides dissolve in water droplets in clouds, making them acidic. They may be carried for hundreds of kilometres.

Plants may be damaged when acid rain falls on them. Acidification of the soil allows toxic elements to dissolve, and be washed into streams, rivers and lakes. Fish and other aquatic organisms may be killed.

Questions

1 a How is sulphur dioxide produced?
   b What damage can be done by sulphur dioxide gas?
   c What is produced when sulphur dioxide dissolves in water?
2 Both sulphur dioxide and nitrogen oxides contribute to the formation of acid rain.
   a What are the major sources of these two gases?
   b Describe the kind of damage which can be done by acid rain.
   c Is normal rain neutral, acidic or alkaline? Explain your answer.
   d What do you think could be done to reduce the damage caused by acid rain?

Questions

1 Copy and complete this table using the suggestions below.

Type of substance		strong acid		neutral	
pH	5				14
Colour with universal indicator			green/blue		

Choose from: 2, yellow, weak acid, weak alkali, purple, green, 7, 8, strong alkali, red.

2 Copy and complete this passage:
Acids are solutions containing ..... ions. Alkalis are solutions containing ..... ions. If an acid and an alkali are mixed, the two types of ion combine to form ..... This process is called ..... .

3 Explain the difference between:
a a strong acid and a concentrated acid.
b a weak acid and a dilute acid.

4 Nitric acid is neutralised by calcium carbonate.
a What three products are formed in this reaction?
b Write a word equation for this reaction.
c Write a molecular equation for this reaction.
d Balance your equation and add state symbols.
e Describe how you could use this reaction to obtain a sample of a salt.

5 Repeat question 4 for the following reactions:
a nitric acid and lead carbonate
b hydrochloric acid and copper oxide
c sulphuric acid and zinc

6 The salt sodium nitrate can be made by titration.
a Which reactants would you use?
b Write a word equation for the reaction.
c Write a balanced molecular equation for the reaction.
d Describe fully how you would make the salt, explaining why you would titrate three times.

— EXTENSION —

7 The table shows the pH range at which certain plants will grow successfully.

Plant	pH range
apples	5.5 - 7.0
blackcurrants	6.0 - 7.5
broad beans	5.5 - 7.0
carrots	6.0 - 7.5
lettuces	6.5 - 7.5
onions	6.5 - 7.5
potatoes	5.5 - 6.5
tomatoes	5.5 - 7.0

a Design a chart or graph to show this information as clearly as possible.
b A market gardener wishes to grow as wide a range of crops as possible. He decides to test the pH of his soil. Describe how he could do this.
c He finds that his soil's pH is 6.0. Is the soil acid or alkaline?
d Which of the above crops could he grow successfully?
e To grow a fuller range of crops, he would like the pH of his soil to be around 6.5. Which crops could he grow then?
f He decides to add lime to the soil. Lime can be calcium carbonate. What effect will this have?

8 Describe how you would prove the following:
a Magnesium and hydrochloric acid react to produce hydrogen.
b Magnesium carbonate and hydrochloric acid react to produce carbon dioxide.
c Sodium hydroxide solution will neutralise vinegar.
d Cola drinks contain stronger acid than orange squash.
e Rhododendrons grow best in acid soils.

9 Explain the following:
a Rubbing sodium bicarbonate on a bee sting helps to stop the pain.
b Rubbing vinegar on a wasp sting helps to stop the pain.
c Lakes in limestone areas are much less affected by acid rain.
d Acid snow affects the young of water animals more than acid rain affects them.

REVERSING THE EFFECTS OF ACID RAIN

Much of the sulphur dioxide and nitrogen oxides which cause acid rain are produced by coal-burning power stations. Efforts are now being made to reduce the amount of acid gases which they produce.

In 1988 the Central Electricity Generating Board in Great Britain was given the go-ahead to build the world's biggest sulphur removal plant at the Drax power station in North Yorkshire. The plant cost £400 million and created 600 construction jobs. It cut Drax's sulphur dioxide emissions by 90%.

The sulphur dioxide is washed out of the gases emitted from the power station by spraying a watery slurry of limestone over the gas. The sulphur dioxide reacts with the limestone to produce gypsum (calcium sulphate). The gypsum can be used for making plasterboard.

The cost of this plant, and others to follow, has to be passed on to consumers in higher electricity prices.

Scientists are very pleased that these sulphur removal plants are being introduced, but are concerned that damage already done by acid rain may be irreversible. A group of environmental scientists in Norway is conducting an experiment to find out if badly damaged soils can recover from acid rainfall. They are investigating the effects of stopping acid rain falling on an area in southern Norway. The area receives rain and snow with a pH of 4.2. It has a thin soil with a sparse covering of pine and birch trees. They have divided their experimental area into three parts. One part, area A, has been left alone. The other two areas, B and C, have been covered with plastic roofs. All the rain falling on the roofs is collected. In area B, the rain which falls on to the roof has the acidity removed from it. It is then sprinkled on to the trees growing under the roof. In area C, the rain which falls on to the roof is sprinkled on to the trees without having its acidity removed.

The scientists have found that in areas A and C the soil has remained very acidic. The water running off from these areas contains a lot of nitrate and sulphate ions. But in area B the amount of nitrate in the runoff water went down by 60% within two weeks of the beginning of the experiment. After three years the sulphate levels in the runoff water had gone down by 50%.

The researchers think that these results are encouraging. They show that the effects of acid rain can be reversed if we stop releasing sulphur dioxide and nitrogen oxide into the air. But it will take a long time for new trees to grow in areas where they have been killed, and for new populations of fish to build up in lakes damaged by acid runoff.

Drax power station before the sulphur removal plant was built. The white clouds from the cooling towers are just water vapour. More harmful gases, such as sulphur dioxide, are coming from the tall, narrow tower.

a Why do you think that the CEGB took so long to introduce sulphur removal plants at its coal-burning power stations?

b Explain how sulphur dioxide can be removed from the waste gases emitted by coal-burning power stations.

c Name one useful by-product from this process.

d One scientist said, of the decision to build the sulphur removal plant at Drax, 'It's welcome, but it's a bit like closing the stable door after the horse has bolted.' What do you think he meant?

e In the experiment in Norway, why did the scientists cover part of their experimental area with plastic roofs?

f The two areas covered with plastic roofs were treated differently. Explain why you think they used these two methods of treatment.

g Briefly summarise the results of the Norwegian experiment, and explain why they are encouraging.

EXOTHERMIC AND ENDOTHERMIC REACTIONS

**Exothermic reactions give out heat.
Endothermic reactions take in heat.**

The reaction between magnesium and sulphuric acid

You will probably have seen how magnesium reacts with sulphuric acid.

The mixture fizzes as hydrogen is given off. The magnesium disappears. The equation for this reaction is:

magnesium + sulphuric acid → magnesium sulphate + hydrogen

$$Mg(s) + H_2SO_4(aq) \rightarrow MgSO_4(aq) + H_2(g)$$

This reaction shows an important property of acids. Acids corrode metals. But there are also two other things happening which have great importance to chemists.

1 *Heat is produced* If you measure the temperature of the acid before and after adding the magnesium, you will find that the temperature goes up.

2 *Electrons move from particle to particle* Before it reacts, the magnesium is in the form of uncharged magnesium atoms. The sulphuric acid is made up of hydrogen ions with a single plus charge – H^+ – and sulphate ions with a double minus charge – SO_4^{2-}.

During the reaction, the magnesium atoms each lose two electrons. They become ions with two plus charges – Mg^{2+}. The electrons are passed to the H^+ ions. They become uncharged H atoms, which join together in pairs to form H_2 molecules.

Chemists are particularly interested in changes in temperature and in electron movement during chemical reactions.

Fig. 79.1 Magnesium reacts vigorously with sulphuric acid, releasing bubbles of hydrogen.

Many reactions give out heat

Most chemical reactions are like the one between magnesium and sulphuric acid. They give out heat to the surroundings. Reactions which give out heat are called **exothermic** reactions. All the reactions we call 'burning' are exothermic reactions. Some examples include:

natural gas burning: $CH_4(g) + 2O_2(g\ ) \rightarrow CO_2(g) + 2H_2O(l)$

coal burning: $C(s) + O_2(g) \rightarrow CO_2(g)$

magnesium burning: $2Mg(s) + O_2(g) \rightarrow 2MgO(s)$

Another very similar reaction occurs in your cells. It is called respiration. It, too, is an exothermic reaction:

respiration: $C_6H_{12}O_6(aq) + 6O_2(aq) \rightarrow 6CO_2(aq) + 6H_2O(l)$

There are many other exothermic reactions which you will be familiar with. For example, as vegetable matter rots on a compost heap, heat is given out. The reactions involved are exothermic. Another exothermic reaction which you have probably performed in the laboratory is the neutralisation of an acid by an alkali.

Fig. 79.2 A flare-out at an oil refinery. This oxidation reaction is exothermic.

Some chemical reactions take in heat

In most chemical reactions, heat is given out. But there are some reactions which take in heat from their surroundings. Reactions which take in heat are called **endothermic reactions**.

One endothermic chemical reaction which is important in industry is the decomposition of calcium carbonate. Calcium carbonate is limestone or chalk. The reaction is:

calcium carbonate → calcium oxide + carbon dioxide

$$CaCO_3(s) \rightarrow CaO(s) + CO_2(g)$$

'Decomposition' means 'breaking down'. A decomposition reaction is one in which a substance breaks down into two or more new substances. In this reaction the calcium carbonate breaks down, or decomposes, to calcium oxide and carbon dioxide. Calcium oxide is called lime, or **quicklime**. It has many uses. Perhaps the most important is in the manufacture of cement. It is also spread on acidic soils to increase their pH, and on clay soils to improve their structure. This decomposition reaction is endothermic. When 1 kg of calcium carbonate decomposes, nearly 3000 kJ of heat are taken in! So the reaction will not happen unless a lot of heat is supplied to it. This is done by putting the calcium carbonate into large ovens called lime kilns. They are kept at around 800 °C.

Fig. 79.4 Limestone is decomposed to form calcium oxide, an important component of cement. The bricklayer is using mortar, a mixture of cement, sand and water.

Another important industrial reaction which is endothermic is the production of sodium and chlorine from salt. This is also a decomposition reaction.

sodium chloride → sodium + chlorine

$$2NaCl(l) \rightarrow 2Na(l) + Cl_2(g)$$

The sodium produced in this reaction has many uses. These include making soap and cooling nuclear power stations. The chlorine can be used to make PVC (polyvinyl chloride) and bleach.

Fig. 79.3 The combustion of magnesium is an exothermic reaction.

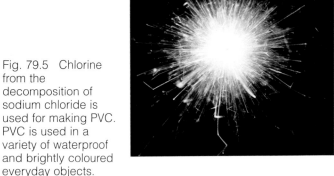

Fig. 79.5 Chlorine from the decomposition of sodium chloride is used for making PVC. PVC is used in a variety of waterproof and brightly coloured everyday objects.

Questions

1 a What is meant by an exothermic reaction?
  b Give one example of an exothermic reaction.
  c What is meant by an endothermic reaction?
  d Give one example of an endothermic reaction.
  e Which type of reaction is most common – exothermic or endothermic?
2 a What is lime?
  b What is lime used for?
  c How is lime produced?

  d Is the reaction which produces lime an exothermic or endothermic reaction?
  e What will be the chief costs of making lime?
  f Lime-making involves a decomposition reaction. Give another example of a decomposition reaction, and explain what decomposition means.

Exothermic and endothermic changes

Carry out the following five experiments. For each one, measure and record the temperature of the liquid. Then add the solid and stir. Finally, measure the temperature of the liquid again.

1 Add two spatula measures of fine iron filings to 25 cm³ of copper sulphate solution.
2 Add two spatula measures of potassium sulphate to 25 cm³ of water.
3 Add two spatula measures of ammonium nitrate to 25 cm³ of water.
4 Add a strip of magnesium ribbon to 25 cm³ of hydrochloric acid.
5 Add two spatula measures of anhydrous copper sulphate to 25 cm³ of water.

Questions

1 Which reactions were exothermic?
2 Which reactions were endothermic?

Bond making is exothermic

When you add water to calcium oxide, the particles in the water and the particles in the calcium oxide form bonds. The bonds hold the five atoms together as a molecule of calcium hydroxide.

$$CaO(s) + H_2O(l) \rightarrow Ca(OH)_2(s)$$

As the bonds are made, heat is released into the surroundings. This reaction is exothermic. Bond making is exothermic.

Bond breaking is endothermic

Calcium carbonate molecules are made up of five atoms. (What are they?) When calcium carbonate is heated, these molecules break up to form a two atom molecule, CaO, and a three atom molecule, CO_2.

$$CaCO_3(s) \rightarrow CaO(s) + CO_2(g)$$

So bonds are broken in this reaction. Bond breaking takes in heat from the surroundings. So this reaction is endothermic. Bond breaking is endothermic.

Decomposition involves breaking bonds. So decomposition reactions are usually endothermic. Some examples include:

$$2NaNO_3(s) \rightarrow 2NaNO_2(s) + O_2(g)$$

$$NH_4Cl(s) \rightarrow NH_3(g) + HCl(g)$$

$$2HgO(s) \rightarrow 2Hg(l) + O_2(g)$$

EXTENSION

Some reactions are exothermic, while others are endothermic

Chemical reactions can be thought of as happening in two stages. (i) Firstly the bonds in the molecules of the reactants have to be broken, forming free atoms. (ii) Secondly these free atoms bond together to form the product molecules. The first stage is endothermic, and the second stage is exothermic. Whether or not the overall energy change is endothermic or exothermic depends on which of the two stages has the largest energy value.

For example, when natural gas (methane) burns, the two stages can be represented by the **energy level diagram** shown in Figure 80.1a. Stage (i) involves the breaking of the bonds in the methane molecule and the two oxygen molecules while stage (ii) involves the bonding together of the nine atoms to form a carbon dioxide molecule and two water molecules. The value for stage (ii) is bigger than the value for stage (i), so the overall energy change is **exothermic**.

When ammonia breaks down to give nitrogen and hydrogen the two stages can be represented by the energy level diagram shown in figure 80.1b. Stage (i) involves the breaking of bonds in two ammonia molecules while stage (ii) involves the bonding together of the eight atoms to form one nitrogen molecule and three hydrogen molecules. The value for stage (i) is bigger than the value for stage (ii), so the overall energy change is **endothermic**.

As these two energy level diagrams suggest, in many chemical reactions the first stage of the reaction is endothermic, so energy is needed in order to get the reaction started. When molecules of the reacting substances meet each other they must have a certain amount of energy or they won't react. This energy amount is called the **activation energy** of the reaction. Reactions with a small activation energy tend to go quickly (because lots of the molecules in the mixture can react), but reactions with a large activation energy tend to go slowly (because few of the molecules in the mixture can react). Scientists have discovered certain substances that, when added to a reaction mixture, decrease the energy values of stages (i) and (ii) by equal amounts. The **overall** energy change for the reaction is therefore not changed, but because the activation energy is decreased the reaction happens more quickly. Substances that do this are called **catalysts**.

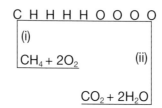

Fig. 80.1a

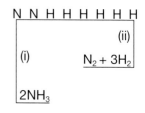

Fig. 80.1b

Questions

1 The energy values of some bonds are as follows:
C–H 412, O=O 496, O–H 463, and C=O 743.

a Work out the energy value of stage (i) for the reaction between methane and oxygen. To do this, add together the energy values of all the bonds in one methane molecule and two oxygen molecules.

b Work out the energy value of stage (ii) for the reaction between methane and oxygen. To do this, add together the energy values of all the bonds in one carbon dioxide molecule and two water molecules.

c Work out the overall energy value for the burning of methane by finding the difference between your value for stage (i) and your value for stage (ii). Will this reaction be exothermic or endothermic?

2 The energy values of some bonds are as follows:
N≡N 944, H–H 436, and N–H 388.

a Work out the energy value of stage (i) for the decomposition of ammonia.

b Work out the energy value of stage (ii) for the decomposition of ammonia.

c Work out the overall energy value for the decomposition of ammonia. Will this reaction be exothermic or endothermic?

A substance which can undergo an exothermic reaction has chemical energy

Stored energy, or potential energy, can be in many different forms. One form of potential energy is chemical energy. A substance which can undergo an exothermic reaction has chemical energy. Petrol has chemical energy. When it burns in an engine, this stored energy is transferred to heat energy and kinetic (movement) energy. The burning of petrol is an exothermic reaction.

In an exothermic reaction, the energy released is usually released as heat. But humans have built machines which cause the energy to be released in other useful forms. A car engine is a good example. If petrol is poured on the ground and set alight, the exothermic reaction produces heat, which disperses into the ground and air. But in a car engine the petrol burns in the cylinder. The hot, fast moving gas molecules which are produced push the piston, which turns the crankshaft. So some of the stored chemical energy in the petrol is transferred to kinetic energy of the car.

chemical energy → heat energy → kinetic energy
  (in the fuel)   (as the   (as the car
                fuel burns)   moves)

In a coal-fired power station, coal burns in an exothermic reaction. The heat produced is used to boil water. The steam produced turns turbines. The turbines turn generators, which generate (produce) electricity.

chemical → internal → kinetic → electrical
energy   energy   energy   energy

(in coal)   (in water   (in turbines
          and steam) and generator)

In your cells, glucose from food combines with oxygen. This is an exothermic reaction. Some of the energy is released as heat. This keeps your body temperature at 37 °C, even when the temperature of your environment is much less. But some of the energy is used for other activities, such as movement.

chemical energy ⟶ movement   and   heat
  (from food)   (in muscles)   (from muscles)

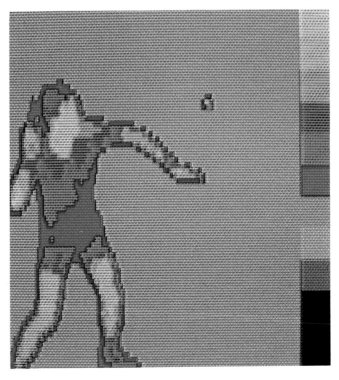

Fig. 80.2 Respiration is an exothermic reaction, which releases heat energy. The heat image, or thermogram, of a man playing squash shows the relative temperatures of different parts of his body. The scale at the side runs from hot (white) at the top to cold (black) at the bottom. Notice that the squash player is hotter than his surroundings. (Why do you think the squash ball is hot?)

81 OXIDATION AND REDUCTION

Many chemical reactions involve movement of electrons.

Chemical reactions often involve movement of electrons

Think again about the reaction between magnesium and sulphuric acid:

magnesium + sulphuric acid → magnesium sulphate + hydrogen

$$Mg(s) + H_2SO_4(aq) \rightarrow MgSO_4(aq) + H_2(g)$$

This equation gives an overall picture of the reaction. But if we include *charges* in the equation, we can see even more about what is happening.

$$Mg(s) + H^+_2SO_4^{2-}(aq) \rightarrow Mg^{2+}SO_4^{2-}(aq) + H_2(g)$$

Look first at what is happening to the magnesium

particles. At the beginning of the reaction, they are uncharged magnesium atoms, Mg(s). But at the end of the reaction, they have become positively charged magnesium ions, Mg^{2+}. The magnesium particles have lost electrons.

Where have these electrons gone? Look at the hydrogen particles. At the beginning of the reaction, they are positively charged hydrogen ions, H^+. At the end of the reaction, they are uncharged hydrogen atoms bonded together in pairs to form hydrogen molecules, H_2. Each hydrogen ion has gained an electron.

So during this reaction, electrons move from magnesium particles to hydrogen particles. Many chemical reactions involve similar electron movements.

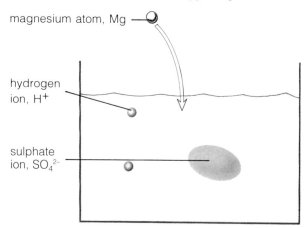

magnesium atom, Mg

hydrogen ion, H^+

sulphate ion, SO_4^{2-}

Fig. 81.1 The reaction between magnesium and sulphuric acid H_2SO_4(aq). The water molecules in the sulphuric acid solution are not shown.
Hydrogen ions have a single positive charge; they have one more proton than electrons. Magnesium atoms have no overall charge; they have equal numbers of electrons and protons.

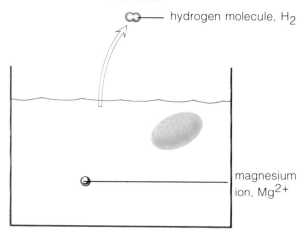

hydrogen molecule, H_2

magnesium ion, Mg^{2+}

When a magnesium atom meets two hydrogen ions in solution, each hydrogen ion takes one electron from the magnesium atom. This leaves the magnesium with a double positive charge. The two hydrogen atoms, now with no net charge, join together to form a molecule of hydrogen.

Loss of electrons is called oxidation

Another example of a chemical reaction which involves movement of electrons is the burning of magnesium.

magnesium + oxygen → magnesium oxide

$$2Mg(s) + O_2(g) \rightarrow 2MgO(s)$$

Putting in the charges, we get:

$$2Mg(s) + O_2(g) \rightarrow 2Mg^{2+}O^{2-}(s)$$

Again, the magnesium particles have lost electrons.
This happens with many metals when they react with oxygen. The metals lose electrons, which are transferred to the oxygen. We say that the metal has been **oxidised**.

Chemists do not only use the word 'oxidised' when oxygen is involved in the reaction. It is used whenever electrons are lost, even if they are transferred to something other than oxygen. **Oxidation is the loss of electrons**. Sometimes the electrons might be taken up by oxygen; sometimes it might be something else. Look back at the magnesium and sulphuric acid reaction. The magnesium particles lost electrons. So we say that the magnesium was oxidised.

Gain of electrons is called reduction

If something loses electrons in a chemical reaction, something else must gain them. When magnesium burns, the electrons from the magnesium are transferred to oxygen. We say that the oxygen is reduced. **Reduction is the gain of electrons**.

Reactions involving electron transfer are called redox reactions

If one substance loses electrons in a reaction, then something else must gain them. So if one substance is oxidised, something else is reduced. Reactions like this are called 'reduction-oxidation' reactions or **redox reactions** for short.

How can you tell if a reaction is a redox reaction? Is the reaction between zinc and hydrochloric acid a redox reaction? You need to follow this method to find out:

1   Write an equation for the reaction.

$$Zn(s) + 2HCl(aq) \rightarrow ZnCl_2(aq) + H_2(g)$$

2   Put charges into the equation.

$$Zn(s) + 2H^+Cl^-(aq) \rightarrow Zn^{2+}Cl^-_2(aq) + H_2(g)$$

3   Look at each type of particle individually.

$$Zn \rightarrow Zn^{2+}$$

$$2H^+ \rightarrow H_2$$

$$2Cl^- \rightarrow 2Cl^-$$

4   Decide whether anything loses or gains electrons.
Zn *loses* electrons. So zinc is *oxidised*.
H+ *gains* electrons. So hydrogen is *reduced*.
So this reaction is a redox reaction.

Fig. 81.2   The reaction between zinc and hydrochloric acid. Which particles are losing electrons? Which are gaining them?

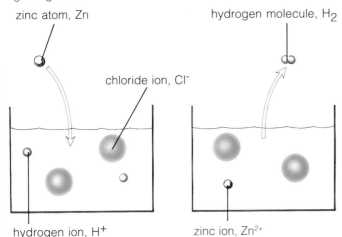

Zinc and copper sulphate solution (on the left) react together to form copper and zinc sulphate. You can see the copper forming in the bottom of the right-hand beaker.

Here is another example. Is the reaction between silver nitrate and hydrochloric acid a redox reaction?

1   Write the equation.

$$AgNO_3(aq) + HCl(aq) \rightarrow AgCl(s) + HNO_3(aq)$$

2   Put in charges.

$$Ag^+NO_3^-(aq) + H^+Cl^-(aq) \rightarrow Ag^+Cl^-(s) + H^+NO_3^-(aq)$$

3   Look at each type of particle individually.

$$Ag^+ \rightarrow Ag^+ \qquad\qquad H^+ \rightarrow H^+$$

$$NO_3^- \rightarrow NO_3^- \qquad\qquad Cl^- \rightarrow Cl^-$$

4   Decide whether anything loses or gains electrons. Nothing does. So this is not a redox reaction.
Notice that a compound ion like the nitrate ion, NO_3^-, is considered as a single particle. You do not have to think about the nitrogen and oxygen atoms in it separately.

Fig. 81.3   As silver nitrate solution is added to hydrochloric acid, a white cloud of silver chloride immediately forms.

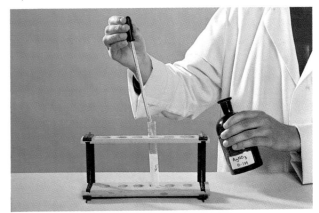

Questions

1   Which of the following are redox reactions?
a $Fe(s) + H_2SO_4(aq) \rightarrow FeSO_4(aq) + H_2(g)$
b $BaCl_2(aq) + H_2SO_4(aq) \rightarrow BaSO_4(s) + 2HCl(aq)$
c $Mg(s) + Cl_2(g) \rightarrow MgCl_2(s)$
d $Zn(s) + CuSO_4(aq) \rightarrow ZnSO_4(aq) + Cu(s)$

2 a   Why is it that when a metal element reacts with a non-metal element, the reaction is always a redox reaction?

b   Which element is oxidised – the metal or the non-metal?
c   Which element is reduced – the metal or the non-metal?

3   Each of the following reactions is a redox reaction. For each one, say which substance is oxidised, and which is reduced.

a $4Na(s) + O_2(g) \rightarrow 2Na_2O(s)$
b $Mg(s) + ZnCl_2(aq) \rightarrow MgCl_2(aq) + Zn(s)$
c $2FeCl_2(s) + Cl_2(g) \rightarrow 2FeCl_3(s)$
d $2NaCl(s) \rightarrow 2Na(s) + Cl_2(g)$
e $Cl_2(g) + 2KI(aq) \rightarrow I_2(aq) + 2KCl(aq)$
f $TiCl_4(l) + 4Na(l) \rightarrow Ti(s) + 4NaCl(s)$

EXTENSION

82 SOME IMPORTANT REDOX REACTIONS

Redox reactions cannot always be identified by looking at electron movement.

The best definition of oxidation is 'the loss of electrons'. The best definition of reduction is 'the gain of electrons'. But in some reactions, these definitions can be hard to apply.

For example, when steam reacts with coal, the following reaction takes place:

$$H_2O(g) + C(s) \rightarrow CO(g) + H_2(g)$$

None of the particles involved in this reaction are ions. So there are no charges to put into the equation. Yet it *is* a redox reaction!

There is another definition of oxidation which can be used. **Oxidation is the gain of oxygen.** Similarly, **reduction is the loss of oxygen.** In this reaction, the steam loses oxygen. The steam is reduced. The coal (carbon) gains oxygen. The coal is oxidised.

There are other reactions where this definition is useful. For example, magnesium reacts with copper (II) oxide to give magnesium oxide and copper:

$$Mg(s) + CuO(s) \rightarrow MgO(s) + Cu(s)$$

Magnesium gains oxygen, so it is oxidised.
Copper loses oxygen, so it is reduced.

In this reaction, we could also use the first definition of oxidation and reduction. Putting charges into the equation, we get:

$$Mg(s) + Cu^{2+}O^{2-}(s) \rightarrow Mg^{2+}O^{2-}(s) + Cu(s)$$

Magnesium loses electrons, so it is oxidised.
Copper gains electrons, so it is reduced.

When you are deciding whether or not a reaction is a redox reaction, use both definitions. Firstly, ask if electrons are being gained or lost. Then ask if oxygen is being gained or lost. If the answer to *either* question is 'yes', then the reaction is a redox reaction. If the answer to *both* questions is 'no', then it is not a redox reaction.

Combustion of fuels

When a fuel burns, it gains oxygen. It is oxidised. Natural gas is a good example. Natural gas is methane, CH_4.

$$CH_4(g) + 2O_2(g) \rightarrow CO_2(g) + 2H_2O(g)$$

Fig. 82.1   Natural gas burning

Respiration

Respiration is the release of energy from food in living cells. Glucose from food reacts with oxygen. The carbon in glucose is oxidised.

glucose + oxygen → carbon dioxide + water

$$C_6H_{12}O_6(aq) + 6O_2(g) \rightarrow 6CO_2(g) + 6H_2O(l)$$

So we get energy from our food by a redox reaction.

Fig. 82.2   A redox reaction in the skier's cells releases energy for muscular movement.

Corrosion of metals

Many metals corrode or rust in air. Iron is a good example. The iron combines with oxygen. It is oxidised.

Iron + oxygen → iron oxide (rust)

$4Fe(s) + 3O_2(g) \rightarrow 2Fe^{3+}_2O^{2-}_3(s)$

The iron loses electrons, so it is oxidised. The oxygen gains electrons, so it is reduced. Rusting is a redox reaction.

Fig. 82.3 Rusting is a redox reaction.

Decomposing sodium chloride

Sodium chloride can be decomposed to sodium and chlorine.

$2Na^+Cl^-(l) \rightarrow 2Na(l) + Cl_2(g)$

The sodium ions gain electrons, so they are reduced. The chloride ions lose electrons, so they are oxidised. This decomposition is a redox reaction.

Making sodium hydroxide

A lot of the sodium which is produced from the decomposition of sodium chloride is used to make sodium hydroxide.

$2Na(s) + 2H_2O(l) \rightarrow 2Na^+OH^-(aq) + H_2(g)$

The sodium atoms lose electrons, so they are oxidised. The water molecules gain electrons, becoming hydroxide ions and hydrogen molecules, so the water molecules are reduced.

Redox reactions

Because scientists can understand redox reactions they have been able to use them in a variety of applications. They can create tiny power cells for quartz watches, control the rusting of iron and explain the role of oxygen in keeping us alive.

Fig. 82.5 Sodium reacting with water. Universal indicator solution has been added to the water. It was green to begin with, but is turning purple as the sodium and water react. Why?

Extraction of metals from ores

Iron is extracted from iron ore in a blast furnace. The iron ore combines with hot carbon monoxide gas.

iron ore + carbon monoxide → iron + carbon dioxide

$Fe_2O_3(s) + 3CO(g) \rightarrow 2Fe(s) + 3CO_2(g)$

The iron ore loses oxygen, so it is reduced. The carbon monoxide gains oxygen, so it is oxidised. Extracting metal from ore is a redox reaction.

Fig. 82.4 A battery of cells used to decompose sodium chloride. The cells contain sodium chloride solution. An electric current is passed through the solution. Sodium and chlorine are produced.

83 RATE OF REACTION

Chemical reactions happen at different rates. You can measure the rate of a reaction by the rate of disappearance of the reactants, or by the rate of appearance of the products.

Reactions go at different rates

By now, you will probably have come across a number of different chemical reactions. A chemical reaction is a process by which one or more substances called **reactants** changes into one or more other substances, called **products**.

Some examples of chemical reactions which you may have met are:

A magnesium + sulphuric acid → magnesium sulphate + hydrogen

B iron + oxygen → iron oxide

C silver nitrate + hydrochloric acid → silver chloride + nitric acid

D copper oxide + sulphuric acid → copper sulphate + water

You may have noticed that these reactions occur at different speeds. C, for example, occurs instantaneously. A is a rapid reaction. D is slow. B is extremely slow, especially if the iron is dry. We say that these reactions have different **rates of reaction**.

INVESTIGATION 83.1

The thiosulphate cross

If you mix sodium thiosulphate solution with hydrochloric acid, a reaction occurs. One product is the element sulphur. This makes the mixture cloudy.
1 Copy the table.

Experiment	1	2	3	4	5	6	7
Volume of sodium thiosulphate (cm^3)	40	30	20	40	40	40	40
Volume of water (cm^3)	0	10	20	5	7.5	0	0
Volume of acid (cm^3)	10	10	10	5	2.5	10	10
Temperature of sodium thiosulphate (°C)	room	room	room	room	room	40	60
Time (s)							

2 Draw a cross, with a dark pencil, on a piece of white paper. Stand a 100 cm^3 beaker on the cross.
3 Put 40 cm^3 of sodium thiosulphate solution into the beaker. Add 10 cm^3 of hydrochloric acid, and start timing. Stop timing when the mixture is so cloudy that you can no longer see the cross. Fill in this result in the table.
4 Rinse the beaker very thoroughly. Perform experiments 2 to 7, rinsing the beaker after each experiment. Record each of your results.

Rate can be measured

Reaction A, above, is quite a rapid reaction. You can tell this because the magnesium disappears quickly. Reaction D is slow, because the copper oxide disappears slowly. Reaction B is very slow. The patches of iron oxide take days to form. Reaction C is instantaneous, because the silver chloride forms immediately when the solutions mix.

So we use two different ways to judge how fast a reaction takes place. Rate of reaction is **the rate of disappearance of a reactant or the rate of appearance of a product**.

Fig. 83.1 As hydrochloric acid is added to sodium thiosulphate solution, a cloudy precipitate of sulphur is formed which obscures the cross. You can measure the speed of the reaction by timing how long it takes for the cross to disappear.

Questions

1 What can you conclude by comparing results for experiments 1, 2 and 3?
2 What can you conclude by comparing results for experiments 1, 4 and 5?
3 What can you conclude by comparing results for experiments 1, 6 and 7?
4 Which are you timing – the disappearance of a reactant, or the appearance of a product?
5 Suggest how you would expect your results to change if you did the experiment in a 250 cm^3 beaker, but used the same quantities of reactants.

Factors affecting the rate of reaction

Hydrochloric acid reacts with marble chips (calcium carbonate). One product is carbon dioxide gas. You can measure the rate of reaction by timing how long it takes to make $10 \, cm^3$ of gas. You are going to try altering several factors, to find out which ones affect the rate of the reaction.

1 Copy the table below.

Experiment	1	2	3	4	5	6	7
Vol. acid (cm^3)	50	50	50	50	100	50	50
Temperature of acid (°C)	room temp	40	60	room temp	room temp	room temp	room temp
Concentration of acid (mol/dm^3)	2	2	2	2	2	1	0.5
Marble chip	whole	whole	whole	crushed	whole	whole	whole
Time for $10 \, cm^3$ of CO_2 to collect							

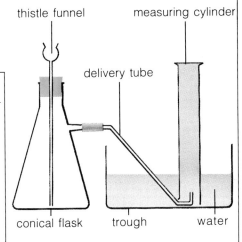

Fig. 83.2 Apparatus for generating carbon dioxide. The thistle funnel is optional. If you do not have one, use a solid bung in the conical flask. Take out the bung to put in the marble chips and acid, then replace it quickly.

thistle funnel   measuring cylinder
delivery tube
conical flask   trough   water

2 Set up the apparatus as in the diagram. The inverted measuring cylinder full of water is to collect the gas. Move it away from the outlet of the delivery tube to begin with.

3 Remove the bung. Put $50 \, cm^3$ of acid and one marble chip into the flask. Replace the bung. 5 s later, place the inverted measuring cylinder over the mouth of the delivery tube. Time how long it takes collect $10 \, cm^3$ of gas.
This is experiment 1. Fill in your result.

4 Perform experiments 2 to 7. Fill in your results for each one. In each case, choose marble chips of similar size.
For experiments 2 and 3, *carefully* warm the acid in a beaker before adding it to the marble chip.
For experiment 4, crush one marble chip in a mortar.
For experiment 6, mix $25 \, cm^3$ of $2 \, mol/dm^3$ acid with $25 \, cm^3$ of water. This makes $1 \, mol/dm^3$ acid.
For experiment 7, mix $12.5 \, cm^3$ of $2 \, mol/dm^3$ acid with $37.5 \, cm^3$ of water. This makes $0.5 \, mol/dm^3$ acid.

Questions

1 Which factor, or factors, increased the rate of reaction?
2 Which factor, or factors, decreased the rate of reaction?
3 Which factor, or factors, did not affect the rate of reaction?
4 Why were marble chips of similar size used?
5 Why was the measuring cylinder not over the delivery tube while the bung was replaced?
6 Which of these conditions would you choose to make the reaction go as fast as possible?
Acid temperature: room temp.   40 °C   60 °C
Acid concentration: $2 \, mol/dm^3$ $1 \, mol/dm^3$ $0.5 \, mol/dm^3$
Marble chip: whole   crushed   whole
7 Which conditions would you choose, from those in question 6, to make the reaction go as slowly as possible?
8 Write a word equation and then a balanced molecular equation, for the reaction involved.

Rate of reaction matters to manufacturers

A slow reaction can be irritating to wait for in a laboratory. But in industry it is much more serious. A slow reaction means slow production. This can mean less profit. Chemical engineers have to understand the factors affecting rate of reaction and use them to make the product form more quickly.

Questions

1 Rank these reactions in order of increasing rate:
a sodium hydroxide solution neutralising hydrochloric acid
b wood rotting
c milk turning sour
d an egg cooking in boiling water
2 Zinc and sulphuric acid react to produce zinc sulphate and hydrogen.

a Write a word equation for this reaction.
b Write a balanced molecular equation for this reaction.
c What are the reactants?
d What are the products?
e List four ways in which you could judge the rate of this reaction.
f How could you speed up this reaction?
g How could you slow down this reaction?

84

CATALYSTS

Catalysts are substances which speed up chemical reactions without being changed themselves.

Catalysts speed up reactions without being changed themselves

One way of making oxygen in the laboratory is by the decomposition of hydrogen peroxide, H_2O_2.

hydrogen peroxide $\longrightarrow$ water + oxygen

$$2H_2O_2(aq) \longrightarrow 2H_2O(l) + O_2(g)$$

Under normal conditions, this reaction is very slow. It can be speeded up by using more concentrated hydrogen peroxide, or by heating it. But a much easier way of speeding it up is to add **manganese (IV) oxide**, MnO_2. Manganese (IV) oxide is a fine, black powder. If you add it to hydrogen peroxide solution – even if it is cold and dilute – the hydrogen peroxide starts to decompose rapidly. Surprisingly, the manganese (IV) oxide is not used up. When the reaction finishes, the black powder is still there.

Manganese (IV) oxide is an example of a **catalyst**. A catalyst is **a substance which speeds up a chemical reaction without being changed or used up.**

Catalysts are important in industry

Catalysts are very important to chemical manufacturers. They allow more product to be made in a given length of time. They are not used up, so they do not have to be continually added to the reactants. But they do have to be cleaned every now and then. This is because the activity of a catalyst involves its surface. Dirt or impurities on a catalyst's surface will stop it acting as efficiently as it should. Industry uses many different catalysts to speed up the reactions it relies on.

Fig. 84.2 Biological washing powders contain enzymes which break down stains such as coffee, egg, blood or sweat.

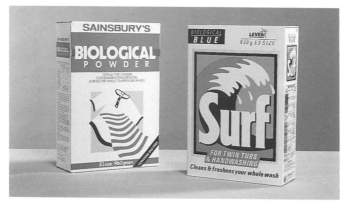

Biological catalysts are called enzymes

Living organisms have hundreds of different chemical reactions going on inside them. Some of these reactions need to happen quickly. The reactions could be speeded up by heating the reactants. But living cells are rapidly killed if their temperature goes too high.

Fig. 84.1 The conical flask contains hydrogen peroxide solution. Without a catalyst nothing happens. But when a small pinch of manganese (IV) oxide is added, the hydrogen peroxide rapidly decomposes to water and oxygen. The manganese (IV) oxide is not used up in this reaction.

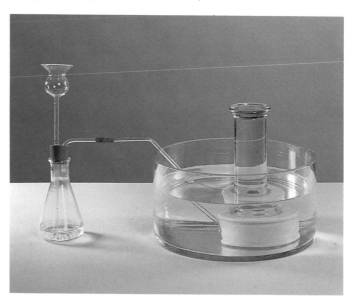

So living organisms use catalysts to speed up their reactions. These catalysts are called **enzymes**. Each different reaction happening in a living organism has a different enzyme catalysing it. So there are hundreds of different kinds of enzyme.

Enzymes are proteins and consist of very large molecules. These molecules have special shapes which are important to their ability to act as catalysts. If enzymes are heated above 45°C, their molecules lose their shapes, and so they don't work well as catalysts at higher temperatures.

Catalysing the decomposition of hydrogen peroxide with an enzyme

You have seen how hydrogen peroxide can be made to decompose quickly by using manganese (IV) oxide as a catalyst. There is also a biological catalyst which will speed up this reaction. It is called **catalase**. Catalase is found in many different kinds of cell. You can find it in apples, yeast, meat (muscle cells) – almost anything! One of the places where there is an especially large amount of catalase is in liver.

Catalase is needed in living cells because hydrogen peroxide is poisonous. Hydrogen peroxide is thought to be made as a by-product of several reactions in cells. It must be decomposed instantly, before it can do any damage. This is what catalase does.

1   Read through this experiment and then design and draw up a results chart.

2   Measure out 20 cm³ of hydrogen peroxide into each of four boiling tubes.

3   Cut four small pieces of fresh liver. They must all be of the same size, around 0.5 cm².

4   Cook two of the pieces of liver in boiling water for about 5 min. Remove them and cool.

5   Grind one raw and one cooked piece of liver to a paste. Keep them separate from one another.

You now have four samples of liver – whole raw, ground raw, whole cooked and ground cooked.

Make sure that you are thoroughly organised before you go any further – things will happen quickly!

6   Start a stopclock. Put your four samples of liver into the four tubes of hydrogen peroxide solution. Try to do them all at the same time. After 2 min, measure the height of the froth in the boiling tube, and record it in your results chart.

7   Light and then blow out a splint.

Push the glowing splint through the froth in one of the tubes, and record what happens.

Questions

1   Write a word equation, and a balanced molecular equation, for the reaction which happened in your boiling tubes.

2   The liver contained a catalyst. What is the name of this catalyst? What type of catalyst is it?

3   Why does liver contain this catalyst?

4   Which type of liver made the reaction happen fastest? Explain why you think this is.

5   Which type of liver was least effective in speeding up the reaction? Explain why you think this is.

Questions

1   This question is about the decomposition of hydrogen peroxide to water and oxygen.

a   Give two ways in which the decomposition of hydrogen peroxide could be speeded up.

b   Give three ways in which the decomposition of hydrogen peroxide could be slowed down.

c   Manganese (IV) oxide acts as a catalyst in this reaction. How could you prove that the manganese (IV) oxide is not used up in the reaction flask?

d   Is manganese (IV) oxide
i   a reactant?
ii   a product in this reaction?

e   Do you think that manganese (IV) oxide would be a better catalyst in lump form, or in powder form? Explain your answer.

2   New cars must by law have catalysts fitted into their exhaust systems.

a   Why do you think that this is done?

b   Find out why using leaded petrol prevents such catalysts being used.

3   Enzymes are biological catalysts. All enzymes are proteins. Protein molecules are easily destroyed by high temperatures.

The graph below shows how a particular biological reaction is affected by temperature. The reaction is catalysed by an enzyme.

a   At which temperatures did the reaction take place most slowly?

b   At which temperature did the reaction take place most quickly?

c   What happened to the reaction as the temperature was raised from 10 °C to 20 °C?

d   What do you think began to happen to the enzyme at 45 °C?

e   Explain why the reaction got slower as the temperature was raised from 45 °C to 60 °C.

f   Human body temperature is normally around 37 °C. If you are very ill, your temperature can go up to 40 °C. Why is this dangerous?

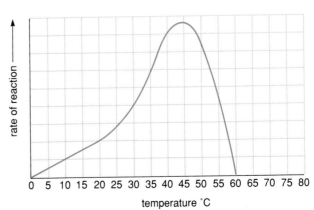

85

SPEEDING UP REACTIONS

Knowing which factors speed up reactions, and why, can be used to make industrial processes more efficient.

Why do some factors affect rate of reaction?

The marble and hydrochloric acid reaction shows you that, when a solid reacts with a solution, the reaction goes faster if:

• the solid is in smaller particles
• the solution is hotter
• the solution is more concentrated.

To understand why this is, you must think about the particles present.

The marble is solid. Its particles are vibrating slightly, but cannot move around freely. The hydrochloric acid is made up of several different sorts of particles. It contains water molecules, H^+ ions and Cl^- ions. These are all in constant motion. They are moving around freely.

For the reaction to occur, the H^+ ions must collide with the marble particles. The harder they hit, the more likely it is that the particles will react together. Anything which makes the acid and marble particles collide more often, or collide more violently, will speed up the reaction.

Finer powdered marble has a larger surface area. So acid particles collide with it *more often*.

Hotter acid has faster moving particles. They therefore collide with the marble *more often* and *more violently*. Only particles that possess a certain minimum amount of energy (called the **activation energy** – see Topic 80) will react when they collide. Heating the acid means that more particles have enough energy to react when they collide, so the reaction happens faster.

More concentrated acid has more acid particles per cubic centimetre. So acid particles will collide with the marble *more often*.

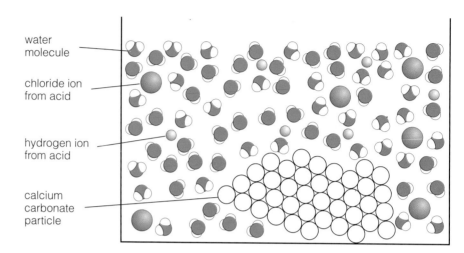

water molecule

chloride ion from acid

hydrogen ion from acid

calcium carbonate particle

Fig. 85.1   In a lump of marble (calcium carbonate) only the particles on the surface can come into contact with the acid. So only they can react. If you break the marble apart, more of it can react.

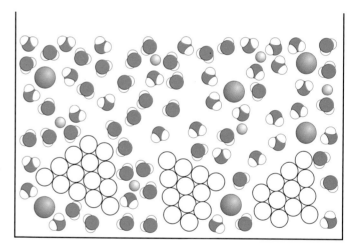

Fig. 85.2   If the lump of marble is broken up into smaller bits, more of its particles can come into contact with the acid, so the reaction will proceed more quickly.

Some reactions are between gases. An example is the reaction between ammonia and oxygen on the next page. How can a gas be made more concentrated? 'More concentrated' means with more particles in a certain volume. The way to force more gas particles into a certain volume is to **pressurise** the gas. Putting gases under greater pressure makes the particles more concentrated, so they collide with each other more often and therefore react faster.

Living organisms speed up their reactions by keeping warm

The reactions inside living organisms are called **metabolic reactions**. The speed at which these reactions take place is known as the **metabolic rate**. The metabolic rate of an organism determines how active it can be. If the metabolic rate is very slow, then the organism is also slow and sluggish. A faster metabolic rate will allow the organism to move more swiftly and be more alert.

One way in which organisms speed up their metabolic rate is by using enzymes as catalysts. In fact, without enzymes, most metabolic reactions would go so slowly that they would never happen at all! All living things rely on enzymes to make their metabolic reactions happen.

Another way of speeding up metabolic rate is to keep warm. Most living things are the same temperature as their surroundings. They are called **poikilothermic** or cold-blooded organisms. When the weather is cold, their metabolism is very slow. Poikilothermic animals, such as insects, are very slow and sluggish in winter. But in warm weather, their reactions happen much faster. They can fly actively.

Two groups of animals go one better. They can keep their temperature warm all the time. These are the birds and mammals. They are **homeothermic** or warm-blooded animals. Mammals keep their temperature around 37°C, even in cold weather. They can stay active all the year round. They use the chemical energy in food to produce the heat energy needed to stay warm.

Fig. 85.3 A smooth newt is poikilothermic whereas a red deer is homeothermic.

Making nitric acid

Ammonia is a very important raw material for many chemical processes. One especially important one is the manufacture of **nitric acid**.

This process involves three stages.

1 Ammonia reacts with oxygen from the air to make nitrogen monoxide.

ammonia + oxygen ⟶ nitrogen monoxide + water

$$4NH_3(g) + 5O_2(g) \longrightarrow 4NO(g) + 6H_2O(g)$$

2 Nitrogen monoxide reacts with more oxygen from the air to give nitrogen dioxide.

$$2NO(g) + O_2(g) \longrightarrow 2NO_2(g)$$

3 Nitrogen dioxide reacts with water to make nitric acid.

$$3NO_2(g) + H_2O(l) \longrightarrow 2HNO_3(aq) + NO(g)$$

Fortunately, reactions **2** and **3** are extremely rapid. They need no help from people controlling the conditions. But reaction **1** is very slow. Conditions must be produced in which the reaction will take place more rapidly. This involves:

a increasing pressure to around seven times atmospheric pressure

b raising the temperature to between 800 and 1000°C

c using a catalyst. This is gauze woven from strands of the precious metals platinum and rhodium.

It is a good job that catalysts are not used up! Each of these pieces of gauze costs around half a million pounds!

The production of nitric acid by this method is a continuous process. The reactants flow into the reactor vessel, and the products flow out continually. This goes on for several weeks or months. The reactor vessel is then shut down and cleaned. But the remains from cleaning are not thrown away. They contain enough tiny fragments of catalyst to make it worthwhile to reclaim them.

Questions

1 a Explain why the reaction between ammonia and oxygen is speeded up by:

   i increased pressure

   ii increased temperature.

   b In this reaction, platinum–rhodium gauzes are used as catalysts.

i State two things about the gauze which justifies its being called a catalyst.

ii Why are the catalyst metals made into a gauze, and not used in lump form?

iii Why do the gauzes need cleaning periodically?

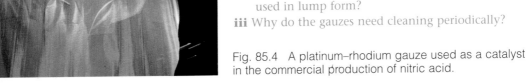

Fig. 85.4 A platinum–rhodium gauze used as a catalyst in the commercial production of nitric acid.

86 THE HABER PROCESS

Ammonia is made from nitrogen and hydrogen by the Haber process. This process is reversible.

Nitrogen and hydrogen react very slowly under normal conditions

In the preceding Topic you have seen how nitric acid is made using ammonia as its main starting material. How is ammonia made? The formula of ammonia is NH_3, so the obvious solution seems to be to react nitrogen and hydrogen together.

$$N_2 + 3H_2 \longrightarrow 2NH_3$$

This sort of reaction is called a **synthesis**. Both the reactants are readily available. The nitrogen can be obtained from the air and the hydrogen is obtained from oil products by a process called **cracking**, or from natural gas.

Unfortunately the reaction between nitrogen and hydrogen is extremely slow under normal conditions, because the triple bond holding the two nitrogen atoms together in an N_2 molecule is very strong. The reaction is slow because it has a high **activation energy**. The problem of how to speed up the reaction, so that a reasonable amount of ammonia is produced in a reasonable time, was solved in Germany by Fritz Haber and Karl Bosch in 1908 and 1909, and was developed into a full scale industrial process in the next few years. You should by now be able to guess the conditions that are used in order to increase the rate of the reaction. Close this book and try to write down the conditions you would choose, and your reasons for choosing them, before you read on.

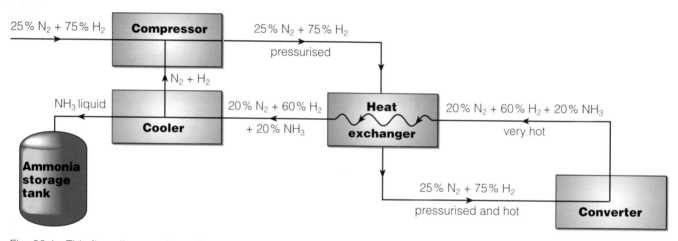

25% N_2 + 75% H_2 — Compressor — 25% N_2 + 75% H_2 pressurised

$N_2 + H_2$

NH_3 liquid

Cooler — 20% N_2 + 60% H_2 + 20% NH_3 — Heat exchanger — 20% N_2 + 60% H_2 + 20% NH_3 very hot

Ammonia storage tank

25% N_2 + 75% H_2 pressurised and hot — Converter

Fig. 86.1 This flow diagram shows the steps in the manufacture of ammonia.

Increasing pressure can speed up gas reactions

The reaction between nitrogen and hydrogen is used to make ammonia. Ammonia is an important material in the fertiliser industry.

nitrogen + hydrogen $\longrightarrow$ ammonia

$$N_2(g) + 3H_2(g) \longrightarrow 2NH_3(g)$$

This is called the **Haber process**.

You know four ways to speed up a reaction:

1 *Use a catalyst* Iron is used as a catalyst in the Haber process. The reactor vessel is filled with short lengths of iron tube.

2 *Use solids in powder form* But in this case both reactants are gases.

3 *Increase the temperature* The gases are heated to 400°–500°C.

4 *Increase the concentration of the reactants* How do you increase the concentration of a gas? The answer is to **pressurise** it. High pressure forces the gas particles closer together. This makes the gas more concentrated. In the Haber process, the pressure used is around 200 times atmospheric pressure. This is very costly, but it is well worth it in terms of increased production of ammonia.

Fig. 86.2 Part of ICI's Number 4 ammonia plant at Billingham. The shortest of the three grey columns at the left is the ammonia converter, where hydrogen and nitrogen combine together to form ammonia. The reaction takes place at 200 times atmospheric pressure. The reaction is exothermic, and the heat from it is used to heat fresh supplies of hydrogen and nitrogen, to speed up their rate of reaction. This happens in the heat exchangers, fed by the yellow pipes at the bottom right of the picture.

The Haber process makes ammonia at an economical rate

The following conditions are used:

1 The nitrogen and hydrogen gases are mixed in the ratio 25% nitrogen to 75% hydrogen. This mixture is then **pressurised** in the **compressor**. This increases the **concentration** of the gases. The pressure used is usually 200 times atmospheric pressure.

2 The nitrogen and hydrogen gases are **heated**. This is done in the **heat exchanger**. The gases are heated to 450°C.

3 A **catalyst** is used. The catalyst is iron metal in the form of short lengths of iron tube. The catalyst is contained in a vessel called the converter. Some of the nitrogen and hydrogen reacts together on the surface of the catalyst in the converter vessel while some remains unreacted. The mixture leaving the converter vessel contains about 20% ammonia, 20% unreacted nitrogen and 60% unreacted hydrogen.

The reaction $N_2 + 3H_2 \longrightarrow 2NH_3$ is **exothermic**. This means that the mixture of gases leaving the converter vessel is hotter than the mixture of gases that went in. This extra heat is used to heat up the gases that are going into the converter vessel. This takes place in the heat exchanger. The next operation is to remove the ammonia from the mixture, which is done by cooling the mixture so the ammonia becomes a liquid. It can then be piped off to storage tanks while the unreacted nitrogen and hydrogen are sent round again giving them another chance to form ammonia. Make sure you can follow all this on Figure 86.1.

The synthesis of ammonia is reversible

20% of ammonia doesn't seem to be a great return for such efforts. The unreacted nitrogen and hydrogen do go round again but perhaps this would be unnecessary if the reaction were made still faster? Unfortunately the answer is no. Some reactions are **reversible** and this is one of them. As soon as ammonia begins to form in the reactor vessel by the reaction

$$N_2 + 3H_2 \longrightarrow 2NH_3$$

some of the ammonia formed begins to break down by the reaction

$$2NH_3 \longrightarrow N_2 + 3H_2$$

So the best way to write the equation is like this

$$N_2 + 3H_2 \rightleftharpoons 2NH_3$$

The double arrow $\rightleftharpoons$ means that the reaction is reversible. No matter what you do you can't get 100% conversion of reactants to products. The reaction always reaches a balanced position where reactants are being converted to products and products are breaking down to reactants at the same rate. This position is called **equilibrium.**

A further factor to be considered is that the proportion of reactants and products at equilibrium is not fixed. It depends on the conditions. In the explanation that follows we will call $N_2 + 3H_2 \longrightarrow 2NH_3$ the **forward reaction** and we will call $2NH_3 \longrightarrow N_2 + 3H_2$ the **backward reaction**.

At higher temperatures the rate increases but less ammonia forms

The forward reaction is exothermic, so the backward reaction is endothermic. For *any* reversible reaction, higher temperatures favour the endothermic reaction. Therefore in the Haber process, the higher the temperature, the more the backward reaction is favoured and the less ammonia there will be in the mixture when equilibrium is reached. 450°C is a compromise temperature – if a higher temperature were used the *amount* of ammonia formed would be too low, but if a lower temperature were used the *rate* of formation of ammonia would be too slow.

At higher pressures the rate increases and more ammonia forms

If a reversible reaction involves one or more gases, the number of molecules involved is important. High pressure favours the side of the equation where there is the **smaller** number of gas molecules. In the Haber process, three hydrogen molecules react with one nitrogen molecule (a total of four reactant molecules) to form two ammonia molecules, so high pressure causes there to be more ammonia in the mixture when equilibrium is reached. High pressure therefore causes more ammonia to be formed *and* increases the *rate* at which it is formed.

The presence of a catalyst doesn't affect the amounts of reactants and products in the mixture when equilibrium is reached, but it does affect how **quickly** equilibrium is reached. The catalyst affects the rate only.

Ammonia is mainly used to make fertilisers

As we have seen, ammonia is used to make nitric acid. If nitric acid is reacted with ammonia, a **neutralisation** reaction takes place and a salt called **ammonium nitrate** is formed.

$$NH_3 + HNO_3 \longrightarrow NH_4NO_3$$

Questions

1 a What are the reactants in the Haber process and how are they obtained?

b What is the product in the Haber process and what is it used for?

c What catalyst is used? Why is a catalyst necessary?

d Why is a temperature as high as 450°C used?

e Why are even higher temperatures not used?

f Give two advantages in running the Haber process at high pressure.

2 What do the terms 'reversible reaction' and 'equilibrium' mean?

87 *MEASURING RATES OF REACTION*

The rate of a reaction can be studied by measuring the amount of product formed every minute, or by measuring the amount of reactants left every minute.

Measuring the rate of reaction of nitric acid with copper carbonate

Often, rates of reaction are studied by measuring the amount of product formed each minute, until the reaction is finished. This method could be used to study the reaction between copper carbonate and nitric acid. This reaction produces carbon dioxide.

This reaction could be carried out by using nitric acid of varying concentration and temperature. An excess of copper carbonate would be used to make sure that all of the acid reacts. The sort of results you might obtain are shown in the table.

Look particularly at the following points:

a Note that the rate of carbon dioxide formation slows down in all the reactions. This is because, as the nitric acid is used up, it becomes less concentrated. This slows the reaction down.

b Compare experiments 1 and 2. They both have the same *concentration* and *temperature* of acid. So they both start at the same rate. But experiment 2 has twice the *amount* of acid. So this reaction carries on for longer, and produces twice the amount of carbon dioxide.

c Compare experiments 1 and 3. Both have the same *amount* of acid, so both produce the same amount of carbon dioxide. But experiment 1 has *more concentrated* acid, so the carbon dioxide forms more quickly.

d Compare experiments 1 and 4. Both have the same *amount* of acid, so both produce the same amount of carbon dioxide. But experiment 4 has *hotter* acid, so the carbon dioxide forms more quickly.

copper carbonate + nitric acid → copper nitrate + water + carbon dioxide

$$CuCO_3(s) + 2HNO_3(aq) \rightarrow Cu(NO_3)_2(aq) + H_2O(l) + CO_2(g)$$

Experiment	1	2	3	4
Volume of acid (cm³)	50	100	50	50
Volume of water (cm³)	0	0	50	0
Temperature (°C)	20	20	20	60
Amount of CO_2 produced (cm³) at:				
1 min	60	60	30	95
2 min	85	115	56	99
3 min	95	155	77	100
4 min	98	180	90	100
5 min	99	192	96	100
6 min	100	196	98	100
7 min	100	198	99	100
8 min	100	200	100	100
9 min	100	200	100	100

—— experiment 1
—— experiment 2
—— experiment 3
—— experiment 4

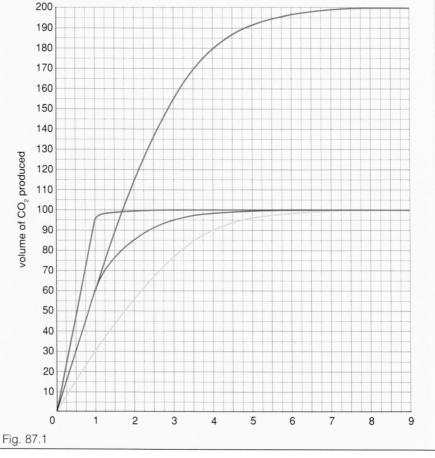

Fig. 87.1

The reaction between magnesium and sulphuric acid

To follow this reaction, you are going to measure the rate of evolution of hydrogen gas. The equation for the reaction is:

magnesium + sulphuric acid $\longrightarrow$ magnesium sulphate + hydrogen

$Mg(s) + H_2SO_4(aq) \longrightarrow MgSO_4(aq) + H_2(g)$

This experiment can be performed with syringe apparatus. You may be shown this. But in the method described here you will collect the hydrogen in a measuring cylinder.

Fig. 87.2   Generating hydrogen

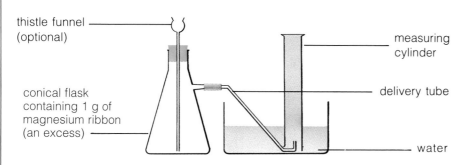

thistle funnel — (optional)

conical flask containing 1 g of magnesium ribbon (an excess) —

measuring cylinder

delivery tube

water

1   Assemble the apparatus as shown in the diagram.
2   Copy the chart. It should continue for 10 min.

Experiment	1	2	3
Volume of acid (cm³)	25	50	25
Volume of water (cm³)	25	0	25
Temperature of liquid (°C)	room	room	50
Volume of gas collected at: 1 min			
2 min			

3   Move the measuring cylinder away from the delivery tube.
4   Remove the bung from the flask. Add 25 cm³ of acid and 25 cm³ of water. Replace the bung. After 2 s, put the measuring cylinder back over the delivery tube. When the first bubble reaches the top of the measuring cylinder, start timing.
5   Record the volume of gas in the measuring cylinder every minute for 10 min.
6   Perform experiments 2 and 3 and fill in the results.
7   Plot these results as a line graph. Put time on the horizontal axis, and volume of gas collected on the vertical axis. Draw all three curves on the same graph.

Questions

1   Why is excess magnesium used?
2   Why does the reaction slow down in each case?
3   Why will the reaction eventually stop?
4   Experiment 2 probably produced hydrogen more rapidly than experiment 1. It will eventually produce twice as much gas in total. Why is this?
5   Experiment 3 probably produced hydrogen more rapidly than experiment 1. But eventually both experiments will produce the same amount of gas. Why is this?
6   Imagine that the conditions of experiment 1 were repeated. However, this time, 1 g of finely powdered magnesium was added. Draw a fourth line on your graph to show the results you would expect. Explain your choice.

The digestion of starch by amylase

Amylase is an enzyme that catalyses the reaction in which starch breaks down to a sugar called maltose. This happens in your mouth, because saliva contains amylase.

1   Label three boiling tubes A, B and C. Measure 10 cm³ of starch solution into each.
2   Stand tube A in a rack in a water bath at 35 °C. Stand tube B in a rack on the bench. Stand tube C in a rack in a refrigerator. Measure and record the temperature on the bench and in the refrigerator. Leave all the tubes for at least 10 min.
3   While you are waiting, put a drop of iodine solution into each space on a spotting tile. Label the rows on the tile A, B and C. Design and draw up a results chart.
4   When the tubes have been left for 10 min, put 10 cm³ of amylase solution into each one. Mix the starch and amylase thoroughly. Try to do this at the same time for all three, and start a stop clock. Put the tubes back into their appropriate places.
5   Every 2 min, take a small sample from each tube, and add it to one of the iodine solution drops on the tile. Record your results as you go along. Keep doing this until at least two of the tubes no longer contain any starch.

Questions

1   Is starch a reactant or a product in this reaction?
2   In which tube did the starch disappear most quickly? Explain why you think this is.
3   What product would you expect to be formed in this tube?
4   In which tube did the starch disappear most slowly? Explain why you think this is.
5   If you had kept one of the tubes at 90 °C, would you expect the starch to disappear quickly, slowly, or not at all? Explain your answer.
6   If you had added water instead of amylase solution to one tube, would you expect the starch to disappear quickly, slowly, or not at all? Explain your answer.

Questions

1 Explain each of the following observations:
  a Hot substances react faster than cold ones.
  b Powdered zinc reacts faster than zinc granules with acid.
  c Zinc reacts faster with fairly concentrated acids than with dilute acids.
  d Zinc reacts faster with strong acids than with weak acids.
  e The chemical industry is constantly searching for new catalysts.
  f Pressuring gaseous reactants affects rate in the same way as concentrating dissolved reactants.
  g The Haber process has been run at pressures of about 1000 atmospheres. However this greatly increases capital expenditure and operating costs compared with running the process at atmospheric pressure.

2 The digestion of starch is a chemical reaction in which large starch molecules are broken down to smaller ones. The equation for the reaction is:

$$\text{starch} + \text{water} \rightarrow \text{maltose}$$

This reaction occurs in your mouth. Saliva contains an enzyme called amylase which catalyses this reaction.
  a What is the temperature inside a person's mouth?
  b What would happen to the rate of this reaction if the temperature was much lower than this? Explain your answer.
  c What would happen to the rate of this reaction if the temperature was much higher than this? Explain your answer.
  d Saliva contains the enzyme amylase, which catalyses this reaction. What other important substance is present in saliva which will help the reaction to proceed?
  e Teeth can also help this reaction to proceed faster. Explain how they can do this.

EXTENSION

3 The graphs show the results of five experiments involving the reaction between hydrochloric acid and excess marble. Details of the experiments are given in the chart beneath the graphs.

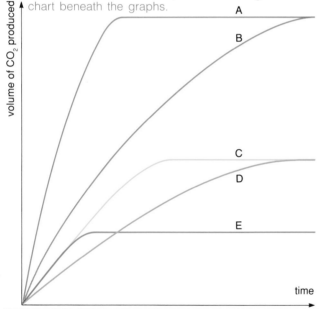

Experiment	1	2	3	4	5
Vol. of acid (cm³)	60	30	30	15	60
Vol. of water (cm³)	60	30	30	15	60
Temp. of liquids (°C)	40	20	40	40	40
Marble (w = whole) (p = powdered)	w	w	w	w	p

  a For each experiment (1–5) identify the correct graph A–E. Give reasons for each of your choices.
  b Which factor is the same in every experiment?
  c Which reaction finished first?

4 An experiment was done to find out what factors affect the rate of decomposition of hydrogen peroxide solution. The pupil had a solution of hydrogen peroxide, distilled water, manganese (IV) oxide powder, some measuring cylinders, a gas generator and a means of collecting gas. She put various mixtures into her gas generator, and measured how much gas was produced each minute for 10 min. The results are shown below.

	Volume of gas collected (cm³)			
Experiment	1	2	3	4
Time (min)				
1	40	80	70	0
2	60	120	77	0
3	70	140	79	0
4	75	150	80	0
5	78	155	80	0
6	79	158	80	0
7	79	159	80	0
8	80	159	80	0
9	80	160	80	0
10	80	160	80	0

In experiment 1, the pupil used 50 cm³ of hydrogen peroxide solution, 50 cm³ of water (both at 20°C) and a pinch of manganese (IV) oxide. Unfortunately, she forgot to note down what she put into experiments 2, 3 and 4. The total volume of liquid in the flasks was always always 100 cm³.
  a What was the gas collected each time?
  b Write an equation for the reaction.
  c What part does manganese (IV) oxide play in this reaction?
  d Draw a diagram of the equipment which might have been used.
  e What might have gone into the flask in experiment 2? Explain your answer.
  f What might have gone into the flask in experiment 3? Explain your answer.
  g What might have caused the result in experiment 4?

CHEMICAL REACTIONS IN LIVING ORGANISMS

88 PHOTOSYNTHESIS

Plant leaves are food factories. All the food in the world is made by plants and microorganisms.

All the energy of living organisms begins as light energy

All living things need energy. They need it to move, to make new cells, and to transport things around their bodies. Even plants, which do not seem to be doing very much, need energy. If you need extra energy, you eat extra food. Animals get their energy from the food they eat. Food contains energy. But where does the energy in food come from? It all begins as **sunlight**.

Plants change sunlight energy into chemical energy in food

If it were not for plants, there would be nothing at all for animals to eat. All the food in the world is made by plants. Plants use the energy in sunlight to make food. The food contains some of this energy. Animals – and plants, too – use the energy from food to stay alive. The name of the process by which plants use sunlight energy to make food is **photosynthesis**.

Fig. 88.1 Green plants harness energy from sunlight to make food. This beech wood is a giant food factory.

Fig. 88.2 A koala bear gets all its energy from that which has been trapped by eucalyptus leaves.

Plants use inorganic substances to make food

What do plants make this food from? They need two chemicals. One is **water**, which they get from the soil. The other is **carbon dioxide**, which they get from the air. They also need **sunlight energy**, which is used to make the water and carbon dioxide react together. Water and carbon dioxide are **inorganic** substances. Inorganic substances are substances which are not made by living things. They usually have small molecules.

These three things – water, carbon dioxide and sunlight – are sometimes called **raw materials**. They are the 'ingredients' from which the plant makes food.

Fig. 88.3   Some examples of inorganic substances

The inorganic raw materials are made into organic substances

The plant uses the energy in sunlight to make the water and carbon dioxide react together. This reaction produces two new substances. One is **glucose**. The other is **oxygen**.

Glucose is an **organic** substance. Organic substances have been made by living organisms. They usually have quite large molecules. They always contain carbon. The formula for one molecule of glucose is $C_6H_{12}O_6$ – so you can see that it *does* contain carbon, and it *does* have quite large molecules. The glucose contains some of the energy which started off in the sunlight. If you suck glucose sweets, you get energy from them. That energy was once sunlight energy and somewhere in the world a plant converted it into chemical energy which your body can now use. Plants are living food factories.

Fig. 88.4   Some examples of organic substances. There is also an inorganic substance here. What is it?

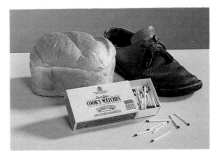

The photosynthesis equation

The raw materials for photosynthesis are water, carbon dioxide and sunlight energy. The products – the things which are made – are glucose and oxygen. This can be written as a word equation:

$$\text{water} + \text{carbon dioxide} \xrightarrow{\text{sunlight energy}} \text{glucose} + \text{oxygen}$$

The balanced molecular equation for this reaction is:

$$6H_2O(l) + 6CO_2(g) \xrightarrow{\text{sunlight energy}} C_6H_{12}O_6(aq) + 6O_2(g)$$

Photosynthesis provides oxygen

Photosynthesis does not only provide all the food in the world. It also provides all the oxygen in the air.

When the Earth was first formed, the air contained gases like methane, ammonia, carbon dioxide and water vapour. There was no oxygen.

Then some tiny organisms appeared which could photosynthesise. The oxygen they made went into the atmosphere. Gradually, over millions of years, the amount of oxygen in the air built up. Now, about 20% of the air is oxygen.

We, and other animals, are completely dependent on plants. They provide us with two of the most important things we need to stay alive – food and oxygen.

Questions

1   Where does all the energy in food originate?
2 a What are the raw materials for photosynthesis?
  b What are the products of photosynthesis?
3 a Explain the differences between *organic* and *inorganic* substances.
  b Which of these substances are organic, and which are inorganic?
    wood, paper, aluminium foil, oil, a cotton shirt, a gold earring, sugar, pepper, salt, lettuce, oxygen

LEAVES

Photosynthesis happens in leaves. Leaves are food-making machines. They are perfectly designed to make food as efficiently as possible.

Chlorophyll absorbs sunlight

As you may have noticed, plants are green! The green colour is **chlorophyll**. All plants contain chlorophyll. Even plants like copper beech trees, which look a reddish-brown colour, have chlorophyll in their leaves.

Chlorophyll is an extremely important substance. Without it, photosynthesis could not happen. Chlorophyll absorbs energy from sunlight. Without this energy, carbon dioxide and water would not react together to make glucose. So chlorophyll is essential for photosynthesis.

Chlorophyll is kept in **chloroplasts**. Chloroplasts are organelles found in plant cells. Only cells which are above ground contain chloroplasts. Cells in roots do not contain them. There would be no point, because there is no light underground.

The cells which contain most chloroplasts are in **leaves**. Most photosynthesis happens in leaves. But other parts of plants above the ground can photosynthesise too. Stems for example may contain chloroplasts.

Fig. 89.1   The green colour of the countryside is caused by chlorophyll.

Leaves have several layers of cells

Leaves are very thin. Yet a leaf is made up of several layers of cells. Figure 89.4 shows these layers. The cells which contain chloroplasts and photosynthesise are in the middle layers. These layers are called the **mesophyll** layers. 'Mesophyll' means 'middle leaf'. There are two mesophyll layers. The one nearest the top of the leaf does most of the photosynthesis. It is called the **palisade** layer. This is where most of the chloroplasts are.

The other mesophyll layer is the **spongy** layer. These cells have big air spaces between them. They also contain chloroplasts, but not as many as the palisade layer. They too can photosynthesise.

The other two layers in the leaf are protective layers on the top and bottom. They are called the **epidermis** layers. The cells in the epidermis make a waxy substance, which spreads out over the surface of the leaf. This layer of wax is called the **cuticle**. The lower epidermis has holes in it, which open directly into the inside of the leaf. These holes are called **stomata**. They are very small, but you can see them with a microscope. Each stoma has a pair of special cells surrounding it, called **guard** cells. The guard cells can open or close the stomata.

Fig. 89.2   Leaves from a pear tree. You should be able to pick out at least five features that you can see in this picture, which help the leaves to carry out photosynthesis efficiently.

The raw materials for photosynthesis are delivered to the palisade layer

The main food-producing part of the leaf is the palisade layer. The raw materials for photosynthesis must be delivered to the cells in the palisade layer as swiftly as possible.

Carbon dioxide gets into the leaf through the stomata. A very small part of the air – less than 0.04% – is carbon dioxide. Carbon dioxide diffuses through the open stomata into the air spaces between the cells in the spongy layer. Because the leaf is so thin, it quickly diffuses all the way to the chloroplasts in the palisade cells.

Water is brought to the leaves in tubes called **xylem vessels**. These are very long tubes, rather like drainpipes. They run all the way up from the root of the plant, through its stem, and into the leaves. The veins of a leaf contain xylem vessels. Branches of xylem vessels run close to every part of a leaf, so each palisade cell is provided with a constant supply of water. The carbon dioxide and water enter the chloroplasts in the palisade cells where chlorophyll is absorbing sunlight. The energy in the sunlight makes the carbon dioxide and water react together. Glucose and oxygen are made.

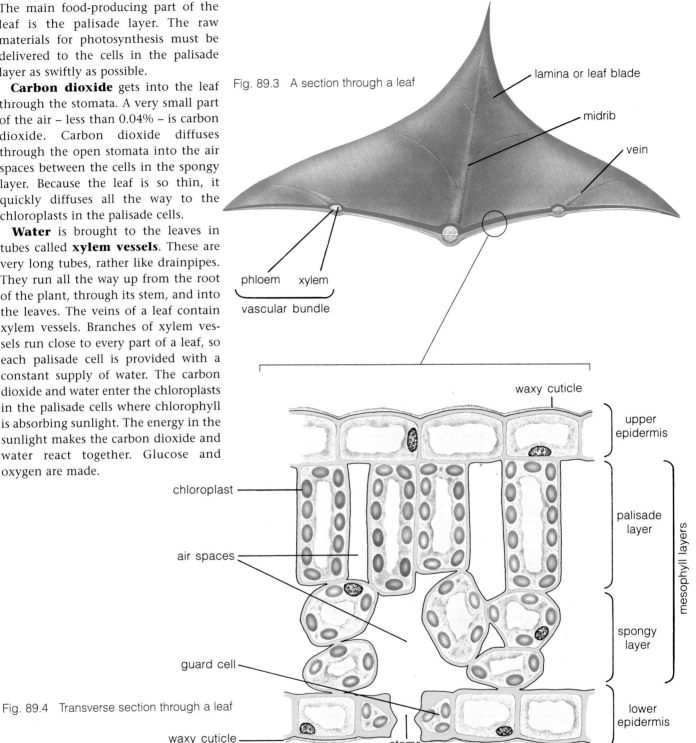

Fig. 89.3   A section through a leaf

lamina or leaf blade

midrib

vein

phloem   xylem

vascular bundle

waxy cuticle

upper epidermis

chloroplast

palisade layer

air spaces

mesophyll layers

spongy layer

guard cell

Fig. 89.4   Transverse section through a leaf

lower epidermis

waxy cuticle

stoma

Questions

1 a Precisely where is chlorophyll found?

b What does chlorophyll do?

2 Which layer(s) of a leaf:

a contain air spaces

b contain small holes called stomata

c contain most chloroplasts

d make a layer of wax

e photosynthesises

f protects the mesophyll cells?

3 Leaves are food factories. They are designed to make sure that the raw materials and energy source are supplied to the production line (the palisade layer) as swiftly as possible. Look at the photographs and diagrams of leaves on these two pages. What features of a leaf do you think:

a help to make sure that plenty of sunlight reaches the palisade layer

b help to provide plenty of carbon dioxide to the palisade layer

c help to provide plenty of water to the palisade layer

d make sure that this water does not evaporate from the leaf too quickly?

90 TESTING LEAVES FOR STARCH

Some of the glucose made by a leaf is turned into starch and stored.

Glucose is turned into starch for storage

When plants photosynthesise they make glucose. Glucose molecules are quite small for organic molecules. They dissolve easily in water. They also react quite easily with other molecules. So they are not very good for keeping in a cell for a long time.

If a plant needs to store glucose molecules, it turns them into **starch** molecules. A starch molecule is very big. It is made of hundreds or thousands of glucose molecules linked together. Natural starch molecules do not dissolve in water. Although they are very long, they curl up tightly, so they fit into a small space. A leaf which has been photosynthesising will have a lot of starch molecules in it. The starch is in the form of **starch grains**, inside the chloroplasts in the mesophyll cells.

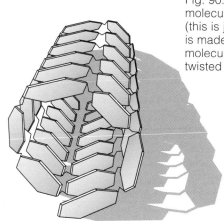

A glucose molecule

A small part of a starch molecule

Fig. 90.1 Glucose and starch molecules. A starch molecule (this is just a small part of one) is made of hundreds of glucose molecules linked together, and twisted into a helix.

Testing a leaf for starch

If you want to find out if a leaf has been photosynthesising, you can test it for starch. Starch turns blue-black when **iodine solution** is added to it. So you can add iodine solution to a leaf and see if goes blue-black. But first, you must break down the cell membranes in the leaf, so that the iodine solution can get into the chloroplasts and reach the starch. You also need to get rid of the green colour in the leaf. If you do not do this, it is very difficult to tell what colour the iodine solution turns.

Fig. 90.3 Testing a leaf for starch

Fig. 90.2 A chloroplast in a pea leaf. The dark green stripes are membranes, on which chlorophyll is spread out. The white area is a starch grain. Glucose made inside the chloroplast can be changed into starch and stored in this way.

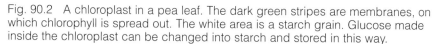

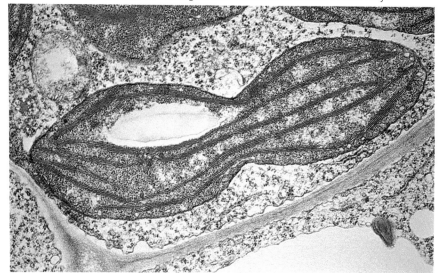

1 Heat the leaf in boiling water until it looks limp.

2 Turn out the Bunsen burner. Stand a tube of alcohol in the hot water, and put the softened leaf into the tube.

3 When the leaf has lost most of its colour, remove it gently from the alcohol. It will be very brittle, so dip it into the water to soften it.

4 Spread the leaf on to a white tile, and cover it with iodine solution. Wait a few minutes for the iodine solution to soak in. If the leaf goes black, it contains starch.

Is light needed for photosynthesis?

You are going to give part of a leaf plenty of sunlight, and keep part of it in the dark. Then you will test the leaf to see which parts of it contain starch.

1 First, remove all starch from a healthy plant. You can do this by leaving it in a dark cupboard for several days.

2 Test a leaf from your plant to check that it does not contain any starch by following these instructions.

a Boil some water in a beaker. When it is boiling, put your leaf into it. 'Cook' it for about 3-5 min. This destroys the cell membranes.

b Turn out your Bunsen burner, and put a boiling tube of ethanol into the hot water in the beaker. Ethanol has a boiling point of around 80 °C, so it will boil. Put your boiled leaf into the ethanol. The chlorophyll will dissolve in the ethanol, making it go green.

c When all the chlorophyll has come out of your leaf, take the leaf out and dip it into the water again. This is to soften it, because the ethanol makes it brittle.

d Now spread your leaf on a white tile. Cover it with iodine solution. If the leaf does not contain starch, the iodine will stay brown.

3 Now go back to your plant. Choose a healthy-looking leaf. *Leave it on the plant!* Cover part of your leaf with black paper. Cut a hole in the paper. Fasten the paper down securely, so that no light can get in round the edges.

4 Put the plant in a sunny window. Leave it for two or three days.

5 Take your leaf off the plant. Take off the black paper. Test your leaf for starch as before.

6 Make a labelled diagram of your leaf after testing it for starch.

Fig. 90.4

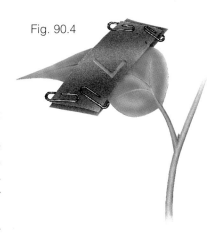

Questions

1 Why is it important to make sure that your plant has no starch in it before you begin the experiment?

2 Why do you need to boil the leaf when testing it for starch?

3 Why do you need to put the leaf into hot ethanol?

4 Why must you not heat ethanol with a Bunsen burner?

5 Do the results of your experiment suggest that plants need light for photosynthesis?

Is carbon dioxide needed for photosynthesis?

1 Take two healthy pot plants, and destarch them by leaving them in the dark for several days. Test a leaf from each one for starch, to make sure that all the starch has gone.

2 Set up the two plants as shown in Figure 90.5. Sodium hydroxide solution will absorb all the carbon dioxide from the air. Leave the two plants in a sunny window for two or three days.

3 Test a leaf from each plant for starch.

Fig. 90.5

tie the bag tightly so that no air can enter or leave

Petri dish – for one plant this contains sodium hydroxide solution, for the other it contains water.

polythene bag

well-watered soil

Questions

1 Why are two plants needed in this experiment?

2 Which plant had carbon dioxide in the air surrounding it?

3 Do your results suggest that carbon dioxide is needed for photosynthesis?

Question

Some leaves have patches which do not contain chlorophyll. Such leaves are called variegated leaves. How could you use a variegated leaf to find out if chlorophyll is needed for photosynthesis? What results would you expect?

Photosynthesis produces glucose and oxygen. The plant can then make the glucose into many other things.

Does light intensity affect the rate of photosynthesis?

1 Set up your apparatus as shown in the diagram. Switch the lamp on.

It is important to have a really healthy piece of pondweed. Cut its stem at an angle. If it does not produce bubbles after a couple of minutes in the tube, tell your teacher – or try another piece of pondweed!

2 When the pondweed has settled down, and is producing bubbles, start your stopclock. Count how many bubbles are produced in 2 min. Without changing anything, do the same thing twice more. Record all your readings, and work out the average number of bubbles per minute.

3 Now switch off the lamp. If the room is still bright, you can arrange some sort of shade around your apparatus. Leave the weed for at least 5 min to settle down. Then count bubbles as before. Repeat it two more times and work out the average number of bubbles per minute.

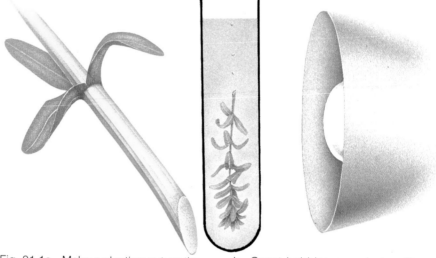

Fig. 91.1a Make a slanting cut on the stem of your piece of weed.

b Count bubbles per minute with and without the lamp.

Questions

1 Why did you take three readings each time?

2 Was there any difference in the rate of bubbling with and without the lamp?

3 Temperature can affect the rate of photosynthesis. Can you see anything wrong with this experiment? How could you solve this problem?

4 How could you collect the gas which the weed produces during this experiment? Draw a diagram to explain your ideas. How could you test it to find out what it is? What would you expect your results to be?

Limiting factors

If you try Investigation 91.1, you will probably find that the plant produces oxygen more rapidly when the light is closer to it. Plants tend to photosynthesise faster in bright light than they do in dim light. We say that light is a **limiting factor** for photosynthesis. Lack of light limits the rate at which the plant can photosynthesise.

Figure 91.2 shows this on a graph. You can see that, at very low light intensity, the plant does not photosynthesise at all. As the light intensity increases, the rate of photosynthesis increases. This is what you would expect to see if light is a limiting factor.

But, as the light intensity gets really high, the curve flattens out. The plant seems not to be able to photosynthesise any faster, no matter how much light it gets. Light is not a limiting factor any more. Something else must be stopping the plant from photosynthesising faster.

What might this be? Plants need carbon dioxide for photosynthesis, and there is only a very little carbon dioxide in the air. So, at high light intensities, it is often a shortage of carbon dioxide which limits the rate at which a plant can photosynthesise. Carbon dioxide can be a limiting factor on a bright summer day.

Temperature can also be a limiting factor. The rate of chemical reactions in a living organism slows down at low temperatures. So plants tend to photosynthesise more slowly at low temperatures. If the temperature is raised, the rate of photosynthesis may increase. Temperature is often a limiting factor on a bright but cold day in winter.

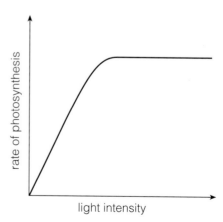

Fig. 91.2 How light intensity affects the rate of photosynthesis

Glucose is used to make other substances

When a plant photosynthesises, it makes sugars such as glucose. You have already seen that the plant changes some of the glucose to starch, which can then be stored inside the cells. But the plant makes many other substances from glucose. With the addition of a few inorganic ions (minerals) which it gets from the soil, it can make all the substances it needs.

Carbohydrates Glucose and starch are carbohydrates. Glucose can easily be changed into other carbohydrates. If the plant needs to transport carbohydrate from one part to another, the glucose or starch is changed into sucrose. Some plants use sucrose for storage, too. Carrots, sugar beet and sugar cane all do this.

Another carbohydrate is cellulose. This is the substance from which plant cell walls are made. Growing plants change a lot of the glucose they make into cellulose.

Fats and oils Glucose can be changed into fats and oils. (Oils are liquid fats.) Cell membranes contain a lot of fat, so a growing plant will need to make quite a lot of fat as it produces new cells. Fat is also used as a storage substance inside seeds. We get most of the oils we use for cooking from seeds and fruits, for example olives, sunflowers and corn.

Proteins Glucose can also be changed into amino acids, which are then used to build up proteins. To do this, the plant needs a source of nitrogen atoms, because glucose contains only carbon, hydrogen and oxygen, whereas proteins contain nitrogen as well. The plant gets its nitrogen in the form of nitrate ions from the soil. These are taken in through the root hairs. You can read about where the nitrate ions come from in Topic 212.

Other substances Glucose can be used to make many other substances, such as chlorophyll. Chlorophyll molecules contain magnesium, so plants need to take in magnesium ions from the soil. Making chlorophyll molecules also requires iron, so plants need this mineral as well. Table 91.1 lists these minerals, and also some others which plants need, in order to be able to make all the substances they need to stay healthy and grow well.

Mineral	Why it is needed	Symptoms shown by plant if mineral is deficient
nitrate	For making proteins, which are used to make cytoplasm, and as enzymes.	Stunted growth, because the plant cannot make proteins and therefore new cytoplasm. Old leaves turn yellow, as proteins are taken from them and used to supply younger leaves.
magnesium	For making chlorophyll; each chlorophyll molecule contains a magnesium atom.	Leaves turn yellow, especially older ones, because not enough chlorophyll can be made.
iron	For making chlorophyll; chlorophyll does not contain iron, but the iron is needed to help to make it.	Similar symptoms to magnesium deficiency.
phosphate	For many of the reactions of photosynthesis. Also, DNA molecules and ATP molecules contain phosphorus atoms. So do cell membranes.	Leaves become an intense green or greenish-purple. Eventually, they become mis-shaped and may develop spots of dead tissue.
potassium	To allow many enzymes to work properly, especially the ones involved in photosynthesis and respiration, making starch and making proteins.	Old leaves develop yellow spots, and then patches of dead tissue around the edges. Stems are shorter and weaker than normal.

Table 91.1 Minerals needed by plants

92 RESPIRATION

Respiration is a process which occurs in all living cells. Respiration releases energy from food.

Every living cell respires

Each living cell in every living organism needs energy. Energy is needed to drive chemical reactions in the cell. It is needed for movement. It is needed for building up large molecules from small ones. If a cell cannot get enough energy it dies. Cells get their energy from organic molecules such as glucose. The chemical process by which energy is released from glucose and other organic molecules is called **respiration**. Every cell needs energy. Each cell must release its own energy from glucose. So each cell must respire. Every cell in your body respires. Every living cell in the world respires.

Fig. 92.1 Florence Griffith-Joyner ('Flo-Jo') winning the 100 m at the Seoul Olympics. The energy which she is using comes from respiration in her muscle cells.

INVESTIGATION 92.1

Getting energy out of a peanut

1   Spear a peanut on the end of a mounted needle. Be careful – it is easy to break the peanut.
2   Put some cold water into a boiling tube. Support the boiling tube in a clamp on a retort stand, with the base of the tube about 30 cm above the bench top. Take the temperature of the water and record it.
3   Set light to the peanut by holding it in a Bunsen burner flame. (Keep the Bunsen burner well away from the boiling tube.) When the peanut is burning hold it under the boiling tube. Keep it there until it stops burning.
4   Immediately take the temperature of the water and record it.

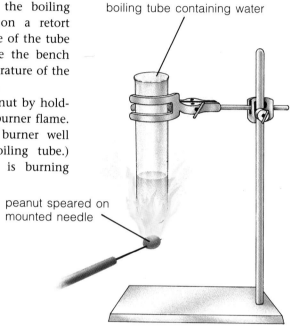

Fig. 92.2   Burning a peanut

boiling tube containing water

peanut speared on mounted needle

Questions

1   Why should you keep the Bunsen burner away from the tube of water?
2   By how much did the temperature of the water rise?
3   What type of chemical reaction was occurring as the peanut burnt?
4   Put these words in the right order, and join them with arrows, to show where the heat energy in the water came from:

    energy in peanut molecules

    energy in glucose in peanut leaf

    sunlight energy

    heat energy in water

5   Put the following words over two of the arrows in your answer to question 4:

photosynthesis   oxidation
6   Do you think all the energy from your peanut went into the water? If not, explain what else might have happened to it.
7   How could you improve the design of this experiment to make sure that more of the peanut energy went into the water?

EXTENSION

8   People often want to know exactly how much energy there is in a particular kind of food. The amount of energy is measured in kilojoules.
    4.18 J of energy will raise the temperature of 1 g of water by 1 °C.
    Design a method for finding out the amount of energy per gram in a particular type of food.

Respiration is an oxidation reaction

You can do an experiment to release energy from food if you do Investigation 92.1. When a peanut burns, the energy in the peanut is released as heat energy. But how do your cells release energy from food such as peanuts?

Obviously, you do not burn peanuts inside your cells! But you do something very similar. The chemical reactions of burning and respiration are very like each other.

First, think about what happens when you burn the peanut. The peanut contains organic molecules, such as fats and sugars, which contain energy. When you set light to the peanut you start off a chemical reaction between these molecules and oxygen in the air. The peanut molecules undergo an **oxidation** reaction. They combine with oxygen. As they do so, the energy in them is released as **heat energy**.

When you eat peanuts, your digestive system breaks the peanut into its individual molecules. Your blood system then takes these molecules to your cells. Inside your cells the molecules combine with oxygen. They undergo an oxidation reaction. This is respiration. The energy in the molecules is released. But, unlike the burning peanut, much of the energy is *not* released as heat energy. It is released much more gently and gradually, and stored in the cell.

So, the burning of a peanut, and the respiration of 'peanut molecules' in your cells are both oxidation reactions. In both of them the peanut molecules combine with oxygen. In both of them the energy in the peanut molecules is released. But in your cells the reaction is much more gentle and controlled.

ATP is the energy currency in cells

Respiration releases energy from food. Each cell must do this for itself. Every living cell respires to release the energy it needs. The energy released in respiration is not used directly for movement or any of the other activities of the cell. It is used to make a chemical called **ATP**. ATP is short for adenosine triphosphate. ATP, like glucose, contains chemical energy.

chemical energy in glucose → chemical energy in ATP

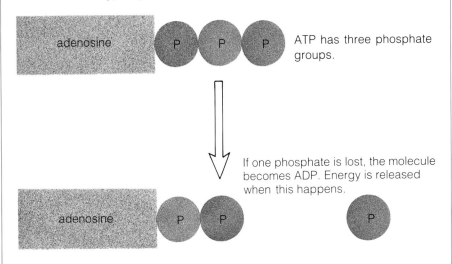

ATP has three phosphate groups.

If one phosphate is lost, the molecule becomes ADP. Energy is released when this happens.

Fig. 92.3   ATP and ADP

ATP is the ideal energy currency in a cell. The energy in an ATP molecule can be released from it very quickly – much more quickly than from a glucose molecule. The energy is released by breaking ATP down to **ADP**. ADP is short for adenosine diphosphate. Another good reason for using ATP as an energy supply is that one ATP molecule contains a much smaller amount of energy than one glucose molecule. If a cell needs just a small amount of energy, then it can break down just the right number of ATP molecules. The amount in a glucose molecule might be too much, and energy would be wasted. Each cell produces its own ATP by the process of respiration. ATP is not transported from cell to cell.

The respiration equation

Respiration is a chemical reaction. The word equation for the reaction with glucose is:

$$\text{glucose} + \text{oxygen} \longrightarrow \text{carbon dioxide} + \text{water} + \text{energy}$$

The balanced molecular equation for the reaction is:

$$C_6H_{12}O_6 + 6O_2 \longrightarrow 6CO_2 + 6H_2O + \text{energy}$$

Questions

1 Respiration is a chemical reaction.
  a Where does it take place?
  b What type of chemical reaction is it?
  c Why is it so important to living cells?
2 List two similarities, and one difference, between the burning of a peanut and the respiration of 'peanut molecules' in your cells.

3 a Write down the word equation for respiration.
  b Write down the balanced molecular equation for respiration.
4 Respiration releases energy from food. Explain how the energy came to be in the food.

5 a What is ATP?
  b Why do cells use ATP as an energy store?
  c A muscle cell uses glucose to provide energy for movement.
    i List all the energy changes involved in this process, beginning with energy in sunlight.
    ii Energy is 'lost' at each transfer. What do you think happens to this 'lost' energy at each stage?

93 THE HUMAN BREATHING SYSTEM

Cells need oxygen for respiration. The human breathing system gets oxygen into the blood.

Every human cell needs oxygen

Every single cell in your body needs oxygen. This is because each cell needs to release energy from glucose. It does this by combining oxygen with glucose. This is respiration.

oxygen + glucose $\longrightarrow$ carbon dioxide + water + energy

So each individual cell needs a constant supply of oxygen and glucose. These are brought to the cell in the blood system. The carbon dioxide which the cell makes is taken away in the blood.

Gas exchange occurs in the lungs

Each cell needs oxygen and produces carbon dioxide as a waste product of respiration. So your body needs to take in oxygen and get rid of carbon dioxide. The process of taking in oxygen and getting rid of carbon dioxide is called **gas exchange**. It happens in your lungs.

Lungs are made up of millions of alveoli

Lungs are like huge pink sponges. They are pink because they contain many tiny capillaries full of blood. The blood collects oxygen from the lungs and takes it to the rest of the cells in the body. The blood also brings carbon dioxide to the lungs to be removed from the body.

Lungs are like sponges because they are full of tiny air spaces. The air spaces are called **alveoli**. Air is drawn in through your nose and mouth and down your throat. It passes through a large tube called the **trachea** which divides into two smaller tubes called **bronchi**. The bronchi divide into a network of even smaller tubes called **bronchioles**. These lead into the alveoli.

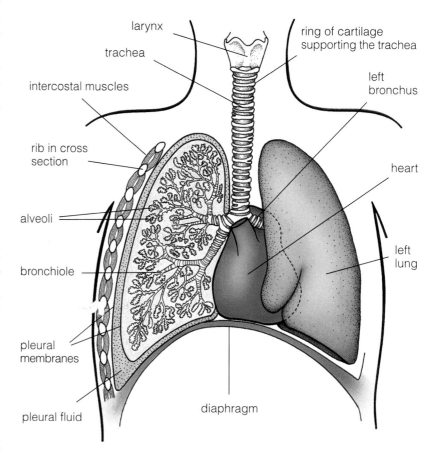

Fig. 93.1 A vertical section through the human thorax.

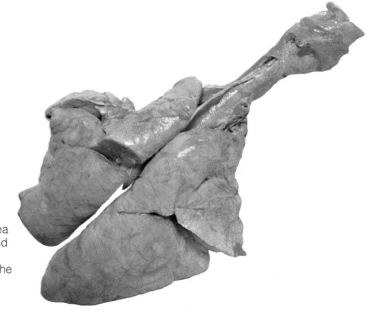

Fig. 93.2 Lungs from a pig. Try to identify: the larynx; epiglottis; trachea (cut open); rings of cartilage; left and right lungs covered with pleural membranes. You can also just see the oesophagus, running alongside the trachea.

224

The structure of sheep lungs

Sets of sheep lungs can be obtained from butchers. They are sometimes sold as 'lights' for dog food.

1   Describe the size, shape and colour of the lungs.

2   You will probably be able to see two tubes leading down to the lungs. One of these is the trachea, or windpipe. It has bands of gristle or **cartilage** around it to hold it open. You can feel these bands on your own trachea in your neck.

    At the top of the trachea there is a wider part. What is the name of this part? What is inside it? What does it do?

3   Covering the top of the trachea is a firm but flexible flap called the **epiglottis**. When you swallow, this flap shuts off the top of the trachea. Why is this necessary?

4   Find the other tube leading down past the lungs. What is it? Where is it leading to? How is it different from the trachea?

5   Now look at the lungs themselves. They are covered with a thin, transparent, slippery skin called the **pleural membranes**. These membranes make a fluid which allows the lungs to slide easily inside your body as they inflate and deflate. They also keep the lungs airtight.

6   Follow the trachea downwards to the lungs. Find where it divides into two tubes. What are these two tubes called? Do they have rings of cartilage, like the trachea?

7   If the lungs and trachea have not been badly cut, you can try inflating them. Put a rubber tube right down inside the trachea, and hold the top of the trachea tightly around it. Blow firmly down the tube. If the lungs do inflate, describe what they look like.

8   Cut a small piece out of one lung. Describe its appearance, weight and texture. Why does it feel like this?

Gas exchange takes place in the alveoli

The function of the lungs is to bring air as close as possible to the blood. Oxygen can then get from the air into the blood. Carbon dioxide can get from the blood into the air. This is gas exchange. Gas exchange happens in the alveoli. It is in the alveoli that the blood and air are brought really close together. The walls of the alveoli are only one cell thick. The walls of the blood capillaries are also only one cell thick. So the air and blood are separated from one another by a thickness of only two cells. The total thickness of this barrier is only about $\frac{1}{1000}$ mm.

The concentration of oxygen in the air is much greater than the concentration of oxygen in the blood in the lungs. So oxygen diffuses from the air into the blood. It diffuses into the red blood cells. Here it combines with a substance called haemoglobin to form oxyhaemoglobin. The red blood cells are swept along in the blood to every part of the body. When they arrive at a part where oxygen levels are low, the oxyhaemoglobin releases its oxygen. This is how your body cells get their supply of oxygen for respiration.

Carbon dioxide is also carried in the blood. It is brought from the body cells to the lungs. The concentration of carbon dioxide in the blood in the lungs is greater than the concentration in the air in the alveoli. So carbon dioxide diffuses into the alveoli.

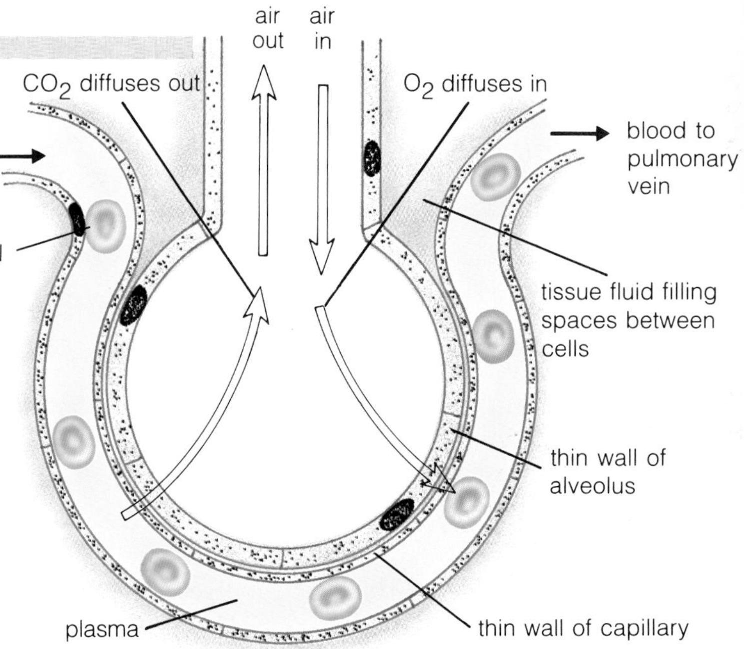

Fig. 93.3 Gas exchange in an alveolus. Notice that the walls of both the alveolus and the capillary are only one cell thick.

DID YOU KNOW?

The combined surface area of all the alveoli in both your lungs is about 70 m² (the area of a tennis court). This enormous surface area means that diffusion between the air and blood can take place very rapidly.

Questions

1 a What is gas exchange?
  b Why is gas exchange necessary?
  c Exactly where does gas exchange occur in humans?

EXTENSION

2   Discuss how the structure of the lungs, including the alveoli, is designed to make gas exchange as efficient as possible.

Breathing movements move air in and out of the lungs. This keeps the concentration of oxygen in the lungs high so that oxygen will keep diffusing into the blood.

Breathing movements supply fresh air to the alveoli

Oxygen gets into your blood by diffusion. It diffuses from the alveoli into the blood. It does this because the concentration of oxygen in the air in the alveoli is greater than the concentration of oxygen in the blood.

As the oxygen diffuses from the alveoli into the blood, the oxygen concentration in the alveoli goes down. So fresh air must be brought to the alveoli to keep the oxygen concentration high. If this did not happen, its concentration would end up the same as the concentration of oxygen in the blood. Then the oxygen would stop diffusing.

You supply fresh air to your alveoli by **breathing**. Breathing means making movements which pull and push air in and out of your lungs.

The diaphragm and intercostal muscles help in breathing

Two sets of muscles produce your breathing movements. One is the muscle in the **diaphragm**. The diaphragm is a sheet of tissue which runs across your body just below the ribs. All around the edge of it are strong muscles. When the diaphragm muscles are relaxed, the diaphragm makes a domed shape. When the diaphragm muscles contract, they pull the diaphragm flat.

The other set of muscles involved in breathing are the muscles between the ribs. They are called the **intercostal muscles**. When they are relaxed, the rib-cage slopes downwards. When the intercostal muscles contract, they pull the rib-cage upwards and outwards.

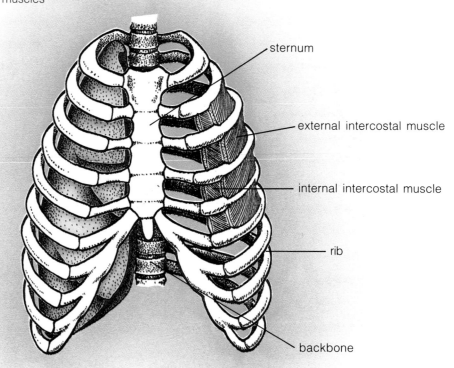

Fig. 94.1 The ribs and intercostal muscles

- sternum
- external intercostal muscle
- internal intercostal muscle
- rib
- backbone

Muscles contract when you breathe in, and relax when you breathe out

How do these muscles help you to breathe? When both set of muscles are relaxed, the volume of the thorax is fairly small. But as your diaphragm and intercostal muscles contract, the shape and size of the thorax changes. The diaphragm is pulled downwards, and the ribs are pulled upwards and outwards. Both these movements increase the volume of the thorax. As the volume of the thorax *increases*, the pressure inside it *decreases*. The pressure inside the thorax becomes *lower* than the air pressure outside. So air rushes into the thorax from the atmosphere. There is only one way in. It is along the trachea and into the lungs. So air rushes into the lungs. You have breathed in.

A few seconds later, your diaphragm and intercostal muscles relax. The diaphragm springs upwards, and the rib-cage drops downwards. Both these movements decrease the volume of the thorax. The pressure inside the thorax is increased. It rises above the pressure of the air outside the body. So air inside the thorax is squeezed out. There is only one way out. Air is squeezed out of the lungs, along the bronchi and trachea, and out through your mouth or nose. You have breathed out.

a. Expiration. Muscles relax, making the thorax volume smaller.

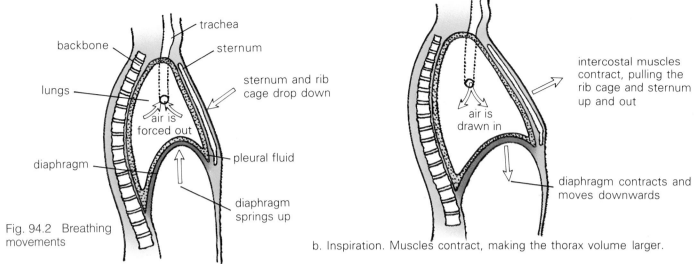

Fig. 94.2 Breathing movements

b. Inspiration. Muscles contract, making the thorax volume larger.

Pressure and volume changes in a model thorax

1 Make a labelled diagram of the apparatus with everything relaxed. On your labels, indicate which parts you think represent the lungs, the trachea, the bronchi, the ribs and the diaphragm.

2 Pull down gently on the plastic or rubber sheet. Watch the balloons carefully. Describe what happens.

Questions

1 When you pull down on the sheet what happens to the volume and pressure inside the bell jar?

2 Use your answer to question 1 to explain why the balloons inflate when you pull down on the sheet.

3 Which important set of muscles is missing from this model of the human thorax?

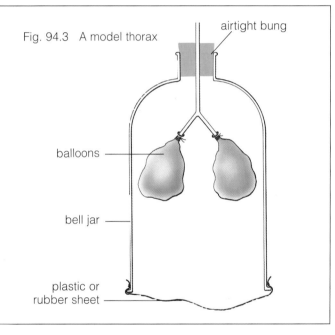

Fig. 94.3 A model thorax

Does breathing rate increase with exercise?

You breathe to fill your alveoli with fresh air. Oxygen from this air diffuses into your blood and is carried to the body cells. The cells use this oxygen in respiration.

When cells are working hard, they need more energy. They get their energy from food, by combining it with oxygen in respiration. So cells need more oxygen when they work hard. Muscle cells work very hard when you do physical exercise.

Design and carry out an experiment to find out if your rate of breathing increases with an increasing amount of exercise. Record and explain your results as clearly as you can.

Questions

1 Explain the difference between breathing, gas exchange and respiration.

2 Copy and complete the following sentences:
When you breathe in, your .... and .... muscles .... This .... the volume and .... the pressure inside the thorax. So air rushes into the lungs.
When you breathe out, your .... and .... muscles .... This .... the volume and .... the pressure inside the thorax. So air rushes out of the lungs.

EXTENSION

3 All living things respire. Do all living things breathe? Use three or four different examples of living organisms in your answer to this question, and describe how they obtain their oxygen.

227

95 INSPIRED AND EXPIRED AIR

The air we breathe out contains less oxygen and more carbon dioxide than the air we breathe in. The level of nitrogen remains unchanged.

INVESTIGATION 95.1

Comparing the carbon dioxide content of inspired and expired air

Inspired air is the air you breathe in. Expired air is the air you breathe out.

You can use either lime water or hydrogencarbonate indicator solution for this experiment. Both indicate the presence of carbon dioxide. Lime water turns milky, and hydrogencarbonate indicator solution becomes yellow if carbon dioxide bubbles through it.

1 Set up the apparatus as in the diagram. The long tubes should reach right into the liquid, but the ends of the short tubes should be in the air space above.

2 Breathe in and out gently through the central tube. Watch to see which tube bubbles when you breathe in, and which bubbles when you breathe out. Make a record of which is which.

3 Keep breathing in and out until the colour of the indicator in one of the tubes changes. Record which it is.

4 Continue breathing in and out for a few more minutes. Watch the colour of the indicator in *both* tubes, and record any further changes.

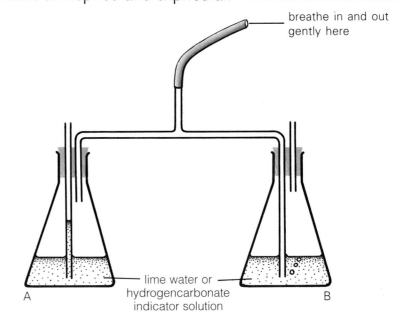

Fig. 95.1   Comparing the carbon dioxide content of inspired and expired air.

Questions

1 Try to explain why your inspired air went into one tube, and your expired air went into the other.

2 Which indicator changed first – the one with inspired air, or the one with expired air bubbling through it?

3 Which contains the most carbon dioxide – inspired air or expired air?

4 Did the indicator in both tubes eventually change colour? If so, what does this tell you?

Table 95.1   The composition of inspired and expired air

	Inspired air %	Expired air %
Nitrogen	78	78
Oxygen	21	18
Carbon dioxide	0.04	3
Noble gases	1	1

INVESTIGATION 95.2

Comparing the temperature and moisture content of inspired and expired air

1 Take the temperature of the air around you. This is inspired air.

2 Breathe out on to the bulb of a thermometer for 2 min. Do not put the thermometer in your mouth! Record the temperature of expired air.

3 Take a piece of dry cobalt chloride paper. Handle it with forceps, because any moisture on your fingers would affect the results. Wave the paper in the air around you for 2 min. Record any colour changes in the paper.

4 Take a second piece of dry cobalt chloride paper, and breathe out on to it for about 2 min. Record any colour changes in the paper.

5 Summarise your findings about the temperature and moisture content of inspired and expired air.

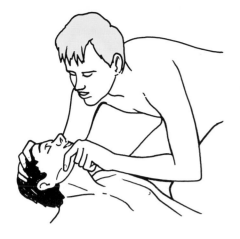

Fig. 95.2 Giving artificial respiration. Having checked that the person really is not breathing, lie them on their back. Check that the airway is not blocked, for example by loose teeth. Pull the head right back to make a clear passage from the mouth, down the trachea, to the lungs.

Either hold the nose tightly closed with one hand or cover both nose and mouth with your own mouth. Breathe firmly and steadily into the person's mouth. Do not rush your breathing – a slow and steady rate is best. Between breaths watch the person's chest to check that you are making it rise and fall. Keep going until help arrives.
If the person begins to breathe on their own, roll them onto their side into the recovery position.

Giving artificial respiration

People may stop breathing if they have an electric shock, are under water for some time or suffer some other type of accident. If this happens their body rapidly becomes short of oxygen. Brain cells are especially likely to be damaged if they get no oxygen. It is very important to help the person to begin breathing again as soon as possible.

The best way to do this is to use your own breath. You should inflate the person's lungs rhythmically. This may start their diaphragm and intercostal muscles working again. As you can see from Table 95.1, even expired air contains quite a lot of oxygen. There is plenty of oxygen in your expired air to supply an unconscious person's brain with the oxygen it needs.

Questions

1   Figure 95.3 shows some apparatus which can be used to compare the air breathed in and out by living organisms. Sodium hydroxide solution absorbs carbon dioxide. For the first experiment a mouse was put into flask C. The pump was then turned on, pulling air through the apparatus from left to right.

a   What would you expect to happen to the lime water in flask B? Explain your answer.

b   What would you expect to happen to the lime water in flask D? Explain your answer.

c   What would happen if the tubes into flask A were put in the other way round, so that the short tube led into it and the long tube led out of it?
For the second experiment a green plant was put into flask C. The experiment was done on a sunny bench top.

d   What would you now expect to happen to the lime water in flasks B and D? Explain your answer.
For the third experiment, the same green plant was left in flask C. However, the laboratory was now blacked out, and the experiment done in darkness.

e   What would you now expect to happen to the lime water in flasks B and D? Explain your answer.

2   Copy and complete. There are only three different words to use, and they are all names of gases.
Expired air contains more .... and less .... than inspired air. This is because body cells use .... in the process of respiration, and produce .... as a waste product.
The percentage of .... in inspired and expired air is the same. Although body cells need this element for making proteins, they cannot use it in this form. So the .... goes into the blood, round the body, and then out again in expired air.

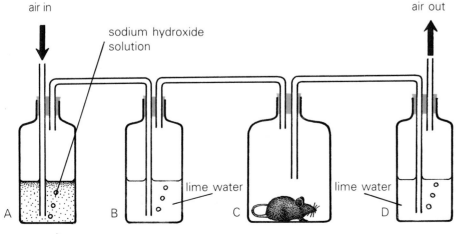

air in   sodium hydroxide solution   air out

lime water   lime water

A   B   C   D

Fig. 95.3

Inspired air contains bacteria, viruses and particles of dust. The body is designed to keep harmful materials away from the lungs.

Cilia and mucus trap dust and bacteria

The air you breathe contains all sorts of things. If you live in a city, there may be carbon particles – soot – in the air. Wherever you live, there are bound to be bacteria and viruses in it.

The alveoli in your lungs are very delicate. They are easily damaged. And your lungs are a warm, moist place for bacteria to live and breed. So your breathing system is designed to protect your lungs from harmful things in the air that you breathe in. There are special cells called **goblet cells** which line the tubes leading to your lungs. They make mucus. The mucus covers the inside of your trachea. Dust and bacteria get trapped in the mucus. In between the goblet cells are other cells. These are covered with a layer of microscopic hairs. The hairs are called **cilia**. They move in a wave-like motion, sweeping the mucus upwards. The mucus, with whatever it has trapped, is swept to the back of your throat. Then you swallow it! The dirt and bacteria are destroyed in your digestive system.

Fig. 96.2 A scanning electron micrograph of the inner surface of a human trachea. The yellow filaments are cilia. The orange parts are goblet cells, which secrete mucus. They are not really this colour!

Fig. 96.1 Cells lining the trachea

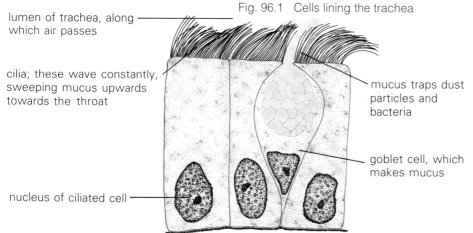

lumen of trachea, along which air passes

cilia; these wave constantly, sweeping mucus upwards towards the throat

mucus traps dust particles and bacteria

goblet cell, which makes mucus

nucleus of ciliated cell

Fig. 96.3 Two white cells patrolling the alveoli in a human lung. The spherical one is in its normal shape. Below it is an elongated white cell, about to engulf the round particle on the left.

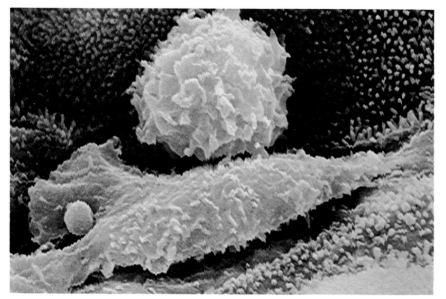

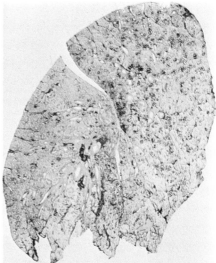

Fig. 96.4 Section through a lung of a smoker. The dark areas are deposits of tar.

White cells patrol the lungs

Despite this well-organised dust removal system, some dust and bacteria do reach the lungs. But a second line of defence now comes into operation. White blood cells constantly move around in the alveoli. They find and destroy bacteria. These white cells are called **phagocytes**.

Cigarette smoke stops cilia working

Cigarette smoke contains carbon dioxide and carbon monoxide. These substances stop the cilia from moving. If a person smokes a lot, their cilia will disappear completely. But the goblet cells go on working. They still make mucus, and the mucus still traps dirt and bacteria. But now the mucus is not swept upwards. It trickles downwards, into the lungs. The lungs now start to fill up with mucus, bacteria and smoke particles. The person coughs to push the mucus upwards. But, very often, the organisms which cause disease stay in the lungs and the tubes leading to them. They may live and breed in the mucus. The person gets **bronchitis**.

The constant coughing of heavy smokers can also damage the delicate alveoli. Their thin walls get broken. Instead of having millions of tiny alveoli in their lungs, the smoker may end up with far fewer bigger ones. These are not very good at letting oxygen get into the blood. The person becomes short of breath. This disease is called **emphysema**.

Fig. 96.5 Emphysema can make it very difficult to get oxygen into the blood. Sufferers may need to breathe oxygen even when resting.

Fig. 96.6 Post-mortem specimens of a normal lung, and a lung destroyed by cancer. The tumour is the white part in the lower part of the lung.

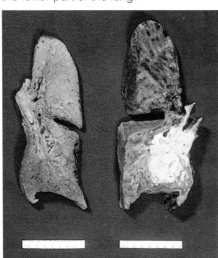

Tars in cigarette smoke may cause lung cancer

Cigarette smoke contains tars. These tars affect the cells lining the tubes leading to the lungs, and cells inside the lungs. They may make them divide more than usual. The cells go on and on dividing. They form a lump of cells called a **tumour**. Quite often, the person's white cells recognise that these dividing cells are not right. They destroy them before they cause any damage. But sometimes the tumour gets quite large. It goes on and on growing. The person now has **lung cancer**.

Many cancers can now be cured. But lung cancer is very difficult to treat. Most people who get lung cancer die from it. But most people who get lung cancer are smokers. So it is quite easy to make it almost certain that you won't get lung cancer. Don't smoke!

DID YOU KNOW?

It has been found that people who smoke are more likely to get *all* types of cancer – not just lung cancer.

Questions

1   A study was made of the incidence of upper respiratory tract infection in US Army recruits in basic training at Fort Benning, Georgia.

1230 soldiers took part in the study. The number of soldiers reporting to the troop medical clinics with upper respiratory infections were recorded over their basic training period.

Overall, it was found that soldiers who smoked were 1.46 times more likely to suffer from upper respiratory tract infection than men who did not smoke. When the smokers were subdivided into different categories, the results below were obtained:

	% of soldiers in group who reported upper respiratory infection
Group 1 - soldiers who smoked throughout basic combat training	25.3
Group 2 - soldiers who gave up smoking part way through training	36.0
Group 3 - soldiers who began smoking during training	21.4
Group 4 - soldiers who did not smoke at all	16.9

a What do you think is meant by 'upper respiratory tract infection'?

b These figures show that soldiers who smoked were more likely to suffer from upper respiratory infection than soldiers who did not smoke. Explain why you think this might be.

c The United States Army has taken a strong position against the use of tobacco by soldiers. Why do you think they have done this?

d Suggest a reason for the high figure of infection in the Group 2 soldiers.

97 RESPIRATION WITHOUT OXYGEN

It is possible to release energy from food without oxygen. This is called anaerobic respiration.

Energy can be released from food without using oxygen

Respiration is a chemical reaction which releases energy from food. The food is usually glucose. The reaction usually uses oxygen. The word equation for this reaction is:

glucose + oxygen → carbon dioxide + water + energy

This reaction is called **aerobic respiration**. 'Aerobic' means 'to do with air'. This reaction uses oxygen from the air.

However, it is possible to release energy from food *without* using oxygen. This method is not as good as aerobic respiration. It does not release as much energy from the food. But it is useful if oxygen is in short supply. Respiration which does not use oxygen is called **anaerobic respiration**. 'Anaerobic' means 'without air'.

Fig. 97.1 Yeast cells. This particular kind of yeast is used in bread-making. Some of the cells are producing buds, which will break off to form new cells.

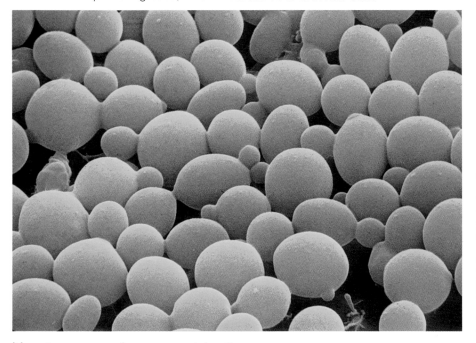

Yeast can respire anaerobically

Yeast is a single-celled fungus. Yeast lives naturally on the surface of fruit. It feeds on the sugars in the fruit. Yeast can respire anaerobically for long periods of time. It breaks down glucose, and releases energy from it, without using oxygen. The word equation for this reaction is:

glucose → carbon dioxide + ethanol + energy

You will see several differences between this reaction and the aerobic respiration reaction. Firstly, this one does not use oxygen. Secondly, this one produces ethanol. Another difference is that anaerobic respiration releases much less energy from the glucose than aerobic respiration does.

INVESTIGATION 97.1

Anaerobic respiration in yeast

1 Set up the apparatus as shown in the diagram.
2 Set up an identical piece of apparatus which does not contain yeast.
3 Leave both pieces of apparatus in a warm place for at least 30 min.

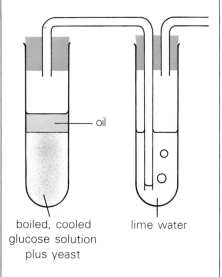

oil

boiled, cooled glucose solution plus yeast

lime water

Fig. 97.2

Questions

1 This experiment is to find out if yeast can respire anaerobically. Why was the water in the tube boiled before use?
2 Why was the boiled water cooled before the yeast was added?
3 Why is a layer of oil floated on the water?
4 Did either of the samples of limewater turn milky? What does this indicate?
5 At the end of the experiment what new substance would you expect to find in the tube containing the yeast?

Yeast is used in brewing and wine making

When yeast respires anaerobically it produces ethanol. This is used in the brewing industry to make beer. Barley grains are allowed to germinate. As they germinate they produce a sugar called maltose. The maltose is dissolved in water. Yeast is added. The yeast respires anaerobically, using the maltose. Ethanol (alcohol) is made. Hops are usually added to the liquid, to give a bitter flavour, and help to preserve it.

Yeast will also use the sugars in grapes. This is how wine is made. Both brewing and wine-making involve anaerobic respiration. Because the reaction makes alcohol, it is sometimes given another name. It is called **alcoholic fermentation**.

Fig. 97.3 In this fermentation tank, yeast is feeding on malt from barley seeds. The yeast respires anaerobically, releasing carbon dioxide and alcohol.

Yeast is used in bread-making

Yeast is also used in bread-making. This time it is the carbon dioxide which is wanted, not the ethanol. Water is added to flour to make a dough. Some of the starch in the flour breaks down to sugar. Yeast is added and respires using the sugar. It makes carbon dioxide. The carbon dioxide forms bubbles in the dough. This makes the bread rise. When the bread is baked, the yeast is killed.

Fig. 97.4 Making bread. Respiration by yeast releases carbon dioxide, which makes the dough rise.

Fig. 97.5 Athletes paying back their oxygen debt after a race.

Human muscles can respire anaerobically

Your muscle cells can respire without oxygen for a short time. They only do this when they are short of oxygen. This might happen if you run a race. Your muscles need a lot of energy, so they respire very fast. But your lungs and heart might not be able to supply enough oxygen to them. So the muscle cells have to manage without oxygen for a short while.

When muscle cells respire anaerobically, they do not make ethanol! Instead, they make **lactic acid**. The word equation for the reaction is:

glucose → lactic acid + energy

When you stop running, you will have lactic acid in your muscles and your blood. This must be broken down. Oxygen is needed to break it down. So you breathe fast to get extra oxygen into your blood to break down the lactic acid. This is why you go on breathing hard, even after you have stopped running. The extra oxygen you need to break down the lactic acid is called an **oxygen debt**.

Questions

1 a List three similarities between the chemical reactions of aerobic respiration and anaerobic respiration.

 b List three differences between them.

2 What is alcoholic fermentation? Give one way in which this reaction is used in industry.

3 Explain why an athlete continues to breathe faster and deeper than usual after finishing a race.

Questions

1 Copy and complete using the words listed on the right:

All the energy in living organisms begins as .... energy. This energy is trapped by the green colouring in plants, called .... The energy is used to drive the chemical reactions of .... This makes glucose and oxygen from .... and ....

The glucose contains some of the energy. The plant may change some of the glucose to .... This may be stored in the plant's seeds. An animal may eat the seeds. The animal's digestive system breaks down the starch molecules in the .... to glucose molecules. They still contain energy. The glucose molecules are taken, in the animal's .... system, to cells which need energy. Inside the cells, the glucose combines with .... This releases .... from the glucose, and produces the gas .... This reaction is called ....

Words to use:

seeds	photosynthesis
sunlight	respiration
starch	oxygen
energy	blood
water	chlorophyll
carbon dioxide (twice)	

2 The diagram shows the result of an experiment to find out how animals and plants affect the air around them.

Carbon dioxide dissolves in water to form a weak acid. Hydrogencarbonate indicator solution detects changes in acidity. It is purple with no carbon dioxide, red with a very little, and yellow with a lot of carbon dioxide present.

a In which of the tubes would you expect respiration to have been occurring?

b In which of the tubes would you expect photosynthesis to have been occurring?

c What gas is
i used
ii produced
in respiration?

d What gas is
i used
ii produced
in photosynthesis?

e In which of the tubes would you expect carbon dioxide to have been used up?

f In which tube would you expect carbon dioxide to have been made?

g Explain, as fully as you can, the reasons for the colour of the indicator in each tube.

h If a fifth tube was set up, containing both animals and plants (but covered with black paper) what results would you expect? Explain your answer.

3 a Describe how a person's lungs are kept clean.

b Explain what happens to this process if a person smokes.

c How can this damage a person's lungs?

d Briefly describe two other ways in which smoking can be harmful.

hydrogencarbonate indicator solution

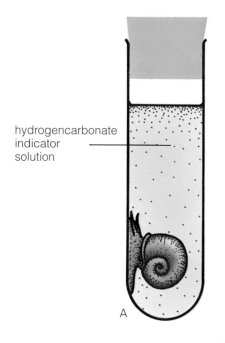

A   B   C   D

FOOD AND DIGESTION

HETEROTROPHIC NUTRITION

Animals cannot make their own food. They rely on food made by plants.

Plants use inorganic substances to make organic substances

One of the things which makes animals different from plants is the way in which they feed. Plants use **inorganic** substances. Inorganic substances are things which have not been made by living things. They often have fairly small molecules.

The inorganic substances which plants use are carbon dioxide and water. These two substances are combined to make glucose. Glucose is an **organic** substance because it contains carbon and is made by a living organism. Plants make glucose by photosynthesis.

All living things are made up of organic substances. Plants can make their own organic substances out of inorganic ones. This way of getting organic food is called **autotrophic nutrition**. 'Auto' means 'self', and 'trophic' means 'feeding'. Plants can make their own food.

Animals must eat organic food made by plants

Animals cannot make organic substances out of inorganic ones. Animals must eat organic food which plants have made. Sometimes they eat plants directly. Sometimes they eat other animals which have eaten plants. But whatever they eat, animals rely on plants to make food for them. Animals cannot make their own food. This way of feeding is called **heterotrophic nutrition**. 'Hetero' means 'other'. Animals rely on other organisms to make their organic food.

Fungi also feed heterotrophically

Fungi are another group of organisms which need organic food. Fungi cannot photosynthesise. Like animals, they rely on food that has been made by plants. Most fungi feed on organic substances that are no longer in a living organism.

The fungus which causes mildew will grow on leather boots or on paper. The leather was once an animal's skin. This

animal will have eaten plants. So the organic substances in the leather were originally made by plants.

Another example of a fungus is yeast. Yeast feeds on sugars made by plants.

The mushrooms you eat are just the part of a fungus above the ground. There is a network of threads underground. These threads feed on organic substances in the soil.

Although fungi and animals both use organic food, they feed in very different ways. Fungi feed by releasing enzymes on to the food on which they grow. The enzymes help to dissolve the food. The dissolved food is absorbed into the fungus cells. This way of feeding is called **saprotrophic nutrition**. It is a special kind of heterotrophic nutrition.

Fig. 98.1 Mushrooms, like all fungi, are saprotrophs. Threads called hyphae grow underground, digesting organic substances and absorbing them.

Fig. 98.2 Mould growing on grapes. The fungus is feeding on sugar and protein from the grapes. It digests them by secreting enzymes, which will eventually reduce the fruit to a liquid.

Animals need several types of food in their diet

At first sight it may seem that the kinds of food that animals eat are very different. You, for example, would probably not be very happy with the diet that an earthworm thrives on. Earthworms eat dead leaves and soil. But there are really more similarities than differences between the diets of different animals. The kinds of food molecules which different sorts of animals need are very much the same. You and an earthworm both need the following substances in your diet:

- carbohydrates
- fats
- proteins
- vitamins
- minerals
- roughage
- water

Throughout the rest of this topic on food and digestion we will consider only humans. But do remember that all other animals use the same kinds of food, even if they eat and digest them in very different ways.

Mineral	Why you need it	Deficiency symptoms	Foods rich in it
Iron	to make haemoglobin, the red pigment in blood cells which carries oxygen	anaemia, caused by a lack of haemo- globin to carry oxygen	liver and red meat
Calcium	for bone and teeth formation	poor growth of bones and teeth	milk and other dairy products; hard water
Iodine	for making the hormone thyroxine	goitre, which is a swollen thyroid gland	sea food; table salt
Fluorine	for bone and teeth formation	more likelihood of tooth decay	milk; water in some areas; toothpaste

Table 98.1 Some minerals needed by humans

Fig. 98.3 The thyroid gland in the neck uses iodine to make the hormone thyroxine. A shortage of iodine in the diet can make the thyroid gland swell, forming a goitre.

Animals do need some inorganic substances

Two of the types of substances in the list of foods are inorganic ones. These are minerals and water.

Minerals are inorganic substances which we need in very small amounts in our diet. There is quite a large number of them. Some of them are shown in Table 98.1.

Water is needed in much larger quantities. You probably need to take in around 2–3 litres of water a day. You will not need to drink all this, because most foods contain some water. A person can live for up to 60 days without food, but will die after only a few days without any water. Water is needed for several different reasons. Firstly, the cytoplasm in your cells is a solution of proteins and other substances in water. The metabolic reactions happening in your cells will only happen in solution. So if your cells lose too much water, metabolism stops and you die.

Secondly, water is used to transport substances around your body. Blood is mostly water. If a person becomes severely dehydrated, their blood gets thicker. It travels more slowly around the body.

There are many other uses for water in the body. It is used in many different metabolic reactions. It helps in digestion. It dissolves waste products so that they can be removed in urine. It evaporates from the skin when you are hot, cooling you down.

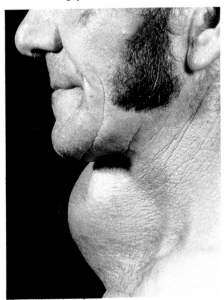

Questions

1 Explain the difference between:
a organic and inorganic substances
b autotrophic and heterotrophic nutrition
2 a List three examples of fungi.
b On what do each of your examples feed?
c Briefly explain how the food gets into the fungus' body.
d What is the name for this kind of feeding?

3 a List the kinds of foods which animals need in their diet.
b Which of these foods are organic?
c Which of these foods are inorganic?
d Give three reasons why water is needed in the diet.
e Why does a lack of iron in the diet cause tiredness and lack of energy?

99 CARBOHYDRATES AND FATS

Carbohydrates and fats are energy foods.

Food provides energy

All the energy that you have comes from the food you eat. If you are 15 years old, you probably use up about 9500 to 12 000 kilojoules (kJ) of energy each day. A lot of this energy is used to keep you warm. The rest is used up in movement, making new cells, and other processes in your body.

Most of your energy comes from **carbohydrates** and **fats** which you eat. **Protein** provides only around 10 % of your energy. So carbohydrates and fats are sometimes called 'energy foods'.

It is important to take in the right amount of energy. If you eat food containing more energy than you need, your body stores the surplus as fat. If you do not eat enough, you lose weight and become tired and ill. Pure carbohydrate contains 17 kJ of energy in every gram. Protein also contains 17 kJ per gram. Fat, however, contains much more energy – around 39 kJ per gram. Many foods are a mixture of carbohydrates, fats, proteins and other substances. To find out how much energy they contain, they can be burnt in a calorimeter.

Fig. 99.1 Taking in more energy than you use up can lead to obesity.

Carbohydrates include sugars and starches

Sugars and starches are carbohydrates. The shape of their molecules is shown in Figure 99.2. Carbohydrates contain three kinds of atoms – carbon, hydrogen and oxygen.

The simplest kinds of carbohydrates are **sugars**. Sugars have quite small molecules. They are soluble in water and they taste sweet. Examples include glucose, sucrose – cane sugar, which you use in tea or coffee – and maltose.

Starches have much bigger molecules. They are made of thousands of glucose molecules linked in a long chain. Because the molecules are so big, they will not dissolve in water. Starch does not taste sweet.

Fig. 99.2 Carbohydrate molecules

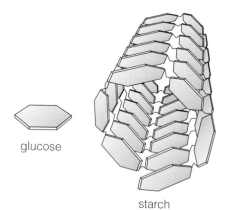

glucose

starch

INVESTIGATION 99.1

Testing foods for carbohydrates

A *Testing for starch*
1. Draw up a results chart. It should have spaces for the name of the food being tested, the colour it goes, and what you can conclude from this.
2. Collect small samples of the foods you are going to test. Take great care not to let any of the foods get mixed up.
3. Put each food in turn on to a white tile. Cover it with iodine solution. This solution is brown. If there is starch in the food, it will turn very dark blue – almost black. Record each result as you go along.

B *Testing for sugars*
1. Draw up another results chart.
2. Collect samples of food as before. This time, each food must be chopped finely, or crushed in a pestle and mortar.
3. Put a sample of the first food you are going to test into a boiling tube. Add Benedict's solution. Shake the tube to mix the food thoroughly with the Benedict's solution. This is so that any sugar molecules in the food will be able to react with the Benedict's solution easily.
4. WEAR SAFETY GLASSES. Holding your tube at an angle, heat it gently over a blue Bunsen burner flame. Shake it gently as you heat it so that each part heats evenly. If it seems to be heating too quickly take it out of the flame for a moment.

If the food contains sugar, a reddish-brown precipitate will form when the Benedict's solution boils. If there is only a little sugar, it may go green or yellowish, but not quite turn red.

Do *not* continue to boil the liquid after it has changed colour.

Questions

1. Make a list of foods which contain starch.
2. Make a list of foods which contain sugar.
3. Why do you need to chop or crush the food for the sugar test?
4. Give two reasons for shaking the solution when doing the sugar test.

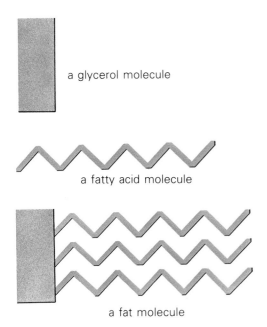

a glycerol molecule

a fatty acid molecule

a fat molecule

Fig. 99.3 Fat molecules

The wrong sort of fats can cause heart disease

Fats, like carbohydrates, contain three sorts of atoms – carbon, hydrogen and oxygen. But the way in which these atoms combine in fats is very different compared with carbohydrates. Figure 99.3 shows the structure of a typical fat molecule. Fats are not soluble in water.

Fats are sometimes called **lipids**.

Fats and oils – oils are liquid fats – should not form too large a part of a diet. Like carbohydrates they are energy foods. Too many make you fat. But you do need to eat *some* fat. Some vitamins are found only in fatty foods. Your body does need fats for making cell membranes, nerve cells and other substances.

Fats are often described as being **saturated** or **unsaturated**. Saturated fats tend to come from animals. Dairy products and meat contain a lot of saturated fat. These foods also contain a fat-like substance called **cholesterol**. People who eat a lot of saturated fats and cholesterol run a higher risk of circulatory problems (blocked blood vessels) and heart disease.

Unsaturated fats tend to come from plants. Sunflower oil, olive oil and many margarines contain unsaturated fats. These fats do not cause heart disease. In fact, there is some evidence that some of them can actually help to *stop* you getting heart disease! Olive oil, especially, is thought to help in this way. Fats from oily fish such as mackerel are also good for you.

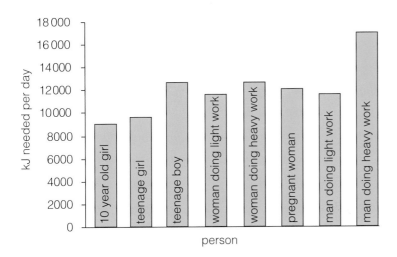

Fig. 99.4 Energy requirements. These figures are only approximate, as the amount of energy you need depends on many things, including your age, sex, weight and how much work you do. It also depends on temperature, as in a cold environment you use a lot of energy to keep warm.

INVESTIGATION 99.2

Testing foods for fats

1 Draw up a results chart.
2 Rub a piece of the food you are testing onto some clean, dry filter paper. If the food is very hard or dry, it will help to warm it first.
3 Hold the paper up to the light. If you can see a translucent mark there may be fat in the food. To check, dry the paper. If the mark stays then it is fat. If it goes, the mark was made by water from the food.

Questions

1 a List four foods which are rich in carbohydrate.
  b List three foods which contain almost no carbohydrate at all.
  c Why do you need carbohydrate in your diet?
  d Describe two properties which distinguish sugars from starches.

2 a List four foods which are rich in fat.
  b List three foods which contain almost no fat at all.

c Why should you avoid too much saturated fat in your diet?
d List two foods which contain a lot of saturated fat.
e List two foods which contain unsaturated fat.
f Give two reasons why everyone should have some fat in their diet.

3 a About how much energy is needed each day by:
  i a 10-year-old girl
  ii a 15-year-old boy

  iii a pregnant woman
  iv a 40-year-old man doing a desk job
  v a 40-year-old man doing heavy work?
b Which types of food provide most of a person's energy?
c What happens if the amount of energy taken into a person's body is more than the amount of energy they use up?

PROTEINS AND VITAMINS

Proteins are needed for making many substances in your body. Vitamins are essential for helping metabolic reactions to take place efficiently.

Proteins are made up of amino acids

Proteins, like fats and carbohydrates, are organic substances. Like fats and carbohydrates, they are made up of carbon, hydrogen and oxygen atoms. But proteins also contain two other types of atoms. These are nitrogen and sulphur. These five types of atoms are bonded together to form **amino acid** molecules. There are about 20 different sorts of amino acids. Amino acids can join together to form chains. A short chain is called a **polypeptide**. A long chain of amino acids is called a **protein**.

There are thousands of different proteins. They are different because of the different order of the amino acids in their chains. As there are about 20 different amino acids, in a chain of 200 there are millions of possible orders in which they could be arranged. Even one amino acid difference in the chain would make a different protein! So there is almost no limit to the number of different proteins which could exist.

Fig. 100.1   Protein molecules. Proteins are long chains of amino acid molecules. There are 20 different amino acids.

a. A single amino acid

b. A short chain of amino acids is called a polypeptide.

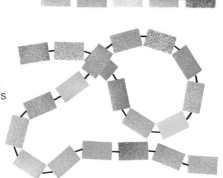

c. A protein molecule is made of several hundred amino acids linked in a chain. The chain coils in a particular shape for each kind of protein.

Proteins are needed for making many substances in the body

Proteins are an important part of the diet. Proteins have many different functions in the body.

The **cytoplasm** from which your cells are made contains a lot of protein dissolved in water. So proteins are needed for making new cells. Muscle cells contain an especially large amount of protein. **Cell membranes** also contain protein.

Haemoglobin, the red pigment in your red blood cells which transports oxygen, is a protein. So is **fibrinogen**, which helps in blood clotting.

Your hair and nails are made of a protein called **keratin**. Your skin is also covered with a layer of keratin. It is tough and insoluble.

The **antibodies** which help to defend you against attacks from bacteria and viruses are proteins. So are **enzymes**. Without enzymes, none of your metabolic reactions would be able to take place.

So you can see that proteins have many important roles to play in your body. When you eat food containing protein, your digestive system breaks down the protein molecules into individual amino acids. The amino acids are taken around the body in your blood and delivered to your cells. Here, the amino acids are linked together again to form whichever proteins that cell needs.

INVESTIGATION 100.1

Testing food for proteins

This test for proteins is called the **biuret test**.

1   Draw up a results chart.
2   WEAR SAFETY GLASSES. Crush or chop a sample of the food you are going to test. Put some into a test tube. Add some potassium hydroxide solution.
   Take care! This is quite a strong alkali and it can burn your skin. If you get it on your hands,

wash it off immediately with cold water.
3   Cork the test tube and shake to mix the food with the potassium hydroxide solution. Then add a little dilute copper sulphate solution. Cork and shake again.

A blue colour means there is no protein in the food. A purple colour means there is protein in the food.

Fig. 100.2   Cheese (on the left) contains protein, and gives a purple colour with the biuret test. The tube on the right shows the blue colour which indicates that no protein is present.

Vitamins are only needed in very small amounts

Vitamins are organic substances. They are only needed in very tiny amounts in your diet. They do not have any energy value at all. But without any one of them, you are likely to get extremely ill.

Vitamin A is a fat-soluble vitamin. Good sources of vitamin A include fish oil, dairy products and carrots. Vitamin A is needed to keep your skin healthy. It is also used to make the substance which helps you to see in dim light. Without vitamin A your skin becomes very dry. The cornea, the covering over the front of the eye, also gets dry. This condition is called **xerophthalmia**. You can no longer see in dim light – you suffer from **night blindness**.

Vitamin B is really lots of different vitamins. The B vitamins are all water-soluble. They are found in wheat germ, brown flour and rice, yeast extract, liver and kidney. The B vitamins help in the chemical reactions of respiration. These reactions take place in every cell in your body. They provide the cells with energy. So without vitamin B your cells run short of energy. One particular disease resulting from lack of vitamin B is beri-beri. People with **beri-beri** have little energy and very weak muscles.

Vitamin C is a water-soluble vitamin. It is found in citrus fruits and many fresh vegetables, including potatoes. But vitamin C is easily destroyed by cooking. This is one reason why you should try to eat fresh, uncooked fruit and vegetables in your diet. Vitamin C helps to keep skin strong and supple. Without it, the skin on the gums begins to crack apart and bleed. Wounds in other parts of the skin cannot heal. This disease is called **scurvy**. Sailors used to suffer from scurvy on long sea voyages when they had no fresh food to eat.

Vitamin D is a fat-soluble vitamin. It is found in fish oil, eggs and dairy products. It can also be made by your skin when sunlight falls on it. Vitamin D is needed to help your body to use calcium to make bones and teeth. Without vitamin D the bones stay soft, and bend. This is called **rickets**.

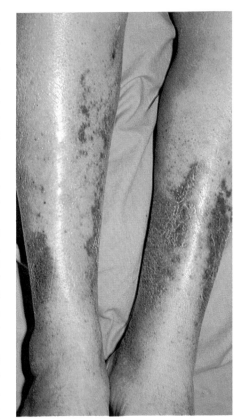

Fig. 100.3 Lack of vitamin C causes scurvy. One symptom of advanced scurvy is bleeding beneath the skin.

INVESTIGATION 100.2

Testing fruit juices for vitamin C

DCPIP is a blue dye which loses its blue colour when vitamin C is added to it. You can find out how much vitamin C a solution contains by adding it to DCPIP. The *more* vitamin C there is in the solution, the *less* of it is needed to make a certain amount of DCPIP lose its blue colour.

You will be given a solution of DCPIP and several different fruit juices. Design and carry out an experiment to find out which of the fruit juices contains the highest concentration of vitamin C.

If you were given a solution containing a known concentration of vitamin C, how could you find out the concentration of vitamin C in each of the fruit juices? Design and carry out an experiment to find out these concentrations.

— E X T E N S I O N —

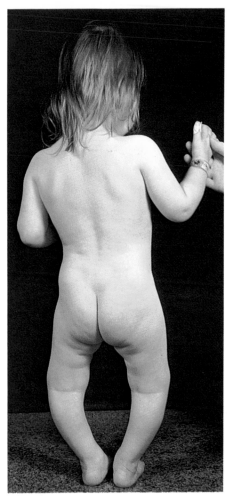

Fig. 100.4 This child is suffering from rickets, a disease in which the bones of growing children do not harden. It can be caused by a lack of vitamin D in the diet.

Questions

1 a Are proteins organic or inorganic substances?
  b Which elements are contained in all proteins?
  c How many different types of amino acids are there?
  d List five foods which contain a lot of protein.
  e List two foods which contain hardly any protein.
2 a Keratin and haemoglobin are two important proteins in your body. Where is each found and what does it do?
  b Even if you do not eat any keratin, your body can still make it. Briefly explain why this is possible.
3 Draw up a chart to show the sources, uses and deficiency diseases for vitamins A, B, C and D.

101 A BALANCED DIET

A balanced diet is one which contains the right amount of food, with enough of all the different substances needed by the body.

Different people need different diets

Different people need different amounts of energy in their diet. Figure 99.4 on page 239 shows a few examples. The more energy a person uses up each day, the more they need in their diet. Different people also need different proportions of the seven types of food in their diet. A growing child, for example, needs plenty of protein for making new cells. An athlete needs protein for building muscles, but also plenty of carbohydrate for energy. A pregnant woman needs extra iron and calcium to build her baby's blood and bones.

A good diet is called a **balanced diet**. A balanced diet contains the right amount of energy to meet a person's needs. It also contains some of all the seven types of food – carbohydrates, fats, proteins, water, roughage, each kind of vitamin and each kind of mineral. It contains enough of all of these, but not too much of any. The right diet for you might not be the right diet for someone else. It all depends on your age, your sex and your lifestyle.

Whatever type of diet you need, there are certain things which are especially good and certain things which should be avoided.

Fig. 101.1 High-fibre foods

Fresh fruit and brown bread are good for you

Fresh fruit and vegetables are a very important part of a balanced diet. They contain many vitamins and minerals. They also contain roughage. Roughage (fibre) helps to keep food moving quickly through the digestive system. This reduces the risk of several different illnesses, including bowel cancer. Brown rice and whole wheat also contain useful vitamins and roughage. So brown bread is much better for you than white.

Vegetarian diets can be very healthy

More and more people are becoming vegetarians. Sometimes they just do not like meat. Some people think that a vegetarian diet is more healthy than one which contains meat. Other people are vegetarians because they do not like the idea of killing and eating animals. Most vegetarians will eat milk and eggs. They are sometimes called lacto-ovo vegetarians. Some vegetarians will not eat any animal products at all. They are called vegans.

A vegetarian diet can be a very healthy one. It is likely to contain plenty of roughage and to be low in saturated fats and cholesterol. But vegetarians do need to be careful to get plenty of proteins. They should eat plenty of pulses (beans, peas and lentils) which are rich in protein. Rice and other grains also contain protein. If they eat milk, eggs and cheese, these are rich in protein too. Vegetarian diets are sometimes low in certain vitamins. Vegetarians must take care to eat as wide a range of different foods as possible, to make sure they get all the amino acids, vitamins and minerals that they need.

Fig. 101.2 Vegetarian food

Too much food causes obesity

Obesity means being too fat. Obesity is caused by eating more food than you need. If the amount of energy you take in is more than you use up, the extra food is stored as fat. Being too fat is not good for you. Obese people are much more likely to suffer from heart disease and circulatory problems than people of normal weight.

The amount of food needed by different people varies widely. Even people of the same age and sex may need very different amounts of food. So do not worry if you seem to eat a lot more than your best friend! So long as you are not obviously overweight, you probably need this extra food.

Anorexia nervosa is a dangerous illness

Anorexia nervosa is a disease in which the sufferer does not eat enough food to keep them healthy. It is commonest in young people. A person suffering from anorexia nervosa does not want to eat food. If they are made to eat, they may be sick afterwards. They lose weight. Eventually, they may lose so much weight that they become ill. They may even die.

No-one is quite sure what causes anorexia nervosa. It is a psychological problem. The person may be really worried about being fat. Or they may not want to grow up and face the world with all its problems. They imagine that if they do not eat, they can avoid these problems. Anorexia can be cured. But there is no magic solution. It takes a long time and a lot of care before an anorexic person can learn to live and eat in a happy, normal way.

Fig. 101.3 This mother and child in Ethiopia are suffering from a lack of food. They were forced to move away from their home because of fighting. Severe lack of food during childhood means that a person will still be small even as an adult.

Malnutrition and starvation

If you have something wrong with you which is caused by a bad diet, you are suffering from malnutrition. Malnutrition means 'bad eating'.

Malnutrition can be caused by eating too much of something. It can also be caused by not eating enough of a particular food. Children who do not get enough protein in their diet, for example, will not grow properly. In some parts of the world this is all too common. It is called **kwashiorkor**.

Starvation is not the same as malnutrition. Starvation means not getting enough food to keep you alive. Starving people do not get enough energy in their diet to balance the energy they use up. They have no fat stores to draw on, so their body begins to break down the proteins in their cells to provide energy. Children with **marasmus** suffer from a general weakness and do not grow properly due to lack of food.

Questions

1 a What is meant by a balanced diet?
  b Why is a balanced diet for one person not necessarily the right diet for another?
2 Below are listed the foods eaten in one day by a 14-year-old boy. It is not difficult to see that this is not a good diet!
  • Breakfast: nothing
  • Morning break: two bags of crisps, can of fizzy drink
  • Lunch: beefburger, chips, baked beans, can of fizzy drink, chocolate
  • Tea: two cups of tea, each with two spoonfuls of sugar; piece of cake, three chocolate biscuits
  • Supper: fish and chips, can of fizzy drink, ice cream
  a Explain, with *reasons*, what is wrong with this diet.
  b Suggest some changes which the boy could make in this diet, without giving up the things that he really likes eating.
3 Read the following passage, and then answer the questions.

Diet and intelligence

A lot of people's diets consist largely of processed food. Children come to school without having eaten any breakfast. They eat crisps and chocolate at break-time and choose cooked, fatty foods at lunchtime. Their diet may contain almost no fresh food.

A science teacher in a comprehensive school in Wales has investigated the diets of some of the children in his school. He found that many of them were not getting enough of up to ten different vitamins and minerals. He wondered if this was affecting their success in school.

In 1986 he gave all children in the second year at school a mental test and recorded the results. For the rest of the year, half of these children were given vitamin pills. The other half were also given a pill, but without any vitamins in it. They did not know which was which. At the end of the school year all the pupils were tested again. On many of the tests, there was no difference between the children who had taken extra vitamins and those who had not. But on one sort of intelligence test, the children who had had vitamin pills did much better than they had before. Other similar tests have been conducted in other parts of the world, with similar results. Does this mean that all children should take vitamin pills? Probably not. But it does mean that we should think carefully about what we eat, and make sure that our diet contains plenty of vitamins.

---

1 What is meant by 'processed foods'? Why are they likely to contain less vitamins than fresh foods?
2 Why did the science teacher decide to try giving children extra vitamins?
3 Why were some children given 'dummy' pills?
4 What does this experiment suggest about the effect of lack of vitamins on intelligence?

102 THE HUMAN DIGESTIVE SYSTEM

The digestive system is made up of the alimentary canal and glands which secrete juices into it.

The alimentary canal is a tube leading from mouth to anus

Food goes into you through your mouth. Eventually, solid remains emerge at the other end as faeces. What happens to the food in between? Food travels through your body along your **alimentary canal**. This a tube which begins at the mouth and ends at the anus. The tube is continuous. It is wide in some parts and narrow in others. It bends back on itself in places but is basically just a single tube running right through your body.

Fig. 102.1 The human alimentary canal. The alimentary canal is the tube along which food passes. The canal, plus the other organs which secrete liquids into it, such as the liver, make up the digestive system.

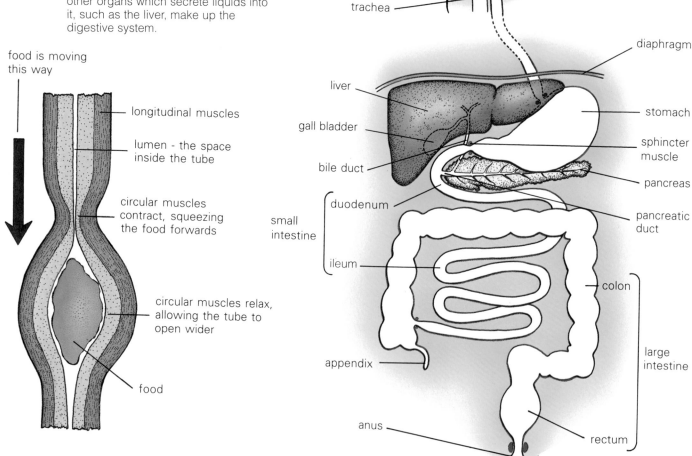

food is moving this way

longitudinal muscles

lumen - the space inside the tube

circular muscles contract, squeezing the food forwards

circular muscles relax, allowing the tube to open wider

food

nasal cavity
palate
tongue
mouth
salivary glands
trachea

food being swallowed
epiglottis
oesophagus

diaphragm
liver
gall bladder
bile duct
duodenum
small intestine
ileum

stomach
sphincter muscle
pancreas
pancreatic duct

colon

large intestine

appendix

anus

rectum

Fig. 102.2 Peristalsis. Muscles in the wall of the alimentary canal contract in waves, pushing food along.

The walls of the alimentary canal contain muscles

All along the alimentary canal there are strong muscles in the walls. These muscles push the food along. You can see this happening in your oesophagus when you swallow. The rippling movements of the muscles are called **peristalsis**. Peristalsis takes place all along your alimentary canal. The muscles work best when you have plenty of fibre in your diet. A lack of fibre can cause constipation, when the muscles do not work so well and the food stops moving along.

Mucus stops food and enzymes damaging the walls of your alimentary canal

The walls of your alimentary canal are made up of soft, living cells. Inside your alimentary canal is a mixture of all sorts of food – some of it quite hard and rough – and enzymes. The living cells must be protected from damage. So scattered amongst them, are goblet cells. These make mucus. The mucus covers the inside of the alimentary canal with a slimy layer, over which the partly digested food can easily slide. And the enzymes inside the alimentary canal cannot get at the living cells and damage them.

The stomach wall needs special protection because the stomach contains hydrochloric acid. This acid helps to kill bacteria in the food. Normally the stomach wall is well protected by its layer of mucus. But sometimes the amount of acid is too great and it begins to digest the stomach wall. This is painful and is called a **stomach ulcer**. The acid may also damage the wall of the next part of the canal – the **duodenum**. This is called a **duodenal ulcer**. If the hole goes right through, the contents of the alimentary canal can leak out. This is called a **perforated ulcer**. It is exceedingly painful and dangerous if not quickly treated.

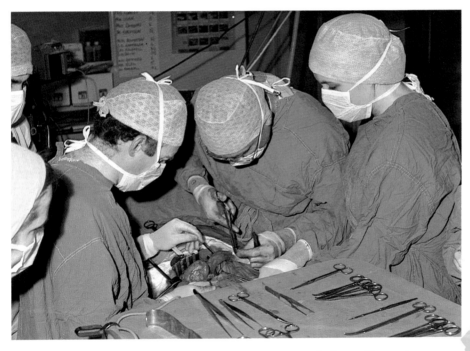

Rings of muscle can close off regions of the alimentary canal

At various places along the alimentary canal there are rings of muscle called **sphincter muscles**. These can contract to squeeze the tube closed. Sphincter muscles at the top and bottom of the stomach close to hold food inside the stomach while it is churned up with enzymes and hydrochloric acid. It may be held there for several hours. Sometimes the sphincter muscles do not work as well as they should. Acid food can then get squeezed out of the top of the stomach into the oesophagus. The acid burns the wall of the oesophagus, which hurts. This is called **heartburn**.

There is also a sphincter muscle right at the end of the alimentary canal. This is the anal sphincter. It stays closed most of the time. It opens to allow faeces to pass out of the rectum.

The liver and pancreas make fluids which help to digest food

The alimentary canal is not the only part of your body which helps with digestion. There are also several glands which pour liquids into the canal. These glands and the alimentary canal make up your **digestive system**.

The largest gland is the **liver**. The liver has many functions besides digestion. But it does help in digestion by producing **bile**. The bile is stored in a small sac inside the liver called the **gall bladder**. A tube called the **bile duct** carries the bile from the gall bladder into the duodenum. It is squirted on to the food as it emerges from the stomach. Bile helps to digest fats.

On the left-hand side of your body, just under your stomach, is a soft, creamy coloured gland called the **pancreas**. Like the liver the pancreas has other functions beside digestion. But most of the cells in the pancreas produce a liquid called **pancreatic juice**. The juice flows along a tube called the **pancreatic duct**. This, like the bile duct, opens into the duodenum. Pancreatic juice contains enzymes which help to digest proteins, fats and carbohydrates.

Fig. 102.3 An operation being performed on the alimentary canal.

Questions

1 a List, in order, the parts of the alimentary canal through which food passes after you have eaten it.
  b What is peristalsis?
  c Why does the wall of the alimentary canal contain goblet cells?
2 a What are sphincter muscles?
  b Name two places in the alimentary canal where sphincter muscles are found.

Large food molecules must be broken down to small ones before they can get into your bloodstream. This is called digestion.

Digestion means breaking down large food molecules into small ones

The walls of the alimentary canal are made up of living cells. Food molecules must pass through these living cells to get into the blood vessels and be taken around the body. Only small molecules can get through the walls of the alimentary canal and into the bloodstream. Water molecules, vitamin molecules and mineral ions are all small enough to get through. But much of the food that you eat is made up of large molecules. These large molecules must be broken down to small ones before they can be absorbed. The process of breaking down large food molecules to small ones is called **digestion**.

Enzymes catalyse the breakdown of large food molecules

The breakdown of large food molecules to small ones is a chemical reaction. Like all reactions which take place in living things, the reactions of digestion are catalysed by enzymes. The enzymes which help in digestion have names ending with 'ase'. Enzymes which catalyse the digestion of carbohydrates are called **carbohydrases**. Enzymes which catalyse the digestion of proteins are called **proteases**. Enzymes which catalyse the digestion of fats are called **lipases**. 'Lipid' is another word for 'fat'.

Like all enzymes, digestive enzymes will only work well at particular temperatures and pH. All of your digestive enzymes have an optimum temperature of around 37–40 °C. But their optimum pHs vary. **Amylase**, a carbohydrase found in your saliva, works best in slightly alkaline conditions at a pH of about 7.5. **Pepsin**, a protease in your stomach, needs acidic conditions around pH 2. And the enzymes in your duodenum and ileum need slightly alkaline conditions. This is provided by sodium hydrogencarbonate, an alkali secreted in bile and pancreatic juice.

Fig. 103.1   How mammals deal with food

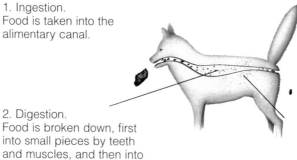

1. Ingestion.
Food is taken into the alimentary canal.

2. Digestion.
Food is broken down, first into small pieces by teeth and muscles, and then into small molecules by enzymes.

4. Egestion.
Any food which could not be broken down into small molecules is passed out of the body.

3. Absorption.
The small molecules pass out of the alimentary canal into the blood, which takes them all around the body.

INVESTIGATION 103.1

The absorption of carbohydrates from a model gut

Visking tubing, like the walls of your digestive system, will only let small molecules pass through it. You are going to give two pieces of visking tubing a 'meal' and see what happens.

1   Take two pieces of visking tubing. Wet them and open them out into tubes. Pushing a pencil inside may help. Tie one end of each, very tightly.

2   Three-quarters fill one piece of tubing with starch solution. Three-quarters fill the other piece with glucose solution. These are the 'meals'.

3   Tie the tops of both pieces of tubing tightly. Rinse them both to remove any spilt starch or glucose solution from their outsides.

4   Put each 'gut' into a beaker. Curl them up to lie in the bottom of their beakers. Pour in enough warm water just to cover each of them. Leave them for at least half an hour — longer if you can.

5   Now take samples of the water in each beaker. Test each sample for starch, and record your results. Test each sample for sugar and record your results.

Questions

1   Which part of your apparatus represented:
a the wall of the digestive system
b the blood?

2   Did you find starch in the water in either beaker? Explain your findings.

3   Did you find sugar in the water in either beaker? Explain your findings.

4   Why did you use *warm* water to cover your pieces of tubing?

5   Why is it best to use only enough water just to cover the pieces of tubing?

6   Starch molecules are made of long strings of glucose molecules, joined in a chain. What does this investigation suggest must happen to starch molecules before they can be absorbed into your blood from your digestive system?

Carbohydrate is digested in the mouth and small intestine

The main types of carbohydrate which you eat are cellulose, starch and sucrose. You cannot digest cellulose at all. Its large molecules cannot be broken down in your digestive system so you cannot absorb them. They pass straight through as fibre and leave your body in the faeces.

Starch, though, is easily digested. The process begins in your mouth. Saliva contains an enzyme called **amylase**, which begins to break down starch to maltose. You do not usually keep food in your mouth for long enough for your amylase to finish digesting any starch in it. When the food is swallowed and enters the stomach the amylase stops working. This is because it cannot work in acid conditions. So no carbohydrate digestion takes place in the stomach.

But when the food leaves the stomach carbohydrate digestion continues. Pancreatic juice flowing into the duodenum contains more amylase to finish changing starch into maltose. Another enzyme, called **maltase**, is found a little further along, in the ileum. Maltase digests maltose, breaking it up into glucose molecules. These are small enough to be absorbed through the ileum wall.

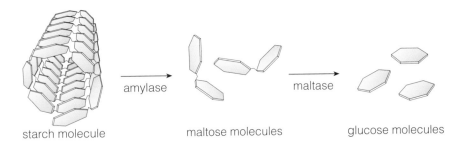
Fig. 103.2 Digestion of carbohydrates

starch molecule → amylase → maltose molecules → maltase → glucose molecules

Protein is digested in the stomach and small intestine

Large protein molecules are digested by proteases. The first time a mouthful of food meets a protease enzyme is when it reaches your stomach. Here the protease pepsin begins to break down the long protein molecules into shorter ones called **polypeptides**.

After a time in the stomach food is allowed through into the duodenum. Pancreatic juice squirts out on to it from the pancreatic duct. This contains a protease called trypsin. **Trypsin** begins to break down the polypeptides into even shorter chains.

The food quickly moves on into the ileum. Enzymes on the walls of the ileum finish breaking down the polypeptides into individual **amino acids**. The amino acid molecules are small enough to get through the wall of the ileum and into the blood capillaries.

Fats are digested in the small intestine

Fats are more difficult to digest than proteins or carbohydrates. This is because they are not soluble in water. They form globules into which enzymes cannot penetrate.

So before enzymes can digest fats the fat globules must be broken up. This is done by **bile salts**. Bile salts are detergent-like substances contained in bile. They emulsify fats, breaking up big globules into very tiny ones. The tiny globules mix in with the digestive juices and can be attacked by enzymes.

The enzyme which digests fats is called **lipase**. There is lipase in pancreatic juice and also on the walls of the ileum. So all fat digestion happens in the small intestine. Lipase breaks up fat molecules into fatty acids and glycerol. These are then absorbed through the walls of the ileum.

Fig. 103.4 Digestion of fat

Fig. 103.3 Digestion of proteins

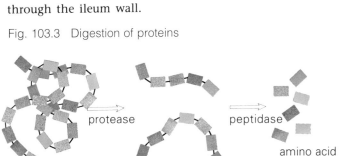

protein molecule → protease → polypeptide molecules → peptidase → amino acid molecules

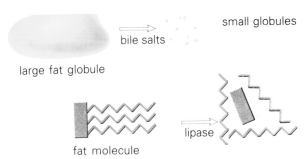

large fat globule → bile salts → small globules

fat molecule → lipase → fatty acid and glycerol molecules

Questions

1 Copy the following chart. In the first three columns, put a tick in the correct line or lines. In the last column write 'acidic' or 'alkaline' in each line.

	carbohydrate digested	protein digested	fat digested	acidic or alkaline
mouth				
stomach				
duodenum				
ileum				

2 a Why do water, vitamins and minerals not need digesting?
  b What type of food does lipase digest?
  c Why is this type of food especially difficult to digest?
  d How does bile help in the digestion of this type of food?

Teeth speed up digestion by increasing the surface area of food particles.

Teeth and muscles help in mechanical digestion

Digestion is the breaking up of large food molecules into smaller ones, so that they can be absorbed through the walls of the alimentary canal. The reactions of digestion are speeded up by enzymes which catalyse the reactions.

The reactions are also speeded up by teeth and muscles. Teeth break up large particles of food increasing their surface area. This allows enzymes to come into contact with the food molecules more quickly. The reactions of digestion are therefore speeded up. Stomach muscles have a similar effect. When your stomach contains food, the muscles in its walls make churning movements. This mixes the food with the enzymes and acid. It also helps to break down lumps of food. The food becomes a liquid mass called **chyme**.

Teeth contain living cells

Teeth are alive. They contain living cells. These cells are supplied with food and oxygen by blood vessels. The blood vessels and also nerves are found in a soft, central area in the tooth. This is called the **pulp cavity**.

Surrounding the pulp cavity is a layer of bone-like material. It is called **dentine**. Dentine, like bone, contains living cells. The cells are embedded in a background material, or **matrix**, containing calcium salts. These make the dentine hard.

The dentine is covered by an even harder material called **enamel**. Enamel is the hardest substance in your body. It is very difficult to scratch the surface of enamel. But it can be cracked and it can be dissolved by acids. Only the part of the tooth which shows above the gum is covered by enamel. This part is called the **crown** of the tooth.

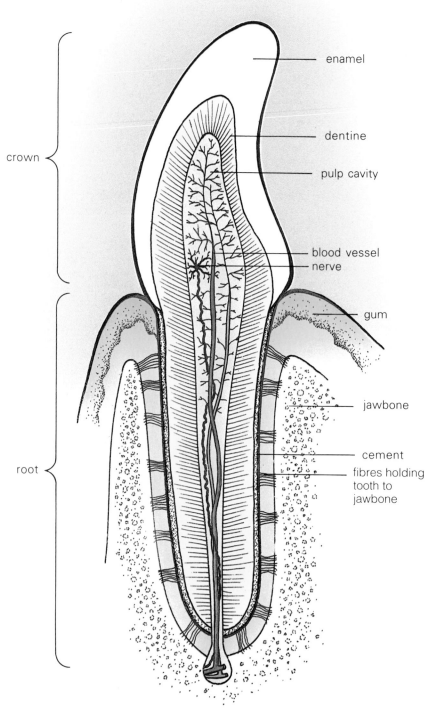

Fig. 104.1 Vertical section through a human incisor

- crown
- enamel
- dentine
- pulp cavity
- blood vessel
- nerve
- gum
- jawbone
- cement
- fibres holding tooth to jawbone
- root

The part which is buried in the gum and jaw bone is called the **root**. The root is covered by **cement**. Cement, like enamel, is very hard.

Connecting the root to the jaw bone are thousands of **fibres**. These are slightly stretchy. If you bite on something very hard, or fall and bang a tooth, these fibres can give a little. They allow your tooth to move slightly without damaging anything.

Tooth decay is caused by acids

Enamel is the hardest substance in your body. It forms an excellent protective covering over your teeth. But it is easily damaged by acids. Acids cause tooth decay. Acids are formed on your teeth by bacteria. Everyone has bacteria living in their mouth. They are not normally harmful. But if you leave sugary foods on your teeth the bacteria will feed on them. The bacteria form a sticky covering called **plaque** on your teeth. As they feed on the sugar, the bacteria produce acid. The acid dissolves the enamel.

If you have plenty of fluoride in your body your teeth are better able to resist this attack by acid. Fluoride toothpastes help a lot. Another way of getting fluoride is in drinking water.

But if the acid *does* get through the enamel the decay spreads quickly. Dentine is much softer than enamel, so the acid dissolves it faster. The tooth may begin to be painful now. Dentine contains living cells, and they are sensitive.

Gum disease is also caused by bacteria

Bacteria often collect along the edges of your teeth where they meet the gum. These bacteria may work their way down between the tooth and the

Fig. 104.2   Tooth decay

1. Bacteria and food are trapped in a groove on or between teeth. The bacteria break down the food, releasing acids. The acids begin to dissolve the enamel.

2. When the acid breaks through to the dentine, the decay spreads faster. Dentine is softer than enamel, and contains living cells.

3. If the decay is not treated, it will probably reach the pulp cavity. This is very painful.

4. Once into the pulp cavity, bacteria can get into the blood vessels. They may cause an infection at the base of the tooth. An abscess forms, which fills with pus.

gum. They can damage the gum and the fibres holding the tooth in position. Over a long period of time the tooth may become loose and may even fall out.

Brushing teeth regularly to remove bacteria from your teeth can help to prevent tooth decay and also gum disease. It is best to use fluoride toothpaste. A regular visit to a dentist can also help to spot problems early so that something can be done about them before it is too late. Making sure that your teeth are not covered with sugar for long periods of time will also help to ensure that you do not lose teeth.

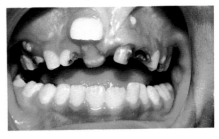

Fig. 104.3   This eight-year-old has very bad tooth decay in his upper jaw. The large tooth at the top is an adult incisor coming through, but it is growing crookedly because of the decay.

Untreated decay may cause an abcess to form

A decaying tooth can be treated with a **filling**. A dentist can drill away all the decayed part, and replace it with a filling material. But if this is not done the decay may spread to the pulp cavity. Bacteria can now get from the surface of the tooth right into the blood vessels and nerves. The tooth becomes very painful. If the bacteria infect it badly, an **abscess** may form. This is an extremely painful swelling, full of bacteria. White blood cells fight the bacteria. The mixture of living and dead white cells and bacteria is called **pus**.

Dentists will not normally take out an abscessed tooth, even though it is very painful. They are worried that if they pull the tooth out the bacteria in the abscess may get into your blood. If this happens they may make you very ill indeed. Instead, the dentist will prescribe antibiotics to kill the bacteria. When the bacteria in the abscess are killed the tooth can safely be removed.

INVESTIGATION 104.1

Human teeth

1   Use a mirror to look at your own teeth. Identify your **incisors, canines, premolars** and **molars**. Count and record how many of each you have on your top jaw. Repeat for your bottom jaw.

2   The full number of teeth in an adult human is 32. The last four teeth to appear are the back molars. They are sometimes called wisdom teeth. Do you have your wisdom teeth yet? Do you have any other teeth missing?

3   Think of how you use your teeth when you eat an apple. (Better still, eat an apple and think hard about how you are using your teeth.) Describe how you use each of your four kinds of teeth.

4   Make two copies of the diagram below. One is for the top jaw, and the other for the lower jaw. Look again at your own teeth. If you are missing any of the teeth on your diagram shade them in. If any of your teeth have fillings put a cross on those teeth on the diagram.

5   Collect the class results for question 4. You will need to design a results chart to do this. Decide on a good way of displaying these results as a graph or chart.

6   Do any particular teeth tend to be missing in your class? If so, can you suggest why this might be?

7   Do any particular teeth in your class tend to have fillings? If so, can you suggest why this might be?

Fig. 104.4

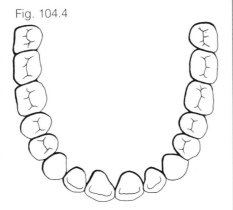

Food molecules move through the walls of the digestive system into the blood. This is called absorption.

Small food molecules are absorbed in the ileum

By the time that food reaches the ileum, digestion is almost complete. Protein molecules have been broken down to amino acids. Starch molecules have been broken down to glucose. Fats have been broken down to fatty acids and glycerol.

These small molecules can now get through the walls of the digestive system into the blood. This is **absorption**. Absorption of digested food happens in the ileum.

Fig. 105.1 A villus. There are thousands upon thousands of these lining your small intestine. They are about 1 mm tall. They contain muscles which shorten and lengthen them rhythmically during digestion, helping to bring their surfaces into contact with fresh supplies of food.

The inside of the ileum has a very large surface area

The ileum is the longest part of your alimentary canal. It may be 5 m long. So it takes a long time for food to pass through it. This gives plenty of time for food to be absorbed into the blood. The ileum is also one of the narrowest parts of the alimentary canal. So food inside it is always quite near to the walls. This makes it easier for the food molecules to pass through the walls into the blood.

The inner surface of the ileum is thrown into folds. These folds increase the surface area. The larger the surface area, the faster absorption can take place. On the folds are thousands of finger-like projections, called **villi**. Villi are about 1 mm high. The villi also help to increase the surface area of the inside of the ileum. Even the villi have little projections on them, called microvilli!

Inside each villus is a blood capillary. The blood capillary absorbs amino acids and glucose. It leads into the **hepatic portal vein**. This vein takes the absorbed food to the liver. Each villus also contains a lacteal. This is a branch of the lymphatic system. The lacteal absorbs fatty acids and glycerol. It leads into a larger lymph vessel which empties its contents into a vein near the heart.

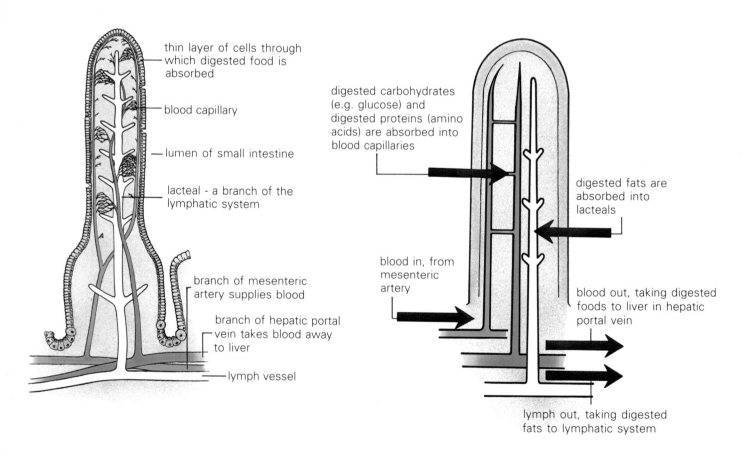

thin layer of cells through which digested food is absorbed

blood capillary

lumen of small intestine

lacteal - a branch of the lymphatic system

branch of mesenteric artery supplies blood

branch of hepatic portal vein takes blood away to liver

lymph vessel

digested carbohydrates (e.g. glucose) and digested proteins (amino acids) are absorbed into blood capillaries

digested fats are absorbed into lacteals

blood in, from mesenteric artery

blood out, taking digested foods to liver in hepatic portal vein

lymph out, taking digested fats to lymphatic system

The colon absorbs water and inorganic ions

When the digested food has been absorbed in the ileum, there is still quite a lot left inside the alimentary canal. This includes water, inorganic (mineral) ions, indigestible food such as cellulose, bacteria, mucus from the lining of the digestive system, and old cells which have worn away from its surface. This mixture passes on into the colon.

In the colon, water is absorbed from the food. The colon also absorbs inorganic ions such as sodium and chloride.

The contents of the colon move on into the rectum. Here they are formed into **faeces**. There is a muscle at the anus called the anal sphincter. Normally this is contracted. When the anal sphincter muscle relaxes, the faeces pass out through the anus.

The food absorbed in the ileum is taken to the liver

The carbohydrates and amino acids that are absorbed into the blood capillaries in the ileum, dissolve in the blood plasma. They are taken in the hepatic portal vein to the liver.

One of the functions of the liver is to regulate (control) the amounts of carbohydrate and amino acids which are allowed into the bloodstream. If you have eaten a meal containing a large amount of carbohydrate, a large amount of glucose will arrive at the liver in the hepatic portal vein. The liver will convert some of this glucose into a polysaccharide called **glycogen**. The glycogen is stored in the liver. Only the right amount of glucose will be allowed out of the liver to be delivered to the body cells.

In the same way, the liver will only allow the right amount of amino acids to be carried to the body cells in the blood. Any extra ones are changed into **urea**. The urea is passed into the blood and taken to the kidneys to be excreted.

Fig. 105.2 A child being given a mixture of glucose, salt and water. This is called 'oral rehydration'. It is a very effective and simple way of replacing fluid lost through illnesses such as diarrhoea.

Assimilation is the use of absorbed food by body cells

The food absorbed by the ileum is eventually used by various cells in the body. The use of this food is called **assimilation**.

Carbohydrate is carried in the blood in the form of **glucose**. Cells take up glucose through their cell surface membranes. They then use the glucose for respiration to produce energy. All cells need a constant supply of glucose.

Some cells can store glucose. They convert it into the polysaccharide **glycogen**. Muscle cells and liver cells are two examples of cells which do this. Excess carbohydrate will be converted into **fat**. This is stored in several places in the body. Much of it is stored under the skin.

Amino acids are also carried to the cells by the blood. They are taken up by the cells through their cell surface membranes and used for building proteins. The DNA in the nucleus of each cell gives instructions about the order in which the amino acids should be joined together. Different orders of amino acids make different proteins. Each cell only follows some of the DNA instructions so each cell only makes certain proteins. Cells in your salivary glands, for example, make the protein amylase. Cells in your hair follicles make the protein keratin.

Questions

1 List the features of the ileum which make it good at absorbing digested food.

2 Match each of the following parts of the alimentary canal with its functions:

Parts:
mouth, oesophagus, stomach, duodenum, ileum, colon, rectum.

Functions:
- passes food to stomach
- absorbs digested food
- begins the digestion of starch by amylase
- absorbs water
- produces hydrochloric acid
- forms undigested food material into faeces
- continues the digestion of starch by amylase
- receives secretions from the gall bladder and pancreas
- begins the digestion of protein by proteases

DID YOU KNOW?

About 500 ml of material enters your colon every day but only about 100 ml is lost in the faeces. The remaining 400 ml is water which the colon absorbs. If you have diarrhoea, the colon does not absorb this water. Diarrhoea can quickly lead to dehydration and is a major cause of death in many developing countries.

Questions

1 Read the following passage, and then answer the questions.

Gastric secretion in humans

The stomach walls secrete a liquid called gastric juice. Gastric juice contains water, hydrochloric acid and the protease pepsin.

Some of the earliest studies on the secretion of gastric juice were performed on people who had been wounded. One of the most famous was Alexis St. Martin, who was wounded by duck shot which made a hole into his stomach in about 1825. The hole never completely closed so that the contents and behaviour of his stomach could be investigated. Another famous patient was an American boy called Tom. Tom swallowed boiling hot clam chowder when he was nine years old, damaging his oesophagus so badly that it closed right over. A hole had to be made directly into his stomach so that he could be fed.

Studies on these people, and more recent experiments, have shown that the secretion of gastric juices depends on all sorts of different factors. It is not surprising that gastric secretion speeds up when a person smells, sees or tastes food. The gastric juices are produced so that they are ready to digest the expected food when it arrives. The presence of food in the stomach also causes gastric juice to be secreted. Food containing protein has a greater effect than pure fat or carbohydrate.

It was also found that gastric juices are secreted when someone is excited or angry. This is thought to be a possible cause of stomach ulcers. In contrast, a depressed, frightened person has very little gastric secretion. Another factor which reduces the secretion of gastric juices is the presence of food in the duodenum.

a What is gastric juice?

b What is 'secretion'?

c In what way did Alexis St. Martin's and Tom's injuries enable research to be done on gastric secretion?

d List three factors which speed up the rate of gastric secretion.

e Why is it sensible for protein-containing foods to have the greatest effect on gastric secretion?

f It is well-known that people who suffer from stress are more likely to suffer from stomach and duodenal ulcers. Explain why this might be.

g Why is it useful for the presence of food in the duodenum to reduce gastric secretion?

2 The following information appeared on the lid of an ice-cream carton.

Ingredients: milk, double cream, sugar, eggs, emulsifier (E471), stabilisers (guar gum, locust bean gum), natural flavouring, malt extract, natural colour (betanin).

Nutritional information;

	typical values per 100 g:
Energy	479 kJ
Protein:	2.1 g
Carbohydrate:	12.7 g
Fat:	6.4 g

a What percentage by weight of the ice cream is protein?

b Which ingredients in the ice cream do you think provide this protein?

c What percentage by weight of the ice cream is probably water?

d A boy had a 250 g serving of ice cream. How much fat did he eat?

e Which ingredients in the ice cream do you think provide this fat?

f The boy needs about 12 000 kJ each day. How much ice cream would he need to eat to provide this much energy?

g Why would this not provide a balanced diet?

h What do you think is the purpose of the emulsifier in the ice cream?

3 A sample of food was tested to find out which food types it contained. The results of the tests are given below.

Test	Colour obtained
A sugar test	blue
B fat test	translucent mark which disappeared when dried
C protein test	purple

a Which food types were present in this food sample?

b What food could it have been? Choose from: lettuce, beef, egg white, sultana, butter.

c Which test would involve heating the food?

d What reagents would be used for Test C?

4 Explain why:

a Taking milk of magnesia tablets, which contain magnesium hydroxide, can help in relieving the pain of indigestion.

b People suffering from diarrhoea should drink plenty of liquid, preferably containing some sugar and salt.

c The lining of the ileum is covered with thousands of villi.

d Both fungi and animals are described as being heterotrophic.

e Lack of iron in your diet makes you tired and listless.

f You should try not to eat too much animal fat.

g Vegetables should not be overcooked.

WAVES

106 **VIBRATIONS**

Many things vibrate when disturbed. Small objects tend to vibrate more quickly than large ones. The number of vibrations in one second is the frequency.

Frequency is the number of vibrations per second

If you twang a ruler against a desk, it vibrates or **oscillates**. A short ruler vibrates more quickly than a long one.

The **frequency** of vibration is the number of vibrations or oscillations in one second.

Frequency is measured in hertz. The abbreviation for hertz is **Hz**. A frequency of 1 Hz is one vibration per second. If the ruler vibrates 100 times in one second, the frequency is 100 Hz.

$$\text{Frequency in Hz} = \frac{\text{Number of oscillations or vibrations}}{\text{time taken in seconds}}$$

If the ruler vibrates 300 times in 2 s, the frequency is:

$$\frac{300}{2} = 150 \text{ Hz}$$

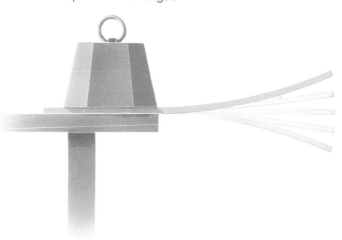

Fig. 106.1   The frequency at which a 'twanged' ruler oscillates depends on its length.

Period is the time taken for one oscillation

If a pendulum swings twice in 1 s, the frequency is 2 Hz. The time taken for one swing is $\frac{1}{2}$ s. This is called the **period.** The period is the length of time it takes to make one oscillation. One oscillation is a swing from one side to the other and **back** to the starting position.

one oscillation

Vibrations can be used to indicate faults in machines

In a complex machine there are many sources of vibration. An engineer who knows what vibrations a machine should produce can predict a failure before it happens. If a study of the machine shows an unusual vibration developing, this could indicate a fault. Regular checks of vibration levels mean that servicing can be carried out less frequently. This wastes less time.

INVESTIGATION 106.1

Investigating a pendulum

1   Set up the apparatus as shown in the diagram. Investigate how the *length* of the string affects the *frequency* of oscillation. (The length of the string should be measured from the point of suspension to the centre of the bob.) Record your results fully. Look for a pattern in them.

2   Using the same apparatus, but with the ruler placed horizontally, investigate how the *size* of the oscillation affects the *frequency*.

Questions

1   A pendulum oscillates twice in one second. If the string is made four times as long, what will be the new frequency of oscillation?

2   What length of pendulum would you use to produce a frequency of 1 Hz?

3   Could you use this for timing something?

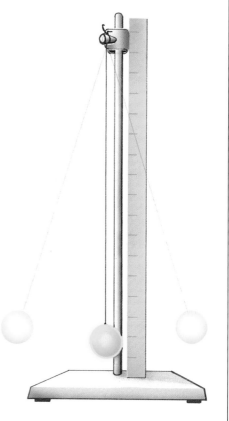

Fig. 106.2   Use this arrangement for Step 1. For Step 2, place the ruler horizontally.

Vibration causes settling

A pile of sand will quickly settle down if you shake what it stands on. When gravel is laid on driveways, it is made to settle by shaking it. You cannot simply shake the whole ground! Instead, a vibrating weight is dragged over the loose gravel. This flattens it, and causes the stones to fit tightly together.

Settling caused by vibration during transport can mean that packets filled in a factory never seem full when you open them at home. Cereal packets usually have an explanation of this written on them.

Vibrations are also used to make sure that concrete has no spaces in it. To make reinforced concrete, wet concrete is poured into moulds containing steel rods. It is important that the concrete should flow around all the rods. A vibrating rod is pushed into the concrete once it has been poured. The rod vibrates at 200 Hz. It is usually driven by a petrol engine, or compressed air. The vibration helps the concrete to flow around the rods.

Fig. 106.4 Vibrations are used to help reinforced concrete to settle without leaving air gaps.

Fig. 106.5 A woman being treated for kidney stones, using ultrasound.
At high sound intensities, ultrasound can be used to break up stones and weld plastic. At lower intensities, the vibrations can be used to shake dirt out of complex shapes like fine jewellery.

Strong vibrations can cause damage

Vibrations can be used to break up objects. A hammer drill is used to drill holes in materials like brick, stone or concrete. The drill not only goes round, but also vibrates backwards and forwards at up to 700 Hz. These vibrations help to break up the material. The drill oscillates 14 times for each revolution. Pneumatic drills do not revolve at all. They rely only on vibration to break up the material.

Kidney stones are hard deposits which can build up inside a person's kidneys. They can be painful and dangerous and are sometimes removed by surgery. Another treatment is to direct ultrasonic (very fast) vibrations at them. This breaks up the stones into tiny pieces which can pass out of the kidney in the urine.

Fig. 106.3 A hammer drill rotates and vibrates backwards and forwards.

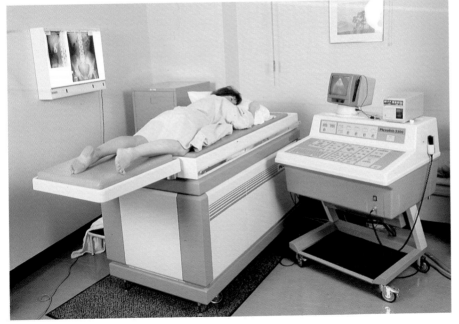

Questions

1 A tall factory chimney sways in the wind. It swings backwards and forwards eight times in 2 s.
   a What is the frequency of oscillation in Hertz?
   b What is the period of oscillation?

————— E X T E N S I O N —

2 The mains electricity delivered to a house in Britain has a frequency of 50 Hz.
   a How many times does the electricity oscillate in 1 s?
   b In America the frequency is 60 Hz. How much longer does one oscillation take in Britain than in America?

A wave carries energy from one place to another. Waves can be mechanical or electromagnetic. A wave can be transverse or longitudinal.

A wave does not carry material with it

If a rope or 'Slinky' spring is shaken from side to side, a series of pulses travels down it. Each point on the rope moves up and down as a wave passes. A single wave pulse makes this easier to see.

The rope or spring is *not* carried along with the wave! After the wave has passed, the rope or spring is still there. What *is* being carried along is **energy**. Waves transfer energy from one place to another.

A wave in a rope or spring is a **mechanical wave**. Mechanical waves disturb material. The particles in the material oscillate to and fro. The oscillating particles transfer energy between themselves. The particles could be in a solid, liquid, or gas.

Fig. 107.1 A wave passing along a rope. The rope itself does not move along; only energy moves in the direction of the wave's travel. This is a transverse wave, in which the direction of oscillation (up and down) is at right angles to the direction of the wave's movement (along).

In a transverse wave, particles move at right angles to the wave

When you send a wave along a rope or spring, you are producing a **transverse wave**. The particles oscillate at right angles to the direction in which the wave is travelling. The wave is travelling *along* the rope, while the particles are moving *up and down*. As the wave passes, each particle moves away from its rest position and then back again. The particles do not move along in the direction of the wave.

In a longitudinal wave, the particles move in the same direction as the wave

If you push a 'Slinky' spring, you can make a single pulse travel along it. A small piece of brightly coloured string makes the movement of a single coil clearer. As the pulse passes, the coils oscillate backwards and forwards along the length of the spring. The coils return to their original position. A ripple passes down the spring, but the individual coils are left where they were in the first place. This is an example of a **longitudinal wave**. In a longitudinal wave the particles oscillate *along* the direction in which the wave is travelling.

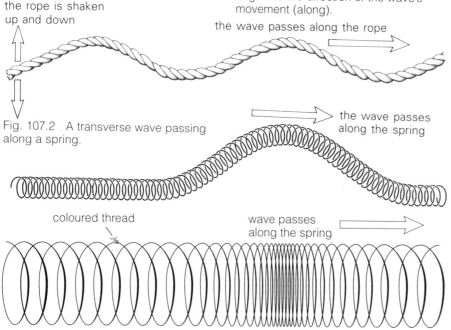

the rope is shaken up and down

the wave passes along the rope

Fig. 107.2 A transverse wave passing along a spring.

the wave passes along the spring

coloured thread

wave passes along the spring

Fig. 107.3 A single pulse passing along a spring. The spring as a whole does not move along, but individual coils move back and forth. This is a longitudinal wave, in which the direction of oscillation (back and forth) is in the same direction as the movement of the wave (along).

Sound is a longitudinal wave

If you push a spring backwards and forwards, you can send a series of pulses down its length. In some parts of the spring the coils are closer together. These areas are **compressions**. As the coils oscillate, the compressions move along the spring. Again, a piece of coloured string may make it easier to see what is happening. Between the

compressions are parts of the spring which are stretched out. These are called **rarefactions**.

Sound is a longitudinal wave. The compressions and rarefactions are regions of high and low pressure. A sound wave moves particles of

materials, so it is a mechanical wave. Sound can only travel when there are particles which can be compressed and rarefied. So sound cannot pass through a vacuum. A sound wave is an example of a longitudinal, mechanical wave.

compression   rarefaction

Fig. 107.4   wave passes along the spring

Electromagnetic waves are not disturbances of particles

Not all waves are mechanical waves. An **electromagnetic wave** is not a disturbance of particles. It is varying electric and magnetic fields. It does not need any particles to pass the energy on. Electromagnetic waves travel more easily when there are no particles. Some examples of electromagnetic waves include X rays, gamma rays, microwaves, light waves and radio waves.

Fig. 107.5 Wavelength and amplitude of a transverse wave. Wavelength is the distance between two crests or two troughs. Amplitude is the height of a crest or depth of a trough from the rest position.

Measuring waves

If you took a photograph of a wave on a rope, or ripples spreading across a pond, it might look something like Figure 107.5. The centre line shows the undisturbed rope, or level of water in the pond. The top of the hump in the water or rope is called a **crest**. The bottom of a dip is a **trough**.

The distance between the peaks of two crests is the **wavelength** of the wave. The maximum distance that a particle moves away from the centre line is the **amplitude** of the wave.

The amplitude is the height of a crest above the centre line.

The number of waves produced per second is called the **frequency**. If you shake your hand backwards and forwards twice per second, the frequency is 2 Hz. You are producing two waves in 1 s, so the wave frequency is 2 Hz.

These measurements also work for longitudinal waves. For these the wavelength is the distance between two compressions.

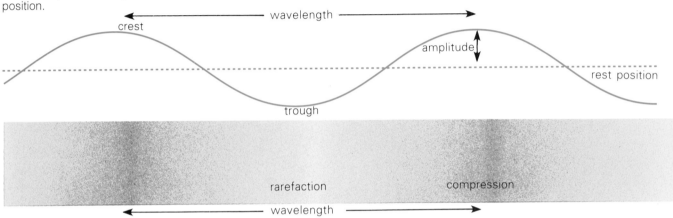

Fig. 107.6 Wavelength and amplitude of a longitudinal wave. Wavelength can be measured as the distance between the centre of two compressions. Amplitude is the maximum distance moved by a particle from its rest position.

INVESTIGATION 107.1

Standing waves

The apparatus shown in the photograph can be used to produce **standing waves**. If you adjust the frequency of the vibrations produced by the electromagnetic vibrator, you will find that you get especially large vibrations at some frequencies. A wave pattern is produced which does not travel down the string.

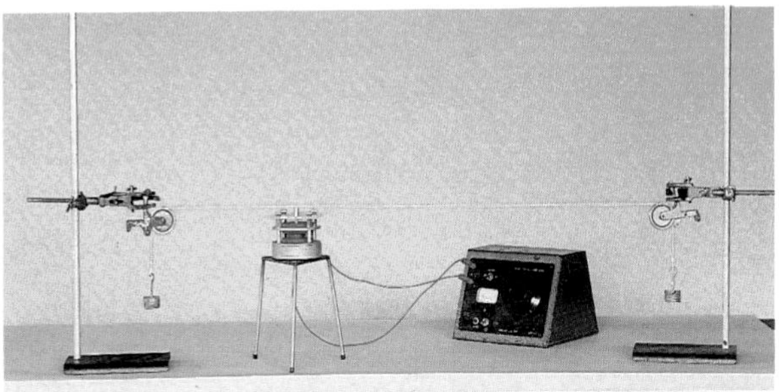

Fig. 107.7 Apparatus for investigating standing waves. This arrangement is using a power pack, which produces vibrations of 50 Hz. If you use a variable frequency oscillator instead, you can investigate the effects produced by many different frequencies.

This is called a standing wave.

A standing wave stores energy in the string. The wave travelling down the string is reflected at the end, and comes back. It combines with the wave from the vibrator to form the standing wave. Standing waves form on the strings of instruments, such as violins, when the string vibrates.

Try the following experiments with this apparatus.

1 Vary the frequency. At what frequencies are standing waves formed on a 1 m length of string? What do you notice about these frequencies?

2 Vary the length of the string. You could begin by trying a 0.5 m length of string. Find the frequencies which produce standing waves again. Can you see a pattern linking the frequencies which produce standing waves in the 1 m and 0.5 m lengths of string?

3 If you have time you could also experiment with different kinds of string, and different weights.

Different types of waves travel at different speeds. Reflected waves are used by humans and other animals to locate objects and measure distances.

INVESTIGATION 108.1

Measuring the speed of sound in air

This experiment will be demonstrated for you, as a starting pistol can be dangerous.

1 Measure the length of your school field in metres, and record it. Mark the two points between which you have measured.
2 Send someone to one of your marked points with a starting pistol. A second person should stand at the other marked point, holding a stopwatch.
3 When everyone is ready, the starting pistol is fired. As soon as the person with the stop watch *sees the smoke* from the pistol, he or she starts the watch. When the person *hears the bang*, he or she stops the watch. Record the time between seeing the smoke and hearing the bang.

You will probably need to practice this a few times before you feel confident that you are getting it right. Collect three sets of reliable results.

Questions

1 Speed is the distance covered per second. In this case, the speed of sound is:

$$\frac{\text{distance travelled}}{\text{time taken}} \quad \text{or}$$

$$\frac{\text{length of field in metres}}{\text{time delay in seconds}}$$

Using an average of your three readings calculate the speed of sound in air.

2 You are assuming that you see the smoke from the pistol immediately.
Light travels one million times faster than sound. Is this assumption a reasonable one?
3 Does your reaction time in starting and stopping the watch produce a significant error in this experiment?
4 How could you reduce the errors in this experiment?

Sound can bounce off surfaces and cause echoes

If you shout at a wall from 340 m away, the sound takes 1 s to reach the wall. The sound reflects from the wall, and takes 1 s to return. So you hear the echo 2 s after you shouted.

The time it takes for an echo to return can be used to find how far away something is. This is how **sonar** works. A ship uses sonar to measure the depth of the seabed below it. Pulses of sound are sent downwards, and the time for them to return is measured. Sound travels at 1500 m/s in water, so the echoes return much sooner than in air. Fishing boats also use sonar to detect shoals of fish in the water below them.

Dolphins use a similar system for locating fish in murky water. They produce ultrasonic waves which bounce off the fish. This is called **echolocation**. Piranha fish use echolocation, and so do bats, rats and some birds. Ultrasonic waves are sound waves of a frequency too high for humans to hear.

You can use an ultrasonic 'tape measure' to measure the size of a room. It sends out ultrasonic waves which bounce off the walls. The time taken for the echo to return gives the distance to the reflecting surface. Some autofocusing cameras use the same method to find the distance between the camera and the object.

In air, over large distances, **radar** is used instead of sound waves. Radar uses the same method, but sound waves are replaced with radio waves. Radio waves travel at the speed of light which is 300 000 000 m/s. An echo from an object 300 km away would take 2 milliseconds (ms) to return.
(1 ms is $\frac{1}{1000}$ s.)

Fig. 108.1

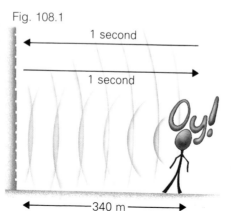

1 second

1 second

Oy!

340 m

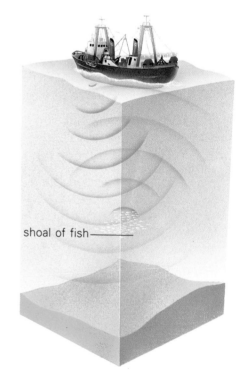

shoal of fish

Fig. 108.2 A sonar system sends ultrasonic waves down into the water, and measures the time taken for the echo to return. Fishing boats can locate shoals of fish in this way.

Fig. 108.3 Bats navigate, and find food, by echolocation. They emit ultrasound squeaks, and pick up the echoes. This horseshoe bat has particularly large ears to pick up the returning vibrations. The closer an object is, the faster the sound bounces back to the bat.

Fig. 108.4 Air traffic control systems use radar. The tower emits radio waves, which travel at 300 000 000 m/s. They are reflected from aircraft, and picked up by receivers in the tower. The time taken for the signal to return can be used to calculate the distance of the aircraft. The signals are fed to a screen, which plots the aircraft's position.

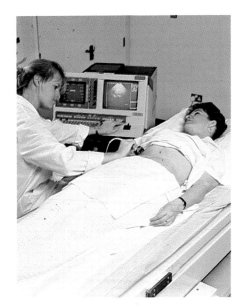

Fig 108.5 An ultrasound scanner beams sound waves into a person's body. Echoes are created when the sound meets a different surface, for example the stomach. The echoes can be used to give an image on a screen. This is thought to be much safer than using X rays to 'see' inside a person's body. Ultrasound scans are used to show the baby inside the uterus. Exactly the same scanning techniques can be used to detect faults in industrial products. A pipe can be tested for leaks by producing ultrasound inside it and seeing where it escapes.

Questions

1 Thunder and lightning occur at the same time. Lightning strikes 1 km away. If you see the flash immediately, how long do you have to wait before your hear the thunder? (Take the speed of sound as 333 m/s.)

2 a An ultrasonic tape measure in a room sends out a sound wave with a frequency of 40 kHz. A returning wave is detected 0.02 s later. How long is the room?

(Assume that the speed of sound is 300 m/s.)

b In a second room, the measure reads 1.5 m. The estate agent thinks that the measure has broken. His scientific customer has another explanation. What might it be?

3 A light beam is bounced off the Moon. It returns 2.5 s later. If the velocity of light is 300 000 000 m/s, how far away is the Moon?

4 A dolphin produces ultrasonic waves. From his first pulse he receives a small echo $1/100$ s later, and another, larger, echo 1 s later. A second pulse from the dolphin produces echoes at $1/100$ s and 2 s later.

a What does this tell the dolphin about the fish and the killer whale?

b On both occasions the dolphin receives an echo at $1/10$ s. Where does this echo come from?

MORE ABOUT WAVE SPEED

Waves travel at different speeds in different materials. Some kinds of waves, such as sound waves, are also affected by temperature.

Waves travel at different speeds in different materials

The speed at which waves travel may be affected by the material through which they are travelling. Ocean waves, for example, travel faster in deep water than in shallow water. As a wave approaches shore, the water gets shallower. The wave slows in the shallow water and gets higher. Eventually the wave collapses and breaks.

The speed of sound is affected by temperature, the type of material through which it is moving, and the density of this material. The speed of sound in air is 331.46 m/s at 0 °C. When the air is warmer, sound travels faster. If you measure the speed of sound on a warm day, you should get an answer of around 340 m/s. The speed also rises as air density falls.

In water, the speed of sound is about 1500 m/s. In aluminium, sound waves travel at 6400 m/s. In steel, the speed is 6000 m/s.

Light travels much faster than this. Nothing travels faster than the speed of light. Light travels at 300 000 000 m/s in a vacuum.

Fig. 109.1 Ocean waves slow down as the water gets shallower. Their wavelength gets shorter, and their amplitude gets larger. Beyond a certain height, the top of the wave 'falls over', or breaks.

Fig. 109.2 Velocity = frequency x wavelength.
The velocity of something is the distance it travels per second. If the rod vibrates at a frequency of 5 Hz, it produces five ripples per second. If each ripple has a wavelength of 2 cm, these five ripples cover a distance of 5 x 2 = 10 cm. This is the distance covered per second = the velocity of the waves.
Velocity = f x λ.

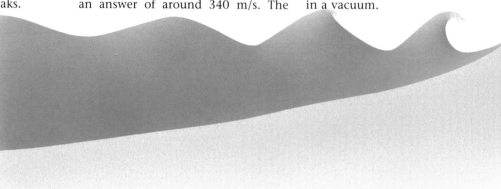

rod vibrates f times per second, so f ripples are produced each second

wavelength

The wave equation

If a vibrating rod is dipped into a pond, ripples spread out from the rod across the surface.

If the frequency of vibration is **f**, then in 1 s there will have been f vibrations. Each vibration produces a ripple. In 1 s, f ripples will have been produced.

The distance between crests is the wavelength, λ. λ is a Greek letter called lambda.

So in 1 s, f ripples, each separated by distance λ, have spread out from the rod. The total distance covered is **f x λ**.

The distance covered in 1 s is the velocity of the wave. The velocity of a wave = frequency x wavelength.

$$v = f \times \lambda.$$

If a sound wave of frequency 330 Hz has a wavelength of 1 m, the velocity is 330 Hz × 1 m = 330 m/s.

This wave equation applies to all waves.

What happens if a wave slows down? Velocity, v, gets smaller. Normally, frequency, f, does not change. This means that wavelength, λ must get smaller. So when a wave slows down its wavelength becomes smaller.

Question

When people are working on a railway line, it is important that they should have as much warning as possible of an approaching train. If a team of workers was supplied with a microphone and detector, where should they place it to give the earliest warning?

EXTENSION

Sonic booms are produced when aeroplanes break the sound barrier

Imagine Concorde travelling at the speed of sound. If the plane is at A, 660 m from the observer, the sound from the engines takes 2 s to reach him. One second later, both the sound and Concorde will reach point B, 330 m from the observer. The pressure waves from A and B will both take 1 s to reach the observer, and arrive together. The noise will be tremendous.

At speeds above the speed of sound, a cone of superimposed pressure waves reaches the ground.

All that sound energy arriving at once creates the 'sonic boom', which can break chimney pots and windows.

This problem has limited Concorde's flight routes. Many countries do not want Concorde to fly over them. Most of Concorde's flight paths lie over water to prevent damage and nuisance from sonic booms.

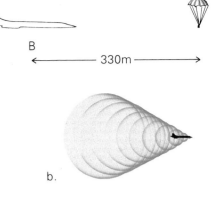

Fig. 109.3 A sonic boom is produced when an aircraft flies faster than the speed of sound. A plane flying very fast creates pressure waves at surfaces which hit the air, such as the nose and the wings. These pressure waves spread out in all directions at the speed of sound. If the plane is flying at the speed of sound, then the waves become concentrated at these points, and combine to produce a very loud noise.

If the plane is flying faster than sound, as shown in (b), then the pressure waves are left behind. As they spread outwards from the points at which they were formed along the plane's flightpath, they become concentrated at the edge of a cone. This reaches right down to the ground, where we hear a sonic boom.

INVESTIGATION 109.1

How is the speed of a wave affected by the depth of water through which it travels?

1 Read through the experiment, and draw up a suitable results chart.
2 Take a shallow rectangular tray and pour water into it to a depth of a few millimetres. Measure the length of the inside of the tray. Measure the depth of the water.
3 Raise one end of the tray 1 cm, and then gently drop it. Observe the ripple which moves across the water. Use a stopwatch to measure the time it takes to travel from one end to the other, and then back again. Repeat this several times, and work out an average reading.
4 Calculate the speed at which the ripple was travelling.

Speed (cm/s) =

$$\frac{\text{distance travelled (twice length of tray, cm)}}{\text{average time taken (s)}}$$

5 Put more water into the tray, and measure the new depth. Repeat steps **3** and **4**.
6 Continue adding more water, and repeating steps **3** and **4**, until you have measurements for at least five different depths of water.

Questions

1 How does the depth of water affect the speed of the wave?
2 What would you expect to happen to the speed of an ocean wave as it approached the shore?
3 Is the wave that you produced a longitudinal wave or a transverse wave?
4 Do you think that your results are really accurate? Discuss any sources of error in your experiment.

Questions

1 a Complete the following table. MHz is short for megahertz, which is 1 000 000 Hz.

velocity m/s	frequency	wavelength	radio station
300 000 000	1089 kHz		Radio 1 MW
300 000 000		330 m	Radio 2 MW
	1215 kHz	247 m	Radio 3 MW
	198 kHz	1515 m	Radio 4 LW
	92.4 MHz		Radio 4 FM

b What do MW, LW and FM stand for?

2 A hi-fi manufacturer produces a system that can make sounds with frequencies between 20 Hz and 20 000 Hz.
a If the velocity of sound is 340 m/s what is the wavelength of the highest and lowest note?
b How many of each of these wavelengths would fit in a room 3.4 m long?

Sound waves are longitudinal waves. They can be displayed on an oscilloscope. Human ears respond to sound waves of a wide range of frequencies.

An oscilloscope can display sound waves

A microphone is made of a diaphragm connected to a coil. Sound waves cause the diaphragm to vibrate and this vibrates the coil. The coil moves in a magnetic field which produces a changing voltage. (This is an **electro-magnetic effect** – you will find much more about it in Topic 235.) The microphone changes sound energy to electrical energy.

The changing voltage caused by sound waves can be displayed on an **oscilloscope**. The oscilloscope shows the changing voltage on the vertical axis. The horizontal axis shows a change of time. Sound waves are longitudinal waves. But the wave on the oscilloscope looks like a transverse wave.

The distance between two crests on the screen shows the time taken for a complete oscillation. This is the **period** of the oscillation. The closer together the crests are, the shorter the period. A short period of oscillation means that the **frequency** is high. Remember – frequency is the number of oscillations in one second.

The height of a crest on the screen shows the **amplitude** of the oscillation. The greater the height of the crests, the greater the amplitude.

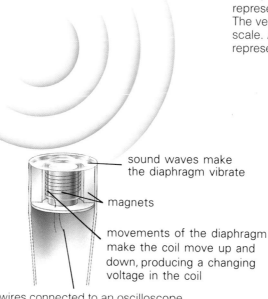

Fig. 110.1 A microphone converts sound waves into electrical signals.

sound waves make the diaphragm vibrate

magnets

movements of the diaphragm make the coil move up and down, producing a changing voltage in the coil

wires connected to an oscilloscope, which detects the changing voltage

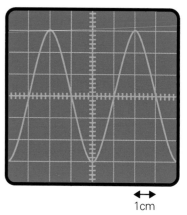

Fig. 110.2 An oscilloscope trace of a pure sound. The grid on the screen is often 1 cm squares. The **time base** setting tells you what time interval each division represents. If the time base is 1 sec/cm, then each large division represents a time of 1 second. The vertical divisions show the voltage scale. At 1 V/cm each large square represents 1 V.

1cm

INVESTIGATION 110.1

Investigating oscilloscope traces of different types of sound

Set up a microphone connected to an oscilloscope. You will also need objects which can produce pure notes, such as tuning forks, a radio – and your voice! Make careful drawings of the traces on the oscilloscope screen produced by different types of notes.

INVESTIGATION 110.2

The range of human hearing

Use the apparatus shown in Figure 110.3. The oscillator produces an oscillating voltage which you can adjust with a dial. The oscillating voltage makes the loudspeaker vibrate, setting up sound waves in the air.

Fig. 110.3

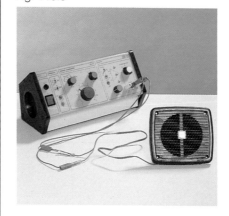

Find out answers to each of the following questions. In each case, you will need to take care that people are not 'cheating'! Devise a method of making sure that people do not say that they can hear a sound when they cannot.

1 What is the lowest frequency produced by your equipment which anyone can hear?
2 What is the highest frequency which anyone can hear?
3 At which frequency do most people find the sound is loudest?
4 Collect class results for the lowest and highest frequencies which people can hear. Draw a block graph to show these results.
5 Does age seem to make a difference to people's range of hearing?

Ears convert sound energy into electrical energy in nerves ■

A human ear has an **eardrum**, which behaves like the diaphragm on the microphone. Sound waves make the eardrum vibrate.

The vibrations of the eardrum set up vibrations in a chain of three small bones. The bones are arranged in a lever-like manner, so that the vibrations in the third bone, the stirrup, are greater than the vibrations set up in the eardrum. The bones **amplify** the vibrations.

The stirrup vibrates against a membrane on the outside of the **cochlea**. This is a coiled tube filled with fluid. The membrane makes the fluid vibrate. Inside it are cells which are sensitive to vibrations. They respond to the vibrations by setting up tiny electrical signals, which travel along the auditory nerve to the brain. The brain interprets these signals as sound.

Fig. 110.4   The human ear

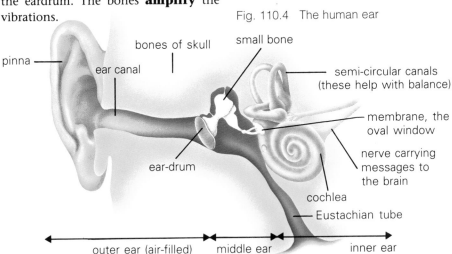

pinna
bones of skull
ear canal
small bone
semi-circular canals (these help with balance)
membrane, the oval window
nerve carrying messages to the brain
cochlea
Eustachian tube
ear-drum

outer ear (air-filled)   middle ear (air-filled)   inner ear (fluid filled)

Deafness has many causes

Anything which stops the passage of vibrations through the eardrum into the cochlea, or which stops the sensitive cells from sending messages to the brain, will make you deaf.

A common cause of deafness is a build-up of **wax** in the outer ear. Wax is made by the ear to stop dust and bacteria entering it. But sometimes such a thick layer builds up that the eardrum cannot vibrate. This sort of deafness is easily cured by dissolving the wax. Because the deafness does not last long, it is called temporary deafness.

If any part of the ear that transmits the vibrations to the inner ear is damaged, this is called a **conductive** hearing loss.

The **eardrum** itself can be damaged by a blow to the head, by very loud sounds or by sudden pressure changes. Any of these can make a hole in the drum. Small tears in an eardrum can often heal themselves, but sometimes the damage is permanent.

Infections in the ear may cause damage to the three small bones. If

these get jammed firmly together, vibrations cannot pass into the cochlea.

Conductive hearing loss reduces the amplitude of the vibrations reaching the inner ear. The hearing loss is similar across a range of frequencies. Some conductive loss can be treated by surgery or drugs.

Infections may also damage the cochlea or the auditory nerve. If the sensitive cells in the cochlea or the auditory nerve cells are damaged this is a **sensorineural** hearing loss. These cells cannot regrow and damage is permanent. Hearing loss is different at different frequencies and sound is **distorted** and reduced.

Long exposure to very loud sound can also cause this type of damage, especially at higher frequencies. People working in noisy environments should always wear ear protectors, or they may find that they become deaf as they get older. Portable hi-fi headphones may also cause damage. The sounds are produced very close to the eardrum, so it is easy to expose the ear to high sound levels for long periods of time.

Hearing aids amplify sound ■

A hearing aid is an electronic amplifier. It has a microphone which collects sound. The sound is amplified, and delivered to the ear by a small receiver, often worn inside the ear canal.

In deafness caused by damage to the eardrum or the small bones, hearing aids often produce very good hearing. But if the deafness is caused by damage to the cochlea, hearing aids may not be as successful. This is because the message sent to the brain by the cochlea is still not clear, even if the sound is loud. Sound may seem distorted, so that it is difficult to pick out important sounds like speech. Background noise is amplified as well, so the person may have problems in sorting out 'useful' noises from unwanted ones.

More advanced hearing aids are designed to amplify the sound selectively, to match more closely the hearing loss. In some cases, it is possible to connect a hearing aid to a **cochlea implant**. The implant is surgically placed inside the head and connects directly to the auditory nerves. It cannot reproduce the same amount of information as a normal ear, because it is not possible to connect up that many nerves. The deaf person can learn to interpret the sounds received. As the technology improves, more life-like hearing becomes possible.

Questions

1   What is meant by:
   a the period of an oscillation?
   b the frequency of an oscillation?
   c How does the sound of a high-frequency sound compare with the sound of a low-frequency sound?

2   Match each of these functions with a structure labelled on the diagram of the human ear in Figure 110.4
   a contains cells which convert vibrations in fluid into tiny electrical signals
   b vibrates when sound waves reach the ear
   c amplify the vibrations as they cross the middle ear
   d transmit the vibrations from the air in the middle ear to the liquid in the cochlea

LOUDNESS, PITCH AND QUALITY

Loudness, pitch and quality are terms we can use to describe the sounds we hear.

The loudness of a sound is related to its amplitude

The amplitude of a wave is the maximum distance that the vibrating particles move from their resting position. Sound waves are longitudinal waves, so the vibrating particles produce areas of compression and rarefaction. This causes pressure changes in the material through which the sound is moving. The further the particles move, the greater the pressure changes. So, in a sound wave, the amplitude can be measured by measuring the pressure changes which are produced.

Sound waves with a large amplitude tend to sound louder to us than ones with a small amplitude. But human ears are more sensitive to some frequencies than others, as you will have found out if you did Investigation 110.2. A sound of a frequency to which your ears are very sensitive will sound loud to you, even if its amplitude is quite small.

Loudness is measured in decibels

The loudness of a sound is often measured in decibels, written **dB**. The softest sound which human ears can hear is said to have a loudness of 0 dB. The sound of a jet aircraft 50 m away is about 10 000 000 000 000 times louder than this, or 10^{13} times louder. The aircraft is said to have a loudness of 130 dB.

A small change in the dB value of a sound represents a large change in its loudness. For example, a noise of 100dB sounds twice as loud as a noise of 90 dB.

Health Inspectors in Britain regularly measure noise levels in people's working environments. Excessive noise is not only unpleasant, but can make people work less safely, suffer stress, or suffer permanent damage to their ears. It is not enough just to measure the physical intensity of the sound, because our sensitivity is different to sounds of different frequencies. The meters the Inspectors use have electronic circuits which compensate for this changing sensitivity.

Fig. 111.1 People working in noisy environments must protect their ears from long-term damage.

Sound intensity is measured in W/m²

The **intensity** of a sound is the amount of energy passing through a square metre every second. It is measured in watts per square metre (W/m²). The intensity of a sound is related to its amplitude. If the amplitude doubles, the intensity is four times greater. Increasing the intensity of a sound increases its loudness. But this also depends on the sensitivity of the person hearing the sound.

— *E X T E N S I O N* —

Table 111.1 Approximate sound levels of different sounds

Sound	Sound level in dB
Quiet countryside	25
People talking quietly	65
Vacuum cleaner 3 m away	70
Lorry 7 m away	90
Very noisy factory	100
Loud music in a disco	110
Jet aircraft 50 m away	130
Rifle firing near ear	160

DID YOU KNOW?

The lowest pitched orchestral instrument is the sub-contrabass clarinet, which can play C at 16.4 cycles/sec.

Pitch depends on frequency

The pitch of a note depends on its frequency. A note of high pitch has a high frequency. On an oscilloscope, a high pitched note has crests close together.

Humans can hear sounds with frequencies of between 20 Hz and 20 000 Hz. As you get older you become less sensitive to high frequencies. This usually begins in the late twenties or thirties, and gradually progresses throughout the rest of a person's life. It is caused by a degeneration of sensitive cells in the cochlea. These cells can also be damaged by exposure to loud noises. Loss of high frequency sensitivity can be a sign of hearing damage and not just old age. People do not normally notice it unless it becomes so bad that they can no longer hear speech.

Quality depends on the mixtures of frequencies in a sound

A tuning fork produces a note of a single frequency. This is a **pure** sound. With practice you can sing a note into

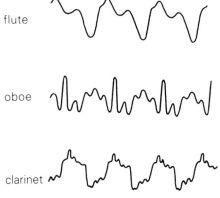

flute

oboe

clarinet

Fig. 111.2   Each type of musical instrument has its own distinctive sound pattern.

a microphone which will produce a single frequency trace on an oscilloscope.

Most sounds, however, contain a mixture of frequencies. If the frequencies are simple multiples of each other they are called **harmonics**. Musical instruments tend to produce notes which consist of a basic, or **fundamental**, frequency and a number of harmonics. Our brains hear these sounds as 'musical'. If a sound contains a mixture of frequencies which are not related to each other, it does not sound musical. A sound with a large number of unrelated frequencies will just sound like 'noise'.

Room acoustics affect how different sounds appear

Sound is reflected from hard surfaces but is absorbed by soft ones. The sound in a gymnasium reflects off the walls and bounces back to the listener. The many echoes make a single clap last longer than the original sound. This is **reverberation**; the sound dies away gradually. If you want to talk or play music, the amount of reverberation is important. The **reverberation time** is the time taken for the sound intensity to fall to one millionth of its original value (a 60 dB drop). Reverberation gives sound 'life', and boosts the sound levels. But if the reverberation is too long, the sound may become confusing as different sounds overlap. For speech,

a reverberation time of a second gives a good balance. For music, the reverberation time can be longer.

Polished stone surfaces reflect nearly all the sound that hits them. Heavy curtains and carpets can absorb most of the sound energy. With careful placing of the different materials in a room, the reverberation time can be controlled. If you have large, heavy curtains at home you may be able to experiment by comparing the sound of some music with them open and closed. You can experiment with a classroom that is empty and one that is full of quiet people. In a concert hall each person in the audience is a very good absorber. It can be difficult to arrange the seating so that the sound properties in rehearsal are not too different from when the hall is

full. In a large enough room, the returning echo could arrive after the sound has finished. This would produce a noticeable echo and would be very confusing. Another problem is if there isn't enough sound coming from a particular direction. With careful design, reflecting materials can be used to fill in any 'gaps' (see Figure 116.5).

Fig. 111.4   A car being tested in an anechoic chamber.

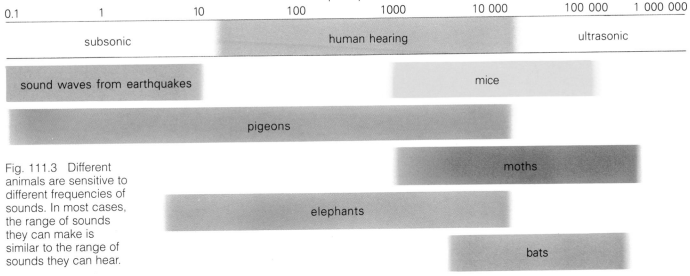

Frequency Hz

| 0.1 | 1 | 10 | 100 | 1000 | 10 000 | 100 000 | 1 000 000 |

subsonic   human hearing   ultrasonic

sound waves from earthquakes   mice

pigeons

moths

elephants

bats

Fig. 111.3   Different animals are sensitive to different frequencies of sounds. In most cases, the range of sounds they can make is similar to the range of sounds they can hear.

We often transfer information in the form of sound. Sound can be recorded on tape or records.

Sound is used for communication

Humans have always used sound for information transfer. Speech is an important way of communicating. Early societies used drums and other sound-producing objects to send sound over long distances.

Using sound for information transfer has its limitations. Sound is instantaneous – it happens and then stops. So if we want to store the signals, we must change them into something else which we can store. Another problem with sound is that the strength of a sound fades as it travels – so it can only be used over fairly short distances.

Both of these problems can be solved if the sound signal can be changed into another kind of signal, which can be stored. The stored signal can then be changed back into sound again whenever we want to hear it. This allows us to 'move' the sound from one place to another, and from one time to another. We can also amplify the sound if we wish.

A microphone is used to convert the sound to electrical signals. An electrical signal can be amplified or recorded. An amplifier takes the electrical signal and makes it larger or **amplifies** it. The amplified signal can then be converted back to a sound signal with a loudspeaker. This is how hearing aids, megaphones and public address (PA) systems work.

Advanced systems can filter and clean up the sounds before amplifying them.

Fig. 112.1 Alpenhorns can be used to transfer information over long distances. In mountainous regions, before modern methods of communication, this was one of the quickest methods of making contact with someone on the other side of the valley.

Fig. 112.2 A microphone, amplifier and loudspeaker can be used to produce a louder sound. The sound signals are converted into electrical signals and then back into sound signals again. The electrical signals change in just the same way as the sound signals. This is an example of an **analogue system**. It processes continuously changing signals.

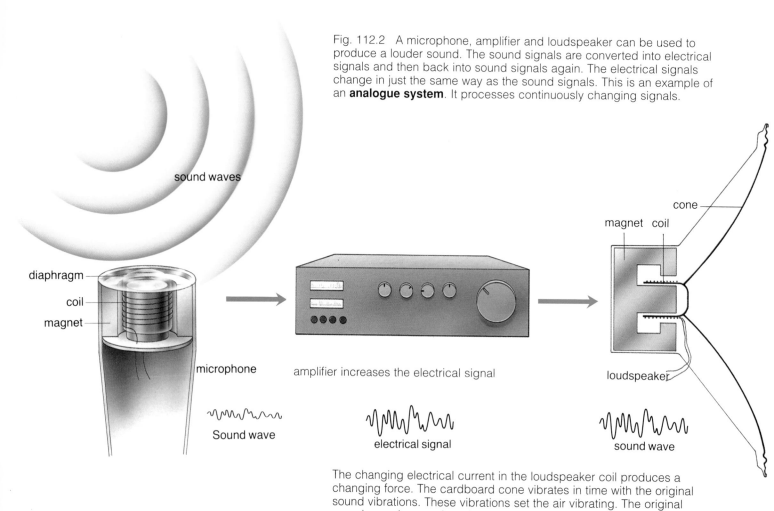

sound waves

diaphragm
coil
magnet

microphone

Sound wave

amplifier increases the electrical signal

electrical signal

cone

magnet coil

loudspeaker

sound wave

The changing electrical current in the loudspeaker coil produces a changing force. The cardboard cone vibrates in time with the original sound vibrations. These vibrations set the air vibrating. The original sound wave is reproduced.

Building a loudspeaker

1 Cut a cardboard ring, as shown in the diagram. Join AB and BC to make a cone. Add a length of cardboard tube, to act as a coil-former.

2 Wind a coil on the former. Leave long ends.

3 Connect the two ends of the coil to an amplifier or a signal generator. If you put a 4Ω resistor in series with the coil, this will prevent damage to the amplifier.

4 Hold one end of a bar magnet inside the coil. You should be able to produce a sound from your loudspeaker.

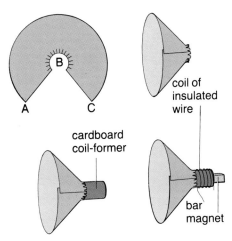

Fig. 112.3

Questions

1 Explain why a sound is produced when you change the current inside the coil.

2 Does your loudspeaker sound equally loud at all frequencies? Which is loudest?

3 Look carefully at a commercially produced loudspeaker. How is it similar to yours? What differences are there between the two loudspeakers?

Tapes store sound as magnetic fields

If a magnetic material is moved past an electromagnet, the magnetic material becomes magnetised. A changing current in the electromagnet will produce a changing magnetic field along the length of the material. This is how a tape or cassette recorder works.

To play back the tape, the tape is moved past a playback head. It moves at the same speed as when it was recorded. The changing magnetic field on the tape generates a changing a.c. signal in the coil as the tape moves past. The signal is then amplified, and passed into a loudspeaker to produce the sound.

The earliest magnetic recorders used iron wire to store the magnetic fields. Modern recorders use magnetic

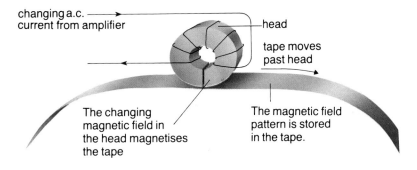

Fig. 112.4 Recording sound onto magnetic tape

particles on a plastic backing tape. The cheapest tapes, called ferric tapes, use particles of iron oxide. Chrome tapes use chromium dioxide. Metal tapes use particles of iron.

The faster the tape moves past the heads, the better the quality of the recording. A cassette tape moves past the heads at 4.7 cm/s. This is quite slow, and it is difficult to eliminate background hiss. The Dolby® system reduces this noise by artificially boosting some frequencies during recording of quiet passages. When the tape is played back, the volume is reduced. This returns the boosted signal to its normal level, but cuts down the volume of the hiss.

Questions

1 Sound is produced by something vibrating. What vibrates to produce sound from the following instruments:
a violin
b guitar
c flute
d piano
e drum?

2 Explain each of the following:
a Sometimes young people can hear the squeaking of bats, but older people cannot.
b It is recommended that no-one should be exposed to steady noise of more than 85 dB in their working environment.
c High pitched notes show a trace with crests close together on an oscilloscope trace.
d Hearing aids help people with deafness caused by damage to the bones in the middle ear, but are less help to people with damage to the cochlea.
e Sound waves with a large amplitude often, but not always, sound louder to us than sound waves with a small amplitude.

3 Figure 110.2 shows an oscilloscope trace of a particular sound. The time base on an oscilloscope tells you the time scale for the trace. The time base was set at 1 ms/cm.
a What is the period of the trace?
b What is the frequency of the trace?
c If this same trace was obtained with the time base set at 5 ms/cm, what would be the period of the trace?
d What would be the frequency of this trace?

EXTENSION

ELECTROMAGNETIC WAVES
Electromagnetic radiation is a disturbance of electric and magnetic fields.

Electromagnetic waves can be grouped according to their wavelength

Energy changes in atoms or electrons may produce **electromagnetic waves.** These are disturbances of electric and magnetic fields. They travel as transverse waves.

The wavelength of electromagnetic waves determines how they behave. So it is useful to group electromagnetic waves according to their wavelength. Figure 113.1 shows these groups. This grouping is called the **electromagnetic spectrum.** Electromagnetic waves of a particular wavelength always have the same frequency in a particular material. Electromagnetic waves with short wavelengths have high frequency. Ones with long wavelengths have low frequency.

Electromagnetic waves are a form of energy. When they pass through a material, some of the energy they carry can be absorbed. This raises the internal energy of the material. The shorter the wavelength of the electromagnetic radiation, the more energy it carries.

All electromagnetic waves travel at the same speed in a vacuum. This speed is 300 000 000 m/s.

Table 113.1 The electromagnetic spectrum

Wavelength in metres	Source	Detector	Uses
10^6 to 10^{-3} Radio waves	Electrons vibrated by electronic circuits, radio and TV transmitters, stars and galaxies including pulsars and quasars	Radio aerial	Communications – radio, television, telephone links, radar, cooking, astronomy
10^{-3} to 10^{-6} Infrared	'Hot' objects, especially the Sun	Electronic detectors, skin, special films	Heating, astronomy, thermography – taking temperature pictures
Visible light	The Sun, very hot objects	The eye, electronic detectors, photographic film	Seeing, photography, information transmission, photosynthesis, astronomy
10^{-6} to 10^{-9} Ultraviolet	The Sun, mercury vapour lamps, electric arcs	Fluorescence of chemicals, photographic film, tanning of skin, electronic detectors	Fluorescent lamps, crack detection, security marking, sterilizing food etc. astronomy
10^{-8} to 10^{-12} X rays	X ray tubes, stars, changes in electron energy	Photographic film, fluorescence of chemicals, ionising effects, electronic detectors	Taking 'X rays' – radiography, astronomy, examining crystal structure, treating cancer, 'CAT scan'
10^{-12} and less Gamma rays	Radioactive materials, nuclear reactions	Geiger-Müller tube, photographic film, electronic detectors	Radiography, treating cancer, measuring thickness

Fig. 113.1 The electromagnetic spectrum

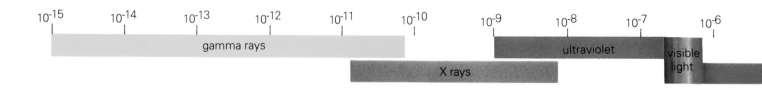

Radio waves have long wavelengths

Electromagnetic waves with wavelengths over 1 mm or so are called **radio waves**. Very short wavelength radio waves are known as **microwaves**.

Radio waves are produced by stars and galaxies. They are also produced by electrons vibrated by electronic circuits.

We use radio waves for communication. Telephone links between cities use microwaves. Communication to satellites is also by means of microwaves. Microwaves are used for cooking. A microwave oven produces electromagnetic waves with a wavelength of about 12 cm. These waves transmit energy to the food. Microwaves can be used to kill insects in grain stores. They are also used to kill bacteria in food, without heating the food so much that it spoils the flavour.

The wavelengths used for radar start in the microwave region of the electromagnetic spectrum and extend into the ultra high frequency radio waves. A short pulse is sent out which bounces back off any object which it hits. The time taken for the 'echo' to return can be used to calculate the distance of the object.

Terrestrial television broadcasts use UHF, or ultra high frequency, waves. Good quality sound broadcasts use VHF, or very high frequency, waves. The signals are sent by slightly altering the frequency of the waves, so this is known as frequency modulation, or FM. Long wavelength radio waves have a greater range in a straight line than short ones. However, short wavelength radio waves (shortwave) reflect off an upper layer of the atmosphere (see

Topic 114). For this reason, very long range sound broadcasts (like the BBC World Service) can use shortwave radio. A great many television and radio broadcasts now use microwaves that are relayed by satellites.

When electromagnetic waves hit materials, they can set the electrons vibrating at the same frequency as the waves. This is how aerials detect radio waves. If the length of the aerial is matched to the frequency of the waves then the aerial will be some multiple of the wavelength. Many aerials have two branches. For an aerial on a car the body of the car counts as one of the branches. To receive radio (or television) waves this length is often set to a quarter of the expected wavelength. The aerial is described as a $\frac{1}{4}$ wave dipole.

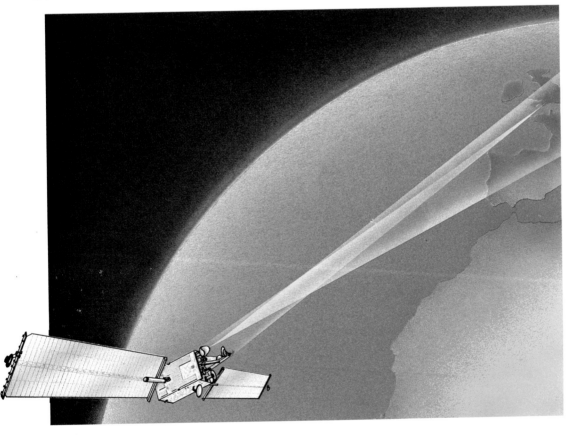

Fig. 113.2 A TV satellite broadcasts television pictures down to Earth from a stationary position 36 000 km above the Earth's surface. Pictures must first be transmitted up to the satellite. It is a relay station in the sky. The large solar panels provide the energy and can span half the width of a football pitch. The energy supply, control and transmission are all provided by electromagnetic radiation.

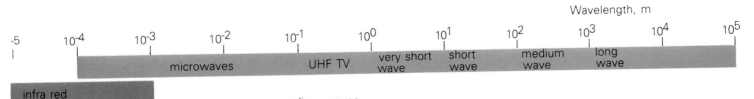

Infrared radiation is felt as heat

Electromagnetic radiation with wavelengths a little longer than that of red light is called **infrared radiation**. It is given off by hot objects. An electric lamp, for example, gives off radiation which we see as light, but also radiation with a wavelength of about $\frac{1}{1000}$ mm, which is infrared radiation. You give off infrared radiation, because you are quite hot! Your body gives off infrared radiation with a wavelength of $\frac{1}{10}$ mm to $\frac{1}{100}$ mm.

The human eye cannot detect infrared radiation, but you can feel it on your skin as warmth. The infrared is absorbed by skin, and the increase in internal energy of the skin is detected by temperature receptors. Some snakes have specially designed sense organs which make good use of this ability. They are situated in pits on either sides of their head, and they can use them to 'see' a warm object, such as a mouse, in the dark.

Infrared cameras can also 'see' hot objects. They collect and focus infrared rays just as an ordinary camera focuses light rays. They then convert the infrared rays to a visible image. These cameras can be used to find people buried in collapsed buildings. They can also be used to take pictures of buildings to find out where heat is being wasted. Infrared satellite pictures of the Earth are used in weather forecasting.

Infrared signals are also used in TV and hi-fi remote controls and for some cordless headphones, keyboards and printers. Infrared signals are also transmitted down fibre optic cables for telecommunications systems.

Light is electromagnetic radiation to which our eyes are sensitive

Light is just the same as all the other types of electromagnetic waves. But it is very important to us because our eyes happen to be sensitive to this particular range of wavelengths. Human eyes contain cells which respond to different wavelengths of light. Some react to red light, some to green, and some to blue light. This is why we can see colour.

Light will affect photographic film in a similar way to our eyes, so we can take pictures which are the same as our view of the world.

Sunlight contains a particular mixture of wavelengths of light. Artificial lighting is designed to give a similar mixture, which results in a 'natural' effect.

Questions

1 **a** What are meant by *electromagnetic waves*?

 b What name is given to electromagnetic waves with each of the following wavelengths?

 i 1 km   **iv** 0.1 µm
 ii 1 cm   **v** 0.0001 µm
 iii 1 µm   **vi** 0.00000001
     (1/1000 mm)   µm?

 c For each of the types of waves in your answer to part **b**, briefly describe *one* source and *one* use.

 d Which of the types of waves in your answer to part **b** carries
 i the most energy?
 ii the least energy?

---

2 Microwaves can be used in surveying. The time taken for them to travel between two points is used to calculate distance.

 a A surveyor transmits microwaves to a second surveyor, whose receiver sends back the signal. The time between the transmission and reception of the signal by the first surveyor's equipment is 0.000033 s. How far apart are the two surveyors?

 b It is known that this method is not completely accurate, and that an error of up to 40 mm might be expected over this distance. What is this as a percentage?

EXTENSION

Fig. 113.3   An infrared image of a house. The brightest parts of the picture show where most heat is being lost.

Fig. 113.4   Animals emit infrared radiation. This picture was taken with an infrared sensitive colour film.

Ultraviolet rays are produced by the Sun

Sunlight contains ultraviolet rays. We cannot see them, because human eyes cannot detect rays with such short wavelengths. But many insects can see ultraviolet light. Many flowers have petals with markings on them which reflect ultraviolet light. Insects can see these markings, which guide them to the nectar at the base of the flower.

Ultraviolet light falling on to human skin enables the skin to produce vitamin D. A lack of sunlight, combined with a lack of vitamin D in the diet, can cause a weakness in the bones, called rickets. In northern Russia, children are exposed to sunlamps for part of the day, to enable their skin to make vitamin D.

But too much ultraviolet light falling on to the skin can be harmful. It damages cells, giving you sunburn. It may even cause skin cancer. Skin cancer is now getting much commoner, as people who are not used to strong sunlight go on holidays to hot countries more often. Your skin can protect itself by making a pigment called **melanin**, which absorbs the ultraviolet rays. This is what happens when you tan. But it takes a while to happen, and until you have built up a deep tan you are in danger of damaging your skin if you expose it to strong ultraviolet radiation.

Ultraviolet radiation is produced by arc welders. An arc welder heats metal with an electric spark. The very hot atoms give out ultraviolet light. So a welder wears protective clothes, and also a dark green filter over his eyes. Eyes are easily damaged by ultraviolet light.

Ultraviolet light entering the Earth's upper atmosphere from the Sun is involved in the formation of the gas ozone. This absorbs a lot of the ultraviolet light reaching the Earth from the Sun. But chemicals such as chlorofluorocarbons, or CFCs, are damaging the ozone layer, so that more ultraviolet light can reach the Earth's surface. This could be dangerous to living things.

Some materials convert ultraviolet radiation into visible light

Inside a fluorescent lamp, electrical energy is converted into ultraviolet radiation. The inside of the tube in a lamp is coated with fluorescent powder. When the ultraviolet light hits this powder, it is absorbed by it. The atoms in the powder absorb some of the energy in the ultraviolet waves, and release it as visible light. This is called **fluorescence**.

Some washing powders contain chemicals which fluoresce. They absorb ultraviolet light in sunlight, and release it as visible light. So your white clothes look even brighter! At a disco an ultraviolet light source makes white clothing which has been washed in these powders glow in the dark.

'Invisible' ink fluoresces in ultraviolet light and is used for security marking. You can write your post code on an object like a video recorder. In normal light, the marking is not visible, but under an ultraviolet lamp the ink fluoresces and is clearly seen.

Fig. 113.5 An evening primrose flower photographed in ultraviolet light. This is how an insect might see it. Notice how the guide-lines show up clearly, directing the insect to the nectar in the base of the flower.

Fig. 113.6 Dark filters on the welder's goggles protect his eyes from electromagnetic radiation.

X rays can penetrate solids

X rays are produced by electrons that are slowed down very quickly. Electrons are accelerated by a voltage of 30 kV. They strike a metal target at 100 000 000 m/s. Some of their kinetic energy is converted into electromagnetic energy as they slow down.

As you can see from the chart on page 268, electromagnetic radiation with wavelengths between about 10^{-12} and 10^{-10} m may be classified either as gamma rays or as X rays. The rays themselves are no different. Whether they are called gamma rays or X rays depends on how they were produced.

X rays are high energy waves. They can be detected with photographic film. X rays can penetrate solids.

Different materials absorb different amounts of X ray radiation. X rays passing through an arm, for example, can pass easily through the skin and muscle, but not so easily through the bone. If you place your arm between an X ray source and a photographic film, you get a shadow picture of the bones in your arm. The film blackens where the X rays hit it, so the areas of skin and muscle show up black. Fewer X rays pass through the bones, so the areas of film behind the bones stay white.

X rays cause materials to fluoresce. If X rays hit a fluorescent screen, they cause the screen to emit visible light. So 'X rays' can also be taken by photographing the visible light emitted from a fluorescent screen which has been bombarded with X rays.

High energy X rays have high frequencies and low wavelengths. They can penetrate metal, and are used to inspect welds.

Fig. 113.7 X ray pictures of luggage can reveal their contents.

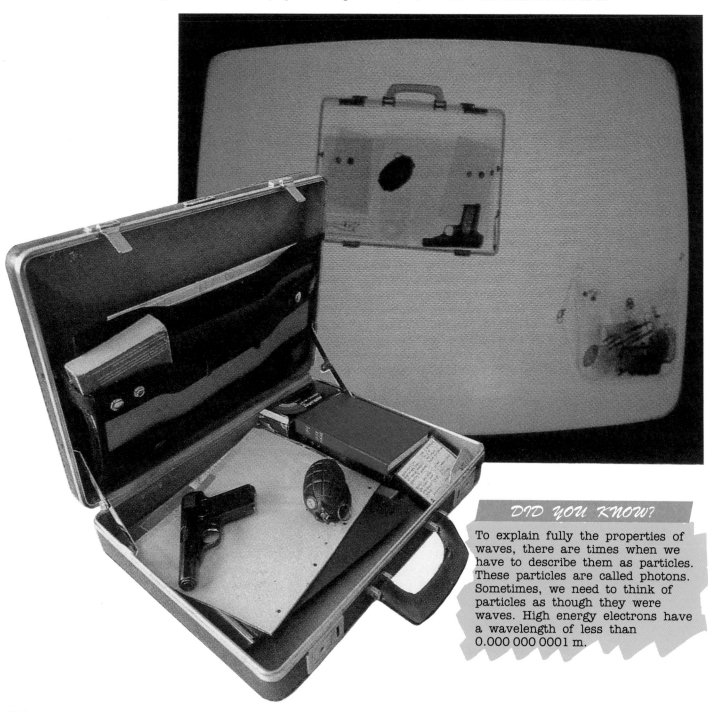

Questions

1 Explain the following:
  a You should always wear protective glasses when using a sunlamp.
  b People with pale skin are more likely to get skin cancer than people with dark skin.
  c X ray pictures show bones as light areas.
  d Snakes can catch mice in the dark.

2 A fluorescent tube produces mainly ultra-violet radiation. A tube using 100 J of electrical energy per second produces 62 J of ultra-violet light, and 3 J of visible light.
  a What is the visible light output power of the tube?
  If the tube is coated with fluorescent powder, 20 J of visible light energy is produced from each 100 J of electrical energy.
  b What is the visible light output power of the coated tube?
  c What is the efficiency of the coated tube?
  d How much more heat will the coated tube produce than the uncoated tube?

3 Iridium 192 is a gamma source. It has a half life of 74 days. An oil company uses iridium 192 to inspect the welds in a pipe line.
  a Explain how iridium 192 could be used for inspecting welds.
  b After a year of using the same iridium 192 sample, it is found that the film used needs to be exposed for much longer than before. Why is this?
  c What happens to the intensity of the radiation from the iridium 192 after 370 days?
  d How could this problem be prevented by the oil company?

4 A company decides to replace the lighting in its offices. In one office, the cost of running three normal light bulbs is £22.50 per year. A lighting consultant points out that fluorescent lighting would produce the same amount of light for only £5 a year.
  a The manager finds this hard to believe. How would you explain to him why a fluorescent lamp saves so much money?
  b If the company had 30 offices, how much money could they save in a year?
  c If each fluorescent light cost £12 to install, how much money would they save in the first year by replacing all their normal lights with fluorescent lights?
  d In houses, people tend only to use fluorescent lighting in the kitchen, if at all. Why is this?

EXTENSION

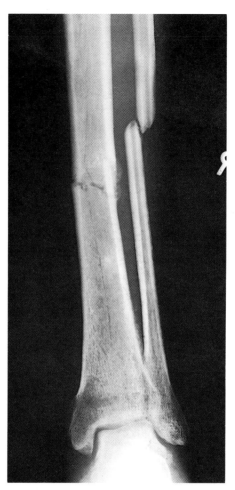
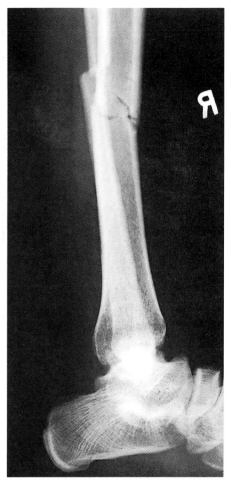

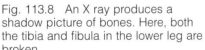

Fig. 113.8   An X ray produces a shadow picture of bones. Here, both the tibia and fibula in the lower leg are broken.

Gamma radiation has very short wavelengths

Electromagnetic waves with the shortest wavelengths and the highest frequencies are called **gamma rays**. Gamma rays are produced by changes in the nuclei of atoms. They carry very large amounts of energy – at least 10 million times more energy than light rays.

Gamma rays are used for measuring thicknesses of metal sheets, and for sterilising medical materials and instruments. The most common industrial sources of gamma rays are iridium 192 and cobalt 90. Gamma rays are harmful to living things because they cause ionisation in their cells. This can cause mutations in the cells. High doses of gamma radiation kill cells. With careful control they can be used to kill cancer cells.

Gamma rays can be detected using a Geiger-Müller tube, or photographic film.

Radio signals are electromagnetic waves

Only 20 years after the first telephone conversation, the Italian scientist Marconi sent a radio signal across a distance of 12 miles. To begin with, signals were sent, like telegraph signals, as a series of pulses. The pulses, however, were of electromagnetic waves passing through the air, and not electrical signals passing along wires. So the system was called 'wireless'.

A radio signal is sent out as a **carrier wave**, on which information about a sound signal is superimposed. There are two ways in which the sound information can be added to the carrier wave. The sound signal can change the **amplitude** of the carrier wave. This method is called **amplitude modulation** or **AM**. Alternatively the sound signal can change the **frequency** of the carrier wave. This is called **frequency modulation** or **FM**.

The modulated radio waves are generated by the oscillation of electrons in wires. This produces an electromagnetic disturbance, or radio wave, which spreads out from the transmitter aerial. The radio waves travel at the speed of light, and can be picked up by a receiver. The receiver sorts out the superimposed sound signal from the carrier wave. This is called **demodulation**.

DID YOU KNOW?

Although we can hear sounds between 20 and 20000Hz, a conversation needs a range of only 3100Hz. An optical fibre can carry only 250 music channels, but the restricted range of frequencies used in speech means the same optical fibre can carry 2000 telephone channels.

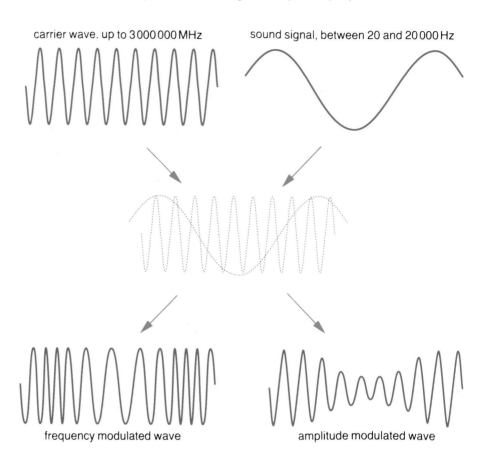

Fig. 114.1 A radio wave can carry information either as changes in frequency (FM) or as changes in amplitude (AM).

carrier wave, up to 3 000 000 MHz

sound signal, between 20 and 20 000 Hz

frequency modulated wave

amplitude modulated wave

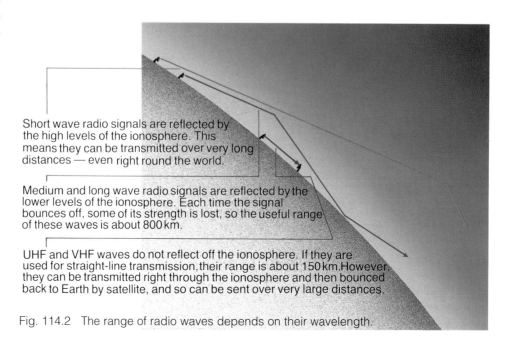

Short wave radio signals are reflected by the high levels of the ionosphere. This means they can be transmitted over very long distances — even right round the world.

Medium and long wave radio signals are reflected by the lower levels of the ionosphere. Each time the signal bounces off, some of its strength is lost, so the useful range of these waves is about 800 km.

UHF and VHF waves do not reflect off the ionosphere. If they are used for straight-line transmission, their range is about 150 km. However, they can be transmitted right through the ionosphere and then bounced back to Earth by satellite, and so can be sent over very large distances.

Fig. 114.2 The range of radio waves depends on their wavelength.

Fig. 114.3 Telecom Tower, London (189 m high) is a microwave relay tower. It can handle over 150 000 telephone messages and over 40 TV channels at the same time.

Fig. 114.4 Telephone conversations are broadcast in straight lines from one relay tower to the next. Away from London's tall buildings, microwave relay towers are much smaller.

Fig. 114.5 To cover large distances, to cross oceans for example, the easiest straight line transmission is up to a geostationary satellite and back down to Earth. Satellite ground stations have large dish antennae.

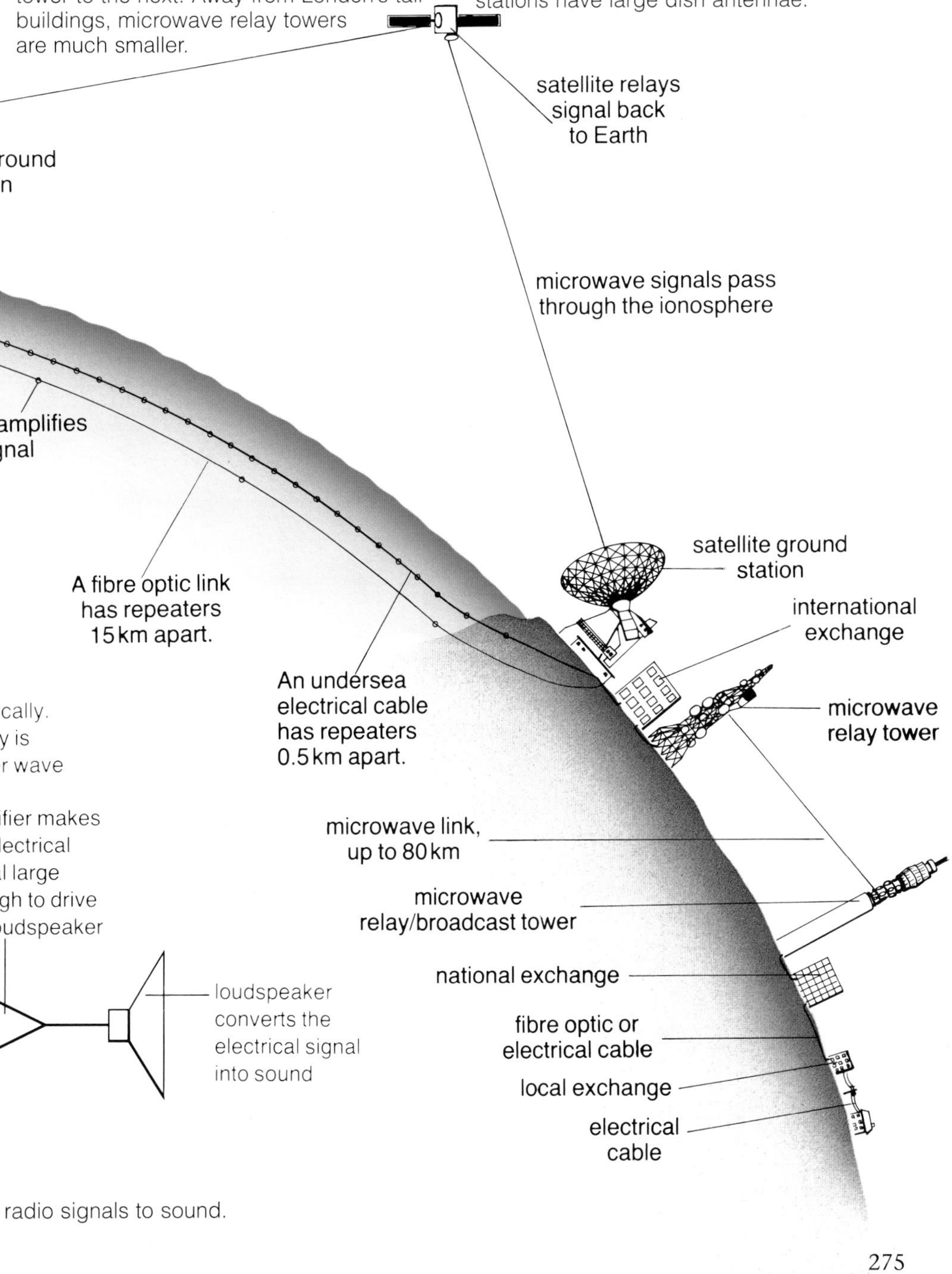

satellite relays signal back to Earth

satellite ground station

microwave signals pass through the ionosphere

Fig. 114.6 The telecommunications network. Sending a telephone message over long distances may involve several methods of communication.

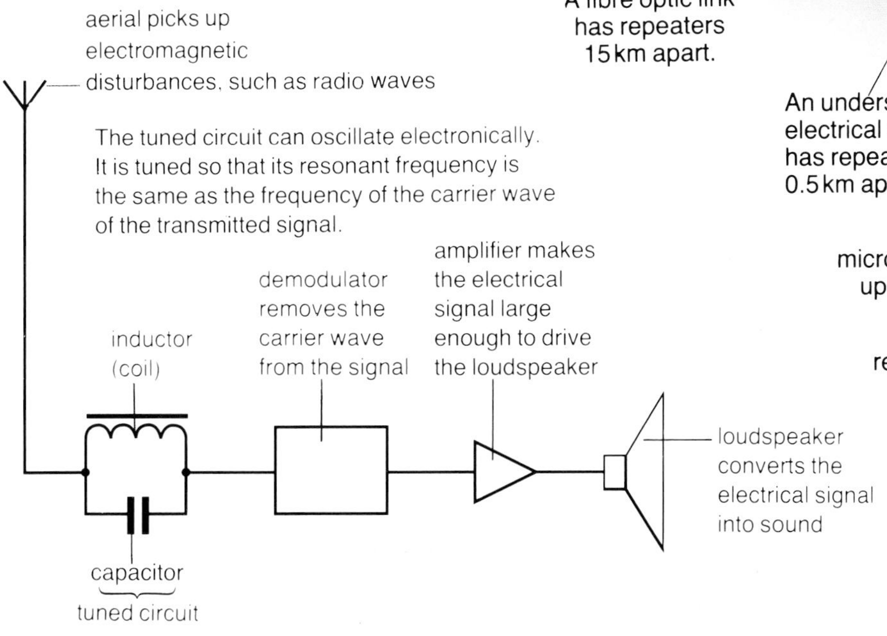

A repeater amplifies the signal

A fibre optic link has repeaters 15 km apart.

An undersea electrical cable has repeaters 0.5 km apart.

satellite ground station

international exchange

microwave relay tower

microwave link, up to 80 km

microwave relay/broadcast tower

national exchange

fibre optic or electrical cable

local exchange

electrical cable

aerial picks up electromagnetic disturbances, such as radio waves

The tuned circuit can oscillate electronically. It is tuned so that its resonant frequency is the same as the frequency of the carrier wave of the transmitted signal.

inductor (coil)

demodulator removes the carrier wave from the signal

amplifier makes the electrical signal large enough to drive the loudspeaker

loudspeaker converts the electrical signal into sound

capacitor

tuned circuit

Fig. 114.7 A radio receiver converts radio signals to sound.

115 LASERS

A laser produces a beam of light in which all the waves are of the same frequency, and exactly in step with each other.

A laser produces a very intense beam of light

If you look back at Topic 9, you will see that atoms emit light when a high-energy electron falls back to a lower energy state. A laser uses ordinary light to stimulate atoms to release light in a special way. The light which they release is concentrated into a very intense beam.

The atoms which produce the laser light may be in the form of a crystal, such as ruby. Or they might be a gas, or even a liquid. If you have a laser in your school it is probably a helium/neon laser, which contains a mixture of helium and neon gas.

One way of putting energy into a laser is by means of bright flashlamps. These produce ordinary light. The light excites some of the atoms in the laser. Their electrons are raised to a higher energy level. Suddenly, an atom returns to its normal energy state. As it does so, it gives out light of a particular wavelength. This emission of light is called **stimulated emission**, because it was stimulated by the light from the flashlamps. The light from the excited atom may stimulate another atom, causing it to produce light as well. There are partly reflecting mirrors at each end of the cavity containing the gas or crystal, so the light bounces back and forth. It stimulates other atoms, so that they, too,

emit light. The light can escape through the partly reflecting mirrors, and an intense flash of light emerges from the laser.

The light pulses emitted by a laser are very short – about a billionth of a second. They are released in a stream from the laser. Some lasers produce their light in pulses, while others emit light continuously. The light emitted from lasers is very special. Ordinary light contains a mixture of frequencies (colours). But laser light is all at exactly the same frequency. Secondly, the light waves produced are all in step, or **in phase**. So the waves reinforce one another, making the light very bright. Thirdly, laser light is produced in a narrow beam, which hardly spreads out at all.

The word 'laser' stands for 'Light Amplification by the Stimulated Emission of Radiation'. A laser uses ordinary light, and amplifies it by stimulating atoms to release light all together and exactly in phase.

1. Energy put into the laser excites electrons in atoms.

Fig. 115.1
How a laser works

2. As an excited electron falls back to its normal energy level, it releases light.

3. The light may cause electrons in other excited atoms to fall back, releasing more light. This is stimulated emission.

4. The light bounces back off the reflectors at each end, stimulating more and more emission from the atoms inside the laser. Some of it escapes through the partly reflecting mirrors. All the light is of the same frequency, and the waves are all perfectly in step.

— *EXTENSION* —

Lasers can be used for cutting

A laser beam concentrates a very large amount of energy into a short time and space. It is very powerful. The power of a laser beam can be used for cutting. Computer controlled laser cutters can cut through forty or more layers of suit fabric at once, with great speed and accuracy.

Lasers are used in surgery. The light-sensitive layer of cells at the back of the eye, the retina, sometimes gets detached. To weld the retina back into position, a laser beam is directed into the eye. The laser passes through the cornea and lens without damaging

them. When it hits the retina, its heat welds the retina back in position. The surgeon can control its narrow beam with great precision, 'spot welding' much more accurately than could ever be done with normal surgical instruments.

Laser beams are also used in surgery to cut through flesh instead of scalpels. Their heat seals blood vessels in the instant that they are cut, so there is much less bleeding. Laser beams have even been used to drill out decayed parts of teeth!

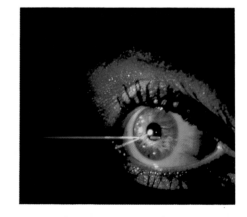

Fig.115.2 A laser beam being used in eye surgery.

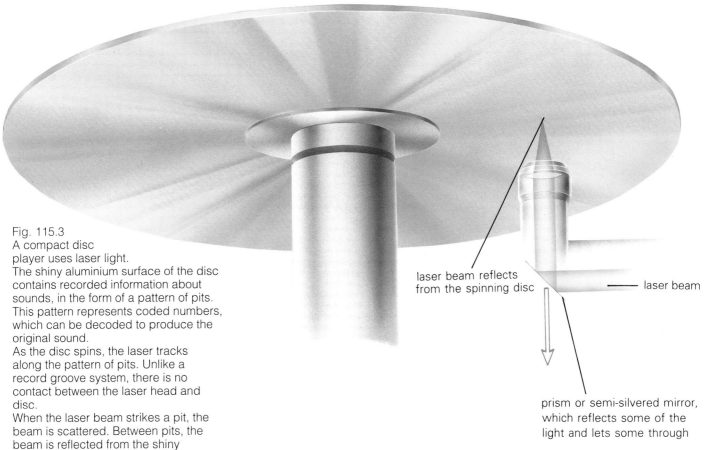

Fig. 115.3
A compact disc
player uses laser light.
The shiny aluminium surface of the disc
contains recorded information about
sounds, in the form of a pattern of pits.
This pattern represents coded numbers,
which can be decoded to produce the
original sound.
As the disc spins, the laser tracks
along the pattern of pits. Unlike a
record groove system, there is no
contact between the laser head and
disc.
When the laser beam strikes a pit, the
beam is scattered. Between pits, the
beam is reflected from the shiny
surface. A detector converts the on and
off flashes of the returning beam into an
electrical signal. The pattern of flashes
represents numbers between 0 and
65536. These are decoded to produce
the sound.

laser beam reflects
from the spinning disc

laser beam

prism or semi-silvered mirror,
which reflects some of the
light and lets some through

Lasers can produce holograms

Perhaps the most impressive applica-
tion of lasers is holography. A hologram is a way of storing a three dimensional picture. When it is reproduced, a
true three dimensional image is formed.
The image looks absolutely real to us,
just like the original object. You can
move around the hologram, and look
at it from different directions – exactly
like the real thing.

Apart from their entertainment value,
holograms have other uses. Dentists
can take holograms of their patients'
mouths. They can then use them to
make measurements of the teeth and
their positions, without bothering the
patient again. Engineers use holograms
of nuclear fuel rods to inspect the
inside of a rod without going near to
it. The complex information contained
in a hologram makes them ideal security aids. Most bank cards now carry a
hologram.

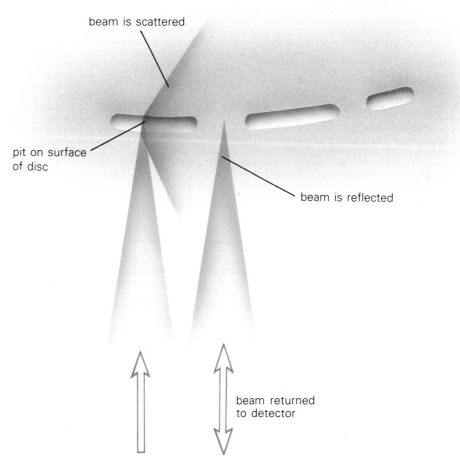

beam is scattered

pit on surface
of disc

beam is reflected

beam returned
to detector

A ripple tank can be used to show how water waves are reflected

A ripple tank is a flat, shallow tank which can be partly filled with water. A vibrator at one end produces ripples, whose shadows can be seen on a screen beneath the tank. The ripples which you see are the **wavefronts**. They are at 90° to the direction in which the waves are moving.

You may be able to use a stroboscope to watch the ripples. The stroboscope produces flashes of light. If the flashes are produced at the same frequency as the ripples, then every time the light flashes on, the ripples seem to be in the same place. Although the waves are really moving, they appear to be stationary.

Figure 116.2 shows what happens if a barrier is placed in the water. The barrier **reflects** the waves.

Water waves are just one kind of wave. There are many other types, including all the different kinds of electromagnetic waves. All waves behave in a similar way. All waves can be reflected.

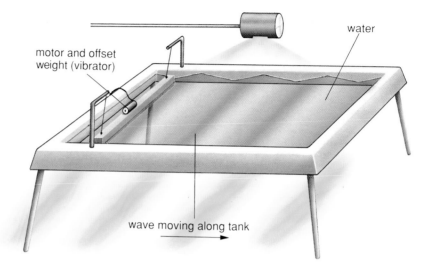

Fig. 116.1 A ripple tank

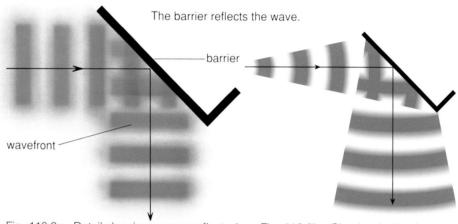

Fig. 116.2a Detail showing a wave reflected in a ripple tank. The arrows show the direction in which the wave is travelling.

Fig. 116.2b Circular ripples from a point continue to spread out after reflection from a flat surface.

We see most objects by reflected light

Most objects do not produce their own light. You can see this page because light is reflected from it. The light originally came from the Sun, or from the electric lights in the room. The light rays hit the page of the book, and are reflected from it into your eyes.

Not all the light which hits an object is reflected. Some of the light is **absorbed**. The brightness of an object depends on how bright the light is which hits the object, and how much light the object reflects. Snow reflects a large proportion of the light which falls on it. If you walk in snow in bright sunlight for a long time, the brightness of the light reflected from the snow can damage your eyes. You may suffer **snow blindness**. Dark glasses will absorb some of this light before it hits your eyes.

Fig. 116.3 An Eskimo in Greenland wears dark glasses to protect his eyes from light reflected from snow.

Reflection of light rays by a plane mirror

A **light ray** is a narrow beam of light. Figure 116.4a shows the apparatus you can use to produce a few parallel rays of light.

A **plane mirror** is a mirror with a flat surface. The mirror you use will probably be made of glass, with a silvered back. Light rays hitting the silvered surface bounce back, or reflect.

1 Set up your apparatus as in the diagram. You may need to partially black out the room. Turn the mirror around at various angles, and watch what happens to the reflected light rays.

2 The **angle of incidence** is the angle at which a light ray hits the mirror. Now make a careful record of exactly what happens to the rays at one particular angle of incidence.

a Place the mirror on a sheet of white paper. Using a pencil, draw a row of dots on the paper to mark the position of the mirror.

b In the same way, mark the position of one of the light rays as it travels to and from the mirror. If your light ray is quite wide, draw along one edge of it. Make sure you draw along the same edge all the time.

c Take your sheet of paper away from the mirror and light rays. Join up the dots, using a ruler. Draw arrows on the lines representing the light rays, to show which way they were travelling.

d Using a set-square, draw a line at right angles to the surface of the mirror, exactly where the light ray hits it. This line is called a **normal**. You should now have a diagram like Figure 116.4b.

3 Repeat step 2 for several other angles of incidence, using a fresh piece of paper each time.

4 On each of your drawings, use a protractor to measure the angle of incidence and the angle of reflection.

Questions

1 What do you notice about the angle of incidence and the angle of reflection in each case?

2 Do you think this also applies to the reflection of water waves? How could you test this, using a ripple tank?

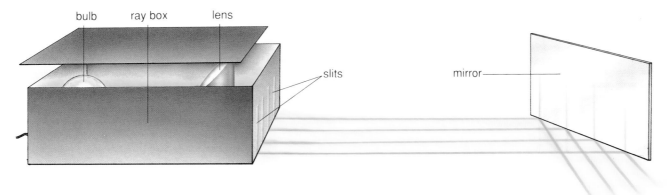

Fig. 116.4a

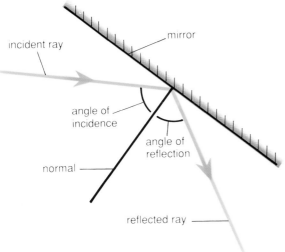

Fig. 116.4b   Detail of the reflection of one ray at the mirror surface.

Fig. 116.5   Sound waves, like all waves, can be reflected from surfaces. Sound reflectors in a concert hall help to provide the best possible sound quality for the audience.

Regular reflection only happens at very smooth surfaces

A mirror has a very smooth, highly polished surface. Under a microscope, its surface might look like Figure 117.1a. Parallel light rays are all reflected to the same new direction. This is called **regular reflection**.

But most surfaces, even if they seem flat, are really quite rough. This page would look very rough under a microscope,

perhaps like Figure 117.1b. Each tiny piece of the surface is angled differently. Parallel light rays falling onto the surface still obey the laws of reflection, and so are reflected to all sorts of new directions. The reflected light is scattered. This is called **diffuse reflection**.

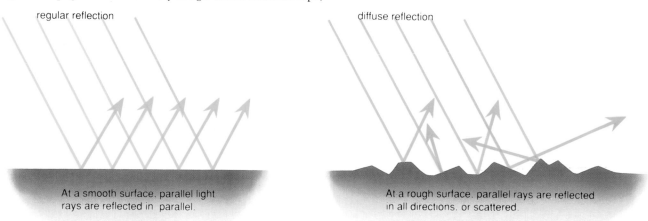

Fig. 117.1a Regular reflection. At a smooth surface, parallel light rays are reflected in parallel.

Fig. 117.1b Diffuse reflection. At a rough surface, parallel rays are reflected in all directions, or scattered.

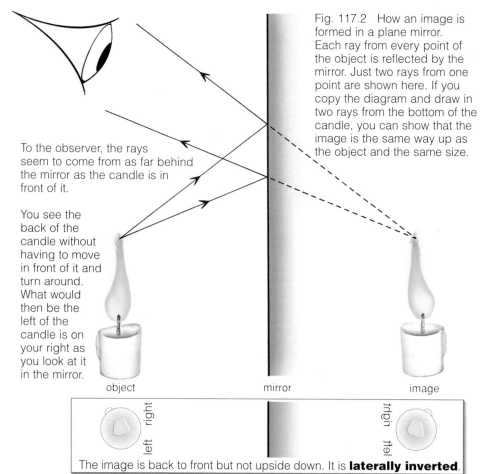

Fig. 117.2 How an image is formed in a plane mirror. Each ray from every point of the object is reflected by the mirror. Just two rays from one point are shown here. If you copy the diagram and draw in two rays from the bottom of the candle, you can show that the image is the same way up as the object and the same size.

To the observer, the rays seem to come from as far behind the mirror as the candle is in front of it.

You see the back of the candle without having to move in front of it and turn around. What would then be the left of the candle is on your right as you look at it in the mirror.

The image is back to front but not upside down. It is **laterally inverted**.

Regular reflection can form an image

A very smooth surface reflects parallel light rays all in exactly the same direction. This is regular reflection. Regular reflection produces reflected light rays in the same arrangement as the incident light rays.

Figure 117.2 shows light rays from the tip of a candle flame being reflected by a mirror. You could draw a similar pattern for all the different points in the candle flame.

The reflected light rays go into the observer's eye. The observer's brain works out where the light rays have come from. The brain has no way of 'knowing' that the light rays have been bent, so it works out that they have come from behind the mirror. So the observer's brain sees an **image** of the candle flame behind the mirror.

If you put a screen behind the mirror, you would not see anything on it. There are not really any light rays behind the mirror. The image is not really there at all. It is called a **virtual image**.

Images in a plane mirror

1 Place a plane mirror in the middle of a piece of white paper. Mark the position of the mirror on the paper. Don't let it move!

2 Make a cross somewhere near the end of the paper, and place a pin exactly in the cross.

3 You are now going to use a ruler as a 'sight' to find the position of the image of the pin in the mirror. Put the ruler edge-on, on the paper, pointing towards the image you can see in the mirror. Look along the ruler, and position it so that it is pointing exactly at the pin. Hold it very still, and draw along its edge.

Alternatively, mark both ends of the ruler, take it away, and then draw a line between your two marked points.

4 Keeping the mirror and pin in exactly the same position, repeat step 3 for two different positions of the ruler.

5 You now have three lines drawn on the 'real pin' side of the mirror. Take away the mirror, and the pin. Carefully continue each of your three lines through to the 'back' of the mirror. If you have lined them up really well, they should all meet at the position where the image of the pin was formed. (If they don't, try again.)

Fig. 117.3

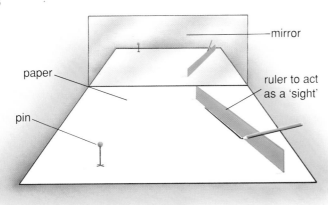

Question

1 What can you say about the position of the pin and its image? Mention both sides of the mirror in your answer.

Questions

1 a Name three objects which produce their own light.

b Name three objects which you see by their reflected light.

2 A mirror and this page both reflect light. Explain why you can see images in the mirror, but not in the page.

3 a A letter P is placed in front of a mirror. Draw a diagram showing what happens to light rays from three different points on the letter as they are reflected by the mirror.

b Extend your three rays behind the mirror, to show where the image is. Join up the three points behind the mirror to show what the image will look like.

c Is the image the same size as the original letter P?

d Is the image the same way round as the original letter P?

4 In a single lens reflex camera, the same lens is used to focus the image onto the film and for

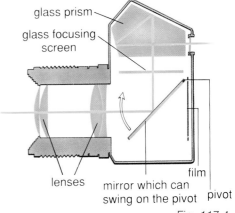

Fig. 117.4

the viewfinder. A hinged mirror can direct light to your eye or allow it to reach the film (see Figure 117.4).

a In which position would the hinged mirror allow light onto the film?

b What would the eye see when the mirror was in this position?

c Some cameras have two separate lenses, one to focus an image onto the film, and one for the viewfinder. What advantages and disadvantages can you suggest between this system, and the single lens system?

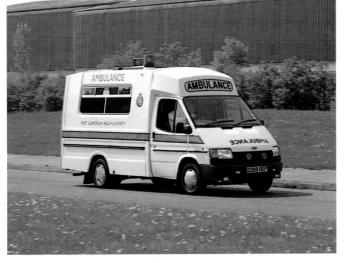

Fig.117.5 Emergency vehicles such as ambulances and breakdown trucks often have their name written in mirror writing, so that it is easily read by drivers looking in their rear-view mirrors.

Reflection from curved surfaces

Fig. 118.1 Reflection of waves at curved surfaces

a water waves

Water waves or light rays reflected from a convex surface seem to come from a point on the other side.

Water waves or light rays reflected from a concave surface converge to a single point.

b light rays

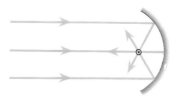

Convex mirrors reflect rays outwards

The rays reflected from a convex mirror appear to come from a single point behind the mirror. An image is formed behind the mirror. The image is smaller than the object.

Convex mirrors can capture rays from a wide area. They are used in shops, so that an image of the whole shop can be seen from one point. Rear view mirrors on cars can also be convex mirrors. They enable drivers to see a wide area of the road behind the car.

Fig. 118.2 Fields of view in convex and plane mirrors

Any light rays entering the yellow area can be reflected into your eye. So you can see objects in the whole of the yellow area.

Fig. 118.3 The convex surface of these mirrors gives a view of the whole road. The image is upright, but smaller than the object.

convex mirror

plane mirror

The field of view in a plane mirror is much smaller.

Concave mirrors reflect light rays inwards

Concave mirrors reflect light rays inwards. The rays are brought to a focus inside the curve of the mirror. The image is larger than the object. It is a magnified image. Dentists use concave mirrors to look at the back of your teeth. The teeth appear larger than they really are.

A concave mirror which is shaped like part of a circle will only give a good focus for a very few rays. One shaped like a **parabola** is much more useful. A parabolic reflector will bring any parallel ray to the same focus. A satellite receiving dish is in the shape of a parabola. It collects radio waves transmitted by a satellite, and reflects them to a focus.

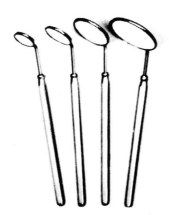

Fig. 118.4 The concave surface of a dentist's mirror forms a magnified image of your teeth.

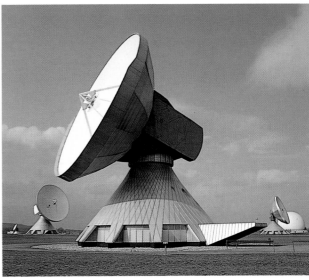

Fig.118.5 A parabolic satellite receiving dish.

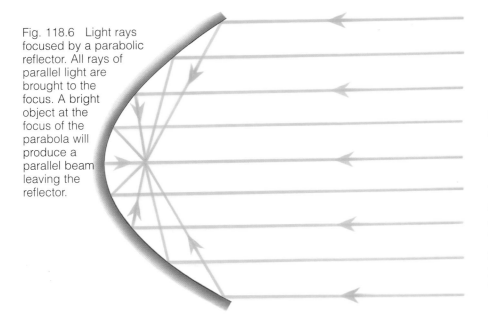

Fig. 118.6 Light rays focused by a parabolic reflector. All rays of parallel light are brought to the focus. A bright object at the focus of the parabola will produce a parallel beam leaving the reflector.

Concave mirrors may be used to spread out radiation into a beam

Waves coming from a point at the focus of a concave mirror will be reflected outwards into a beam of parallel rays. The bars of an electric fire, and the bulb of a torch, are at the focus of the parabola-shaped metal reflector behind them.

Fig. 118.7 The concave reflector of an electric fire directs infrared radiation into the room. The stretched image of the elements almost fills the whole mirror.

Questions

1 a Give three uses for each of the following:
  i a plane mirror
  ii a convex mirror
  iii a concave mirror.

  b For one of your examples for each type of mirror, explain why you would choose this type of mirror.

2 Images may be real or virtual. A real image is produced by light rays which really are in the position in which your eye sees them. A screen placed at this point would have the image on it.

A virtual image is not really there at all. You see it because your brain 'thinks' that that is where the light rays are coming from. A screen placed at the point where you see the image would not have an image on it.

Which type of image is formed by:
a a plane mirror?
b a convex mirror?
c a concave mirror?

119 REFRACTION

When a wave travels from one material to another, its speed may change. This alters the wavelength of the wave, and can make the wave change direction.

Wave speed changes in different materials

As sea waves approach the shore, the shallower water causes them to slow down. Their wavelength also changes. The wavelength becomes shorter.

You can see this happening to waves in a ripple tank. A piece of Perspex can be placed in the bottom of the tank, to make the water shallower at this point. As the waves pass over this shallower area, the wavelength becomes shorter.

If the piece of Perspex is at an angle to the direction of travel of the waves, you can see something else happening. The waves *change direction* as they pass through the shallow area. As they pass back into deeper water, they go back to their original direction of travel.

This change of direction as waves pass from one material to another is called **refraction.**

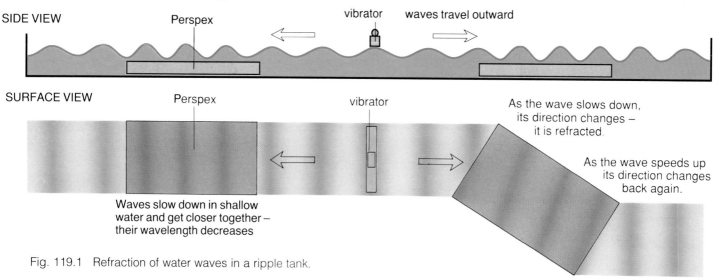

Fig. 119.1 Refraction of water waves in a ripple tank.

Light rays are refracted as they pass from one material to another

Light rays passing from air into a Perspex block behave like the water waves passing from deep water into shallow water. The rays slow down. You cannot see their wavelength change, but you can see how they change direction.

Figure 119.2 shows what happens when a light ray enters and leaves a Perspex block. If it hits the face of the block at right angles, it passes straight through. But if it hits the face of the block at an angle, the ray is bent, or refracted. It is bent towards the normal.

As the ray leaves the block, it bends again. This time, it is refracted away from the normal. It ends up travelling in exactly the same direction as the one in which it began. But the ray has been displaced sideways. The ray leaving the block is parallel with the ray entering the block.

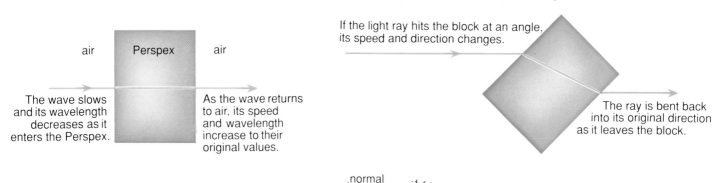

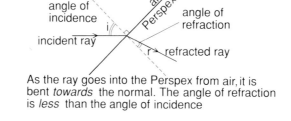

Fig. 119.2 Refraction of light through a rectangular glass or Perspex block.

Refraction of light rays in a semicircular Perspex block

1 Put a Perspex block onto a sheet of white paper. Move the block around, and watch what happens to light rays as they enter and leave it.

2 Now make a careful record of what happens to a light ray with the block in one particular position. Place the block so that a light ray falls onto its straight edge. If you aim the incident ray to hit the centre of this edge, as in Figure 119.3a below, it will hit the curved face at 90°, and not be deflected.

This will make your measurements easier.

Draw a diagram to show what happens to the light ray, using the same technique as for Investigation 116.1.

3 Repeat step 2 for several other positions of the block. Make sure the light ray hits the centre of the flat face each time.

4 Now turn the block around so that the light ray hits the rounded surface, as in Figure 119.3b below. The light

ray will pass straight into the block without bending, but will change direction as it passes from the block back into the air. Make drawings to show what happens for several different positions of the block.

5 Measure the angles i and r for each of your drawings. Angle i is the angle of incidence. Angle r is the angle of refraction.

Fig. 119.3   Refraction of light rays through a semicircular block. In both examples, the incident ray is aimed at the centre of the flat surface.

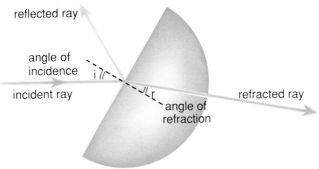

a refraction from air to glass or Perspex

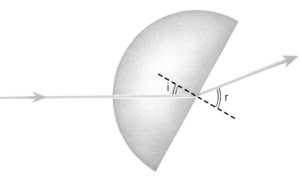

b refraction from glass or Perspex to air

Questions

1 As light rays pass from air into Perspex, are they refracted towards or away from the normal?

2 As light rays pass from Perspex into air, are they refracted towards or away from the normal?

3 In which material do you think light rays travel faster – Perspex or air?

4 Some of the light which hits the block does not pass through it, but is reflected. What do you notice about the strength of this reflected light as the angle of incidence changes?

5 What do you notice when the angle between the normal and the light ray in the Perspex is about 42°?

Light rays are refracted as they enter and leave water

Light rays travel more slowly in water than in air. As light rays leave water, they speed up and bend away from the normal.

Light rays from a brick on the bottom of a swimming pool spread outwards as they travel up through the water. When they reach the water surface, they are refracted. Light rays hitting the surface at a large angle of incidence are refracted more than rays hitting the surface at a small angle of incidence. This makes the rays spread apart even more.

When these rays hit your eye, your brain works out where it thinks they have come from. The brick seems to be much closer to you than it really is. If the pool was 3 m deep, for example, it would look as though it was only 2.25 m deep!

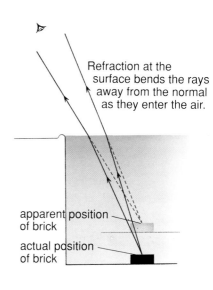

Refraction at the surface bends the rays away from the normal as they enter the air.

apparent position of brick

actual position of brick

Fig. 119.4   Real and apparent depth

Light rays may be unable to escape from a dense material if they hit its boundary with a less dense material at a large angle.

Fig. 120.1 Light rays from glass or Perspex to air. As the angle of incidence increases, the refracted ray becomes weaker and the reflected ray becomes stronger. Eventually, all the light is reflected back into the glass.

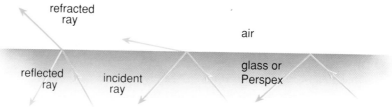

Light rays may not be able to escape from a dense material

As a light ray leaves a dense material, its path is bent away from the normal. But not all the light travels along this path. Some of the light is reflected. You will have seen this happening if you investigated what happens when light rays pass through a Perspex block. Some of the light is reflected back into the block.

As the angle of incidence from the block to the air becomes larger, the reflected ray becomes brighter. If you go on increasing the angle of incidence, there comes a point where the refracted ray is very dim, and travels along the surface of the block.

If you increase the angle of incidence a little more, the refracted ray disappears altogether. All the light is reflected back into the block. This is **total internal reflection**.

The angle of incidence at which total internal reflection happens is different for different materials. For a light ray passing from glass or Perspex into air, the angle is about 42°. At angles greater than this, all the light will be reflected back into the material.

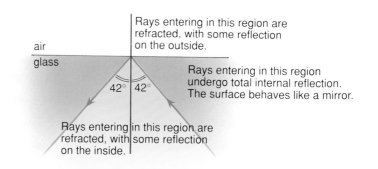

Fig. 120.2 Total internal reflection

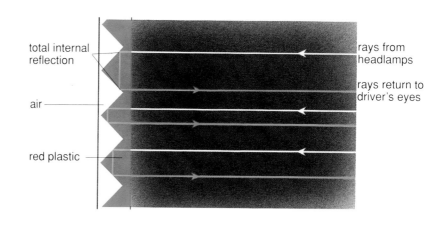

Fig. 120.3 A rear reflector on a bicycle returns light rays to their source.

Total internal reflection can replace mirrors

The red plastic reflector on the back of a bicycle uses total internal reflection.

Figure 120.3 shows what the surface of a rear reflector looks like. The plastic is shaped so that light from a car's headlights hits the front surface at a very small angle of incidence. But the back of the plastic is angled. The light hits this surface at a high angle of incidence. Total internal reflection

occurs. All the light bounces back, and is returned in the direction from which it came. So car drivers see the reflection of their own headlights in the reflector.

No matter what direction the car's headlights come from, this reflector returns them to their source. A mirror only does this for rays hitting it at 90°.

This also works in three dimensions. If you construct a corner out of three mirror tiles, you can see your image in the centre of the corner from many different angles. A reflector like this was

used to measure the distance to the Moon. It was placed on the Moon's surface by American astronauts. A laser beam was then sent from Earth to hit the reflector and bounce back to Earth. The time taken by the laser beam to travel to the Moon and back was used to calculate the distance between the Earth and the Moon. If a flat mirror had been used, a tiny mistake in its alignment would have made the returning beam miss the Earth completely.

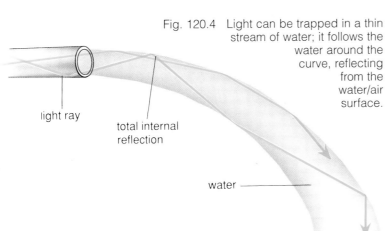

Fig. 120.4 Light can be trapped in a thin stream of water; it follows the water around the curve, reflecting from the water/air surface.

light ray

total internal reflection

water

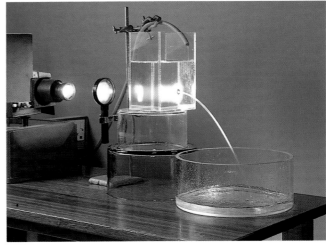

Total internal reflection is used for fibre optics

If a beam of light is sent down a thin glass rod, total internal reflection traps the light inside the rod. Light can go round corners! This technique is called **fibre optics**. The glass rods are so thin that they are called fibres. They may be only as thick as a human hair. The thinner they are, the more they can be coiled without the light 'escaping'.

Any scratches on the surface of the rod might allow a beam to pass through and escape. Optical fibres are made with an outer glass coating and a protective layer of plastic, so that this cannot happen.

Fibre optics are very important in communications, where they are replacing wires in the telephone system. Your voice is transmitted as pulses of light along such fibres. Fibre optics are also used for sending light into, and getting pictures from, inaccessible places. For example, a patient can swallow a tube containing a fine glass fibre, through which a doctor can examine the inside of their stomach without having to perform surgery. This tube is called an **endoscope**.

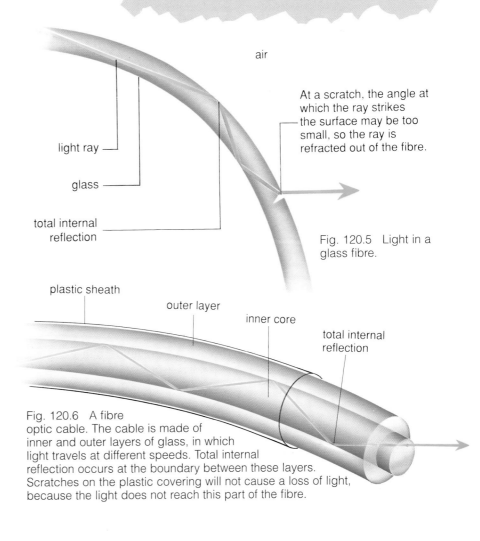

air

light ray

glass

total internal reflection

At a scratch, the angle at which the ray strikes the surface may be too small, so the ray is refracted out of the fibre.

Fig. 120.5 Light in a glass fibre.

plastic sheath

outer layer

inner core

total internal reflection

Fig. 120.6 A fibre optic cable. The cable is made of inner and outer layers of glass, in which light travels at different speeds. Total internal reflection occurs at the boundary between these layers. Scratches on the plastic covering will not cause a loss of light, because the light does not reach this part of the fibre.

121 DIFFRACTION

When waves pass the edge of an object they can spread out around the corner. This happens for sound waves and for electromagnetic radiation.

Diffraction bends waves around edges

As a wave moves forward it keeps going in a straight line. The ripples in water show the **wavefront**. The wavefront is at a right angle to the direction in which the wave is travelling. Fig 121.1 shows how the wavefront moves forward. You can think of each point on the wavefront as a new source of ripples. With no objects in the way, all the small ripples (wavelets) add together to form the new wavefront. If something gets in the way, the wavelets can't add together to make a straight wavefront.

Some of the wave can then move behind the object. This process is called **diffraction**. You can see this in a ripple tank if you set up a barrier. You can also try it out with sound waves. Find a large building and see if you can stand just round the corner from someone and have a conversation with them. Experience tells us that you can. But isn't light a wave as well? If the sound can be diffracted around the corner, why can't you see the person as well as talk to them?

Fig. 121.2 The amount of diffraction depends on the size of the object or hole and the wavelength of the waves. When the size of the object is about the same as the wavelength, diffraction becomes important. If the gap is large compared to the wavelength the waves don't spread out much.

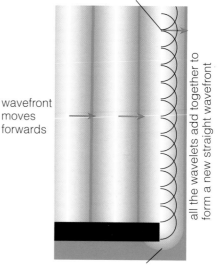

Fig. 121.1 each point sends the disturbance forwards

wavefront moves forwards

all the wavelets add together to form a new straight wavefront

An obstruction allows the wavelets to expand into the shadow area.

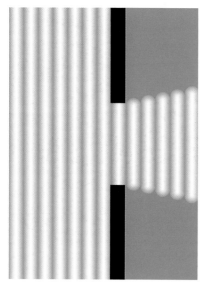

When the gap is the same width as one wavelength the waves spread out completely.

Diffraction increases with wavelength

Figure 121.2 shows two examples of waves being diffracted by a gap in a barrier. In the first diagram the gap is about eight times the wavelength. There is not much diffraction. This is like a sound wave travelling through a double garage doorway. In the second diagram the gap and wavelength are the same. The diffraction bends the wavefront completely into the area you would expect to be in shadow. Instead of making the gap smaller, you could increase the wavelength to get the same effect. Because sound waves have large wavelengths they are easily diffracted. To get the same diffraction for light you would need a hole 0.0000005 m across. This doesn't just work for holes. A small object will diffract light into the shadow region. You can see a small object distorting light waves if you try the investigation.

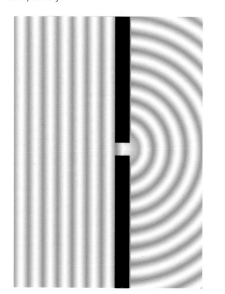

INVESTIGATION 121.1

Bending light

Take a sheet of paper and a pin. Use the pin to push a hole into the paper. Stand at least 2 or 3 metres from an electric light. Hold the paper at arm's length and look at the light through the pinhole. Do not look at **any** light that would be too bright to look at normally. You may be able to see a halo around the bright pinhole. Now move the pin in front of the pinhole and focus your eye on the pin – be careful not to poke your eye! With a bit of practice you can get the bright light to look as if it's shining right through the pin. Try different sizes of pinhole and widths of pin.

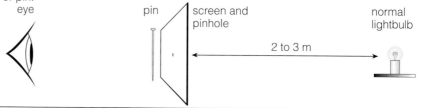

eye — pin — screen and pinhole — normal lightbulb — 2 to 3 m

Radio waves have different wavelengths

Long wave transmissions have wavelengths around 1 km and are diffracted by geographical features like mountains. This means that wherever you are you can probably pick up long wave transmissions. FM broadcasts have wavelengths of around 3 m.

Reception is best when you have a clear view in a straight line to the transmitter – as motorists often find when they are driving around! There is more about the range of radio waves in Topic 114.

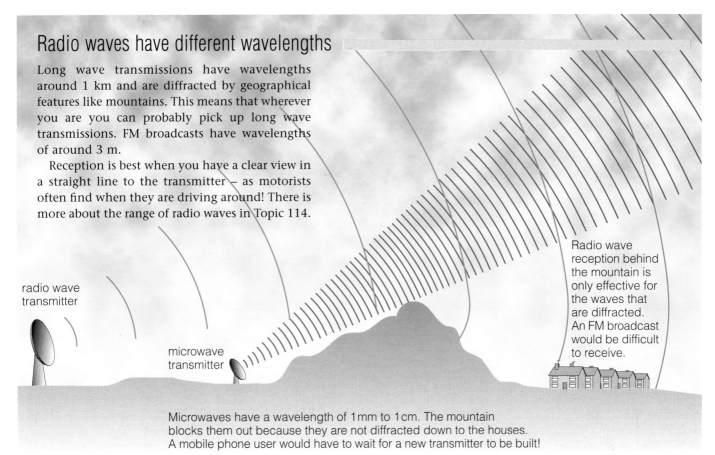

radio wave transmitter

microwave transmitter

Radio wave reception behind the mountain is only effective for the waves that are diffracted. An FM broadcast would be difficult to receive.

Microwaves have a wavelength of 1 mm to 1 cm. The mountain blocks them out because they are not diffracted down to the houses. A mobile phone user would have to wait for a new transmitter to be built!

Fig. 121.4

Diffraction happens when waves add or cancel each other

When waves meet, the two disturbances add together. You can work out the new amplitude by combining the two values. This means that if two waves are in step or **in phase** then the new wave is twice as big. If the two waves are out of phase then they will cancel each other out. This effect is called **interference**. When waves add and get bigger this is **constructive interference**. When they cancel it is **destructive interference**. Diffraction happens because in some directions the wavelets arrive in step. In other directions they don't arrive in step and cancel out. In some directions they add together and produce a disturbance where you wouldn't expect it. This means a small hole produces a pattern of light, not just a single spot. How far over you have to move before the pattern gets dark depends on how far it takes for the waves to get out of phase. This is affected by how wide the hole is and the wavelength of the light. Figure 121.6 shows the pattern you see on a screen if you shine a red laser at a small hole. The central bright spot is bigger than the hole. There are other directions in which the waves also add constructively to make the rings.

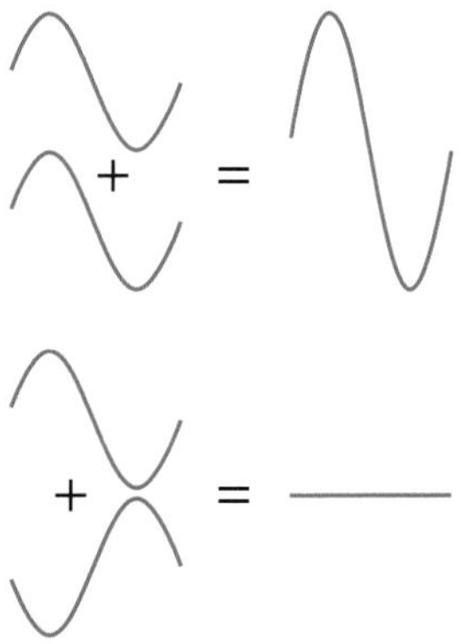

Fig. 121.5

Fig. 121.6

Questions

1 A high pitched shriek may have a frequency of 3000 Hz, whereas a low bellow might have a frequency of only 60 Hz. For each frequency, calculate the wavelength. Use this information to explain why

   a in woods it is easier to locate the direction of a scream rather than a bellow

   b in cities you can often hear low traffic noise wherever you are.

2 Estimate the size of the hole needed to produce the diffraction pattern in Figure 121.6.

Light is electromagnetic radiation with wavelengths of between 0.0004 mm and 0.0007 mm. We see light of different wavelengths as different colours.

The shape of a glass block affects the way in which a ray passes through it

When a light ray passes through a Perspex block, the ray is refracted as it enters and leaves the block. If the two sides of the block are parallel, the ray emerges parallel to the direction in which it entered. But if the two sides through which the ray enters are not parallel, then the direction of the ray is changed.

Fig. 122.1

The two opposite sides of a rectangular block are parallel. A light ray passing through them is **displaced**, but its direction is not changed.

There are no parallel sides on a triangular prism. A light ray passing through it is not returned to its original direction. The ray is **deviated**.

Fig. 122.2 A rainbow is produced when sunlight is refracted by water droplets in the air.

A prism splits light into different colours

When a beam of light hits the surface of a triangular prism, it slows down. Its wavelength changes. If the light is made up of a number of different wavelengths, each wavelength is altered by a different amount, so each wavelength is bent by a different amount. The shorter the wavelength, the more it is bent as it enters the prism.

As the light leaves the prism, it is bent again, back towards its original path. But, because this face of the prism is not parallel with the first face, the light is not bent back onto its original path.

White light contains light of many different wavelengths. A prism splits up the light into all these different wavelengths. We see the different wavelengths as different colours. Each colour leaves the prism at a slightly different angle. The pattern of colours is called a **spectrum.**

Fig. 122.3 As white light passes through a triangular prism, different wavelengths are deviated through different angles. A spectrum is produced.

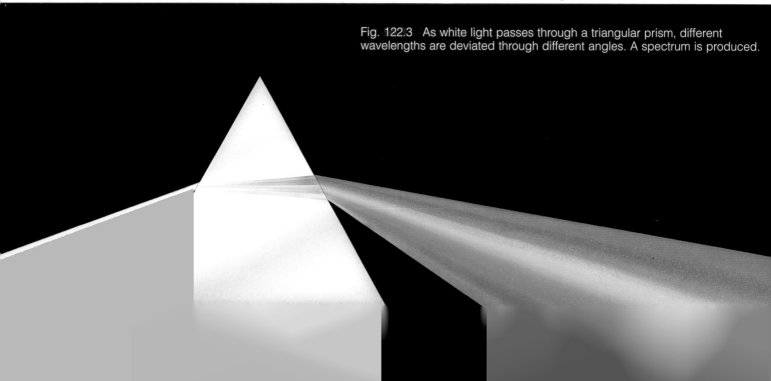

Objects absorb and reflect the light which falls onto them

Sunlight, and the light from electric lighting, is white light. White light contains most of the different wavelengths from 0.0004 mm (violet) to 0.0007 mm (red). But some of the objects around you look coloured. Why is this?

Look at something red – a book perhaps. You can see the book because light from it is going into your eyes. The light from the book is reflected light. The light has come from the Sun, or from the electric lights in the room.

It is white light. When the white light hits the book, some wavelengths are absorbed by the book. The green, blue, yellow and violet wavelengths are absorbed. But the red wavelengths are reflected. This is why the book looks red. It reflects only red light into your eyes.

Objects which look white to us reflect light of all wavelengths. Objects which look black absorb light of all wavelengths.

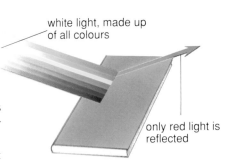

white light, made up of all colours

only red light is reflected

Fig. 122.4   A red book reflects red light, and absorbs all other colours.

Fig. 122.5

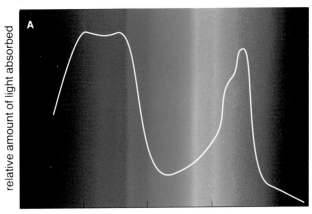

A — relative amount of light absorbed

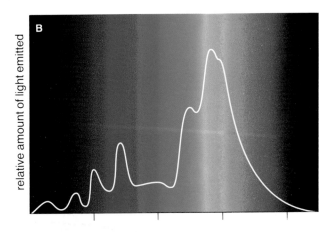

B — relative amount of light emitted

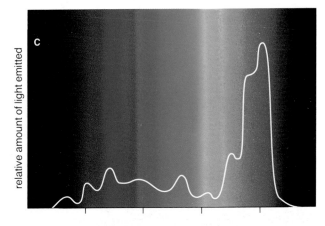

C — relative amount of light emitted

Chlorophyll does not absorb green light

Plants are green because they contain a pigment called **chlorophyll**. Chlorophyll is used in photosynthesis. Chlorophyll absorbs light, and transfers the light energy into organic molecules such as glucose.

However, chlorophyll cannot absorb all the different wavelengths in the sunlight which hits it. It can absorb red and blue light, but it cannot absorb green light. This is why chlorophyll is green. All the green light which hits it is reflected from it, or passes through it.

Some plants, such as copper beech trees, or red seaweeds, do not look green. They do have chlorophyll, and the chlorophyll reflects green light, just as in other plants. But copper beech trees and red seaweeds also contain other pigments which absorb green light and reflect red light. The mixture of green light from the chlorophyll, and red light from these other pigments, looks to us like a reddish-brown colour.

Questions

1 Look at Figure 122.5. Graph A shows the wavelengths of light which are absorbed by chlorophyll. Graph B shows the wavelengths of light which are emitted by an ordinary tungsten filament light bulb. Graph C shows the wavelengths of light which are emitted by a special type of fluorescent light.
   a What colours of light are absorbed by chlorophyll?
   b Why does chlorophyll look green?
   c Why do plants need chlorophyll?
   d Plants growing in a room lit only by tungsten filament lights will normally only survive for a few weeks. Plants growing under the special type of fluorescent light used in graph C thrive. Explain why you think this might be.

We see colour because we have cells in our eyes sensitive to different wavelengths of light. The colour we see depends on what combination of these cells is stimulated.

The human retina contains cells sensitive to different colours of light

At the back of each of your eyes is a layer of cells called the **retina.** The cells in the retina are sensitive to light. When light hits one of these cells, it sets up a tiny electrical impulse which travels along the **optic nerve** to the brain. The brain sorts out the pattern of impulses coming from all the hundreds of thousands of different cells in your retinas, and makes them into an image.

There are two types of sensitive cells in the retina. One type are called **rods.** Rods respond in the same way to all wavelengths of light. No matter what colour light falls onto them, all the rods in your eyes send the same message to the brain. So rods cannot help you to tell what colour anything is. If only your rods are working, you just see in black, white and shades of grey.

The other type of sensitive cells in the retina are called **cones**. Most people have three types of cones. One type is sensitive to red light, one type to blue light, and one type to green light. If red light falls onto a 'red-sensitive' cone, it will send an impulse to the brain. But if green light falls onto this cone, it will not send an impulse. By analysing the messages from all the cones in your retinas, your brain can work out exactly what colour light is falling on which part of the retina. It can build up a colour image of whatever you are looking at.

Cones do, however, have one big disadvantage over rods. Cones will only respond to bright light. Rods will respond to quite dim light. So cones are useless in the dark, or even at dusk. Many night-active, or **nocturnal**, animals do not have any cones at all. They just have rods. They do' not have colour vision.

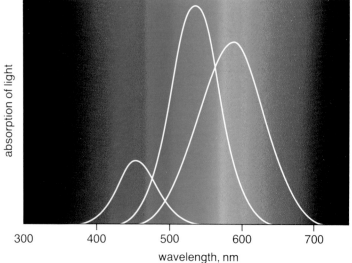

Fig. 123.1 There are three types of cone cell in the retina of the eye, each sensitive to different wavelengths (colours) of light. The three lines on the graph show the colours absorbed by the three types of cone.

Fig. 123.2 The huge eyes of a bush baby collect as much light as possible, because it hunts at night. Like many nocturnal mammals, it has few cones; the retina contains a high density of rod cells, which are sensitive even at low light intensities.

Any colour can be made from red, green and blue light

Any colour may be made by adding together red, green and blue light in the correct amounts. If a mixture of red and green light hits your eye, your red-sensitive and green-sensitive cones send impulses to your brain. Your brain interprets this as yellow light.

A colour television works in this way. The picture on the screen is made up of dots of light. The colours are made up of red, green and blue dots, in different combinations and of different intensities. If you look closely at a television screen, you can see these dots.

Red, green and blue are called the **primary colours of light.** You can make any colour from red, green and blue light. But you cannot make red, green or blue light from any other coloured light.

Colours which can be made by adding any two of the primary colours of light are called **secondary colours of light**. Figure 123.3 shows how the three secondary colours – yellow, magenta and cyan – are made.

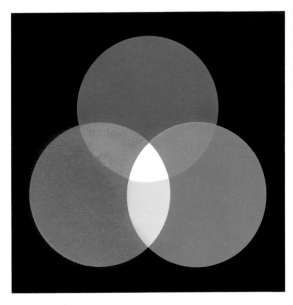

Fig. 123.3 Mixing red, green and blue light produces white light. Which coloured lights produce cyan (turquoise), magenta and yellow?

Objects look different in different colours of light

If you shine white light onto a red book, the book looks red because it reflects only the red light into your eyes. If you look at the book in red light, it still looks red, because it reflects the red light. But if you look at the book in green light it looks black. There is no red light for it to reflect, so it does not reflect any light at all, and it looks black.

What happens if you shine yellow light onto the red book? Yellow light is a mixture of red and green light. The book will absorb the green part of the yellow light, and reflect the red part. So it still reflects red light into your eye, and still looks red.

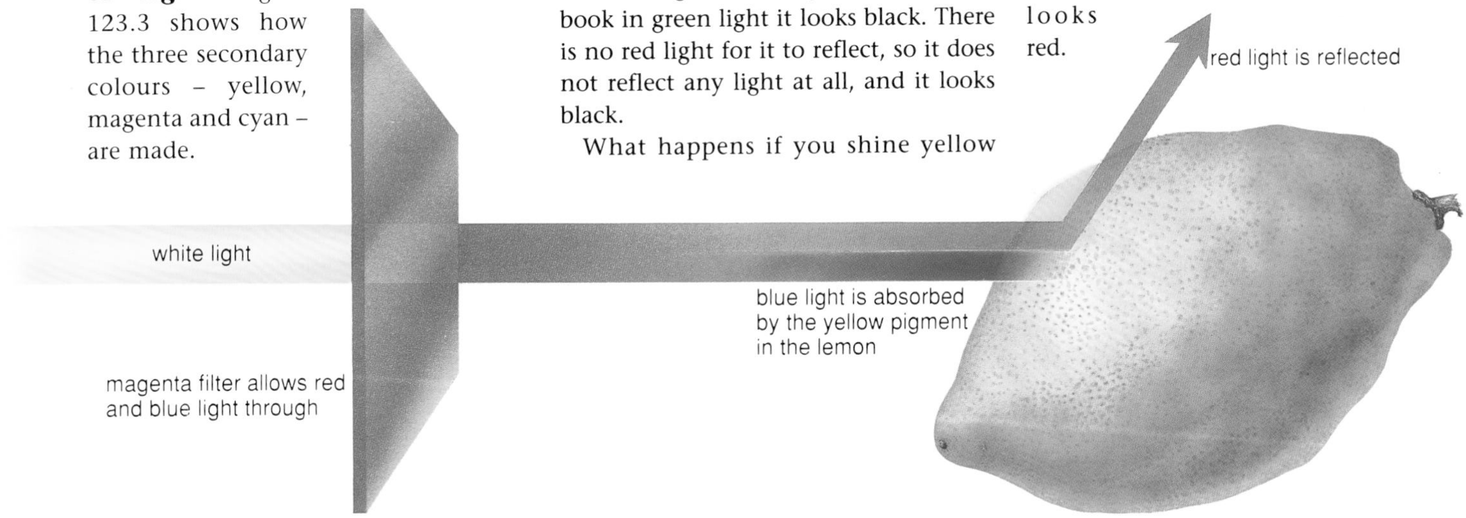

white light

magenta filter allows red and blue light through

blue light is absorbed by the yellow pigment in the lemon

red light is reflected

Fig. 123.4 A lemon appears red under magenta light. What colour would it appear if you used a cyan filter?

Questions

1 Pigeons can be trained to peck at a panel to make it open and allow them to reach food inside. If different patterns – for example, a circle or a square – are drawn on the panels, the pigeons can learn which pattern to peck in order to find food.

Design an experiment, using a similar method to that described above, to find out if pigeons have colour vision.

2 A car accident happened in a street lit by sodium lamps. A witness reported that a green car pulled out of a side road. It caused a red car to swerve into the path of a black car.

The following day, the three drivers were interviewed. Driver A has a blue-green car. Driver B has a blue car, and driver C has a magenta car. Could the witness' report have been accurate? Explain your answer.

3 a What colours does a magenta filter allow through?

b What colours would a cyan filter allow through?

c What colour would be produced if a floodlight producing white light had cyan and magenta filters placed together in front of it?

124 PIGMENTS

Pigments are substances which absorb some wavelengths of light and reflect others. We use them for colouring all sorts of different materials.

We use pigments to colour objects, material, and skin

The chemicals used to colour objects are called **pigments**. The first pigments people used were naturally occurring ones. Ancient Britons, for example, used a blue dye extracted from a plant to colour their skin. The dye was called woad. The purple pigment from a marine mollusc called *Murex* was much prized by the Romans for dyeing cloth. Many paints are also made from naturally occurring pigments. Cobalt Blue, for example, is made from cobalt phosphate. Carmine is made from cochineal, which comes from an insect.

Many of these natural pigments are still used, but some of them have been replaced by synthetic ones. Aniline, an organic chemical which can be extracted from coal or oil, is the basis of many modern, bright pigments.

We can also use dyes to colour food. Food colourings may be natural or artificial. Natural ones include carotene, a yellow colour from carrots, and caramel, which is a brown colour. But there are about twenty permitted food colourings which are artificial. Food colourings are used to make food look more appetising, but many people now prefer naturally coloured food, without unnecessary additives.

Fig. 124.1 Modern dyes can produce an almost infinite range of colours.

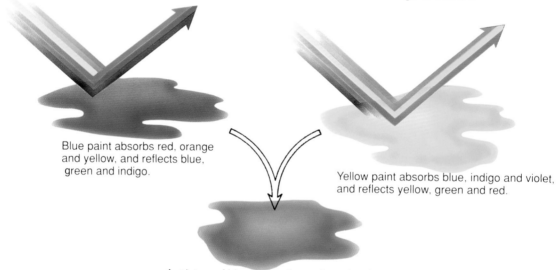

Blue paint absorbs red, orange and yellow, and reflects blue, green and indigo.

Yellow paint absorbs blue, indigo and violet, and reflects yellow, green and red.

A mixture of blue and yellow paints absorbs red, orange, yellow, blue, indigo and violet. Only green is reflected.

Fig. 124.2 The reason why blue and yellow paint make green.

Pigments produce colours by subtraction

Pigments are coloured because they absorb some colours and reflect others. A red pigment absorbs all light except red. In fact, most pigments reflect a mixture of different wavelengths of light. The red pigment used to colour a red book cover, for example, might reflect a lot of red light, and also some yellow light.

We say that pigments produce colour by **subtraction.** A red pigment absorbs, or subtracts, all the green and blue light from white light. Only the red light is left to be reflected and enter your eyes.

A yellow pigment subtracts all the colours except red, green and yellow from white light. A blue pigment subtracts all the colours except blue, green and indigo.

What happens if you mix yellow and blue pigments? The only colour which is not subtracted by the two of them together is green. So the new pigment looks green.

An experienced artist can produce many different colours just by mixing red, yellow and blue paints. In theory, if you mixed these together, they should, between them, absorb all the colours which fall on them. The mixture should look black. But the pigments are, in fact, quite impure, so you actually get a dirty brown colour when they are all mixed.

Colour printing uses cyan, magenta and yellow

Colour printing uses just three colours – cyan, magenta and yellow – to produce the entire range of colours. All the colours in this book have been made from just these three colours.

- Cyan reflects green and blue light, and subtracts red.
- Magenta reflects red and blue light, and subtracts green.
- Yellow reflects red and green light, and subtracts blue.

To produce a green colour on a page, a mixture of cyan and yellow is used. This mixture subtracts red and blue light from the white light falling onto the page. So the reflected light is green. Different shades of green can be made by using different proportions of cyan and yellow.

To print a colour picture, four prints are made. One is made in pure cyan, one in magenta, one in yellow, and one in black to heighten contrast between light and dark areas. The four prints are superimposed to build up the full colour picture. This process is much more expensive than black-and-white printing, which can be done with just one print.

Fig. 124.3 Four single colour images are combined to make a full-colour print. You can see the final picture on the cover of this book.

Fig. 124.4 Printing colours

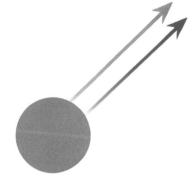

Cyan absorbs red, but reflects green and blue.

Magenta absorbs green, but reflects red and blue.

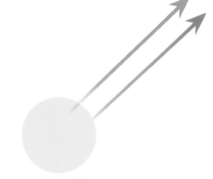

Yellow absorbs blue, but reflects red and green.

Questions

1. A dress designer wishes to produce a green pattern on a cyan background. The fabric is first dyed one colour, then the pattern is added. What colours should be used for:
   a the first dye?
   b the pattern colour?
   (Choose from red, green, blue, cyan, magenta, and yellow.)

2. 'Newton's disc' is a circle, divided into segments, coloured with all the colours of the spectrum. (You can easily make one with some white card and felt-tip pens or paints.) If the disc is spun very fast, it looks almost white. Explain why.

3. A colour photograph is made from three layers printed in magenta, cyan and yellow, on white card.
   a What colour will appear on the photograph where both magenta and cyan are printed? Why?
   b What colour will appear where both yellow and cyan are printed? Why?
   c What colour will appear where magenta, cyan and yellow are printed? Why?
   d What colour will appear where none of these colours are printed?

A convex lens bends light rays inwards, and brings them to a focus, so it is a converging lens. Concave lenses are diverging lenses.

A lens can be thought of as a series of prisms

You have seen how a triangular prism changes the direction of a beam of light passing through it. Figure 125.1 shows what would happen if three rays of light passed through two triangular prisms and a rectangular block. If you used prisms with carefully chosen angles, you could get all three rays to cross at one point. You would have brought the rays to a **focus**.

A **lens** can be thought of as a series of tiny prisms. In Figure 125.1, the prisms at the edge of the lens have sharply angled edges. The prism in the middle has straight edges. Light rays can pass straight through the prism in the middle, but are sharply bent by the prisms at each end. The lens can bring the light rays to a focus.

The lens in this diagram is a **convex** lens. It bends light rays inwards, and brings them to a focus. The light rays are brought closer together, or made to **converge**. A convex lens is a **converging lens**.

Figure 125.2 shows how a **concave** lens bends light rays. The light rays are spread outwards. A concave lens is a **diverging** lens.

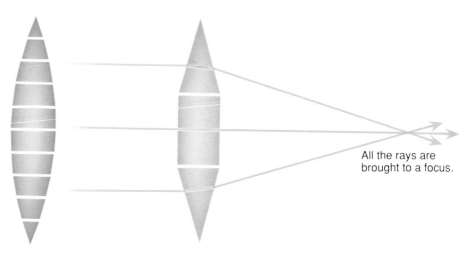

Fig. 125.1   A converging lens can be thought of as a series of prisms. The 'prisms' at the edge have more sharply angled sides and bend the light rays a lot. A ray passing through the centre is not deviated at all.

All the rays are brought to a focus.

Fig. 125.2   A similar model can be used for a diverging lens. The outer parts of the lens bend the rays more than the central parts.

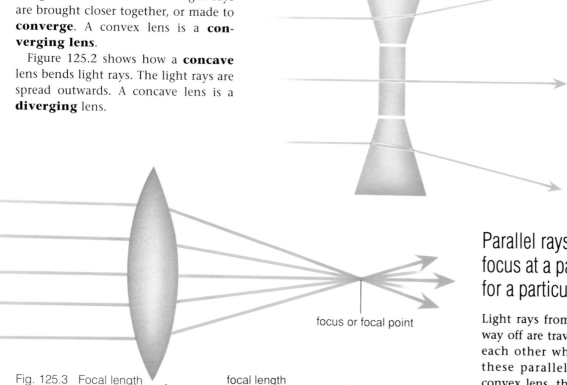

Fig. 125.3   Focal length of a convex lens.

focus or focal point

focal length

Parallel rays are brought to a focus at a particular distance for a particular lens

Light rays from an object a very long way off are travelling almost parallel to each other when they reach you. If these parallel rays pass through a convex lens, they will be brought to a focus. The distance between this focusing point and the centre of the lens is called the **focal length** of the lens.

Investigating a convex lens

TAKE CARE! It is dangerous to look through a lens at a bright object.

1 Find the **focal length** of your lens. Choose an object a long way away – the further away the better – such as a tree outside the window or the horizon. Hold a piece of paper (as a screen) behind the lens, and move it around until you get a sharp image. Measure the focal length.

For these experiments, a lens with a focal length of between 10 cm and 20 cm will work best. You may need to work in a darkened room.

2 Place an object (such as a candle or a lit light bulb) more than two focal lengths away from the lens. Move the screen backwards and forwards until you have a focused image.

a Measure the distance of the image from the lens.

b Which is larger – the image or the object?

c Which is larger – the image distance, or the object distance?

3 Move the object closer to the lens. Again, move the screen until you get a focused image.

a What happens to the image distance as the object distance gets smaller?

b What happens to the image size as the object distance gets smaller?

4 Place the object at a distance of exactly twice the focal length from the lens.

a Compare the image and object distances.

b Compare the image and object sizes.

5 Place the object somewhere between the focal length and twice the focal length.

a Where must the object be to produce the largest image?

b Where is this image?

6 Place the object at a distance less than the focal length of the lens.

a Can you get a focused image on the screen?

b If you are using a light bulb reduce its brightness by turning down the voltage on the power pack. Look *through* the lens. Can you see an image now? What sort of image is it? In what ways is it different from all the other images you have seen?

7 There are several patterns in the information you have collected about image and object sizes and distances. Try to summarise them.

Fig. 125.4

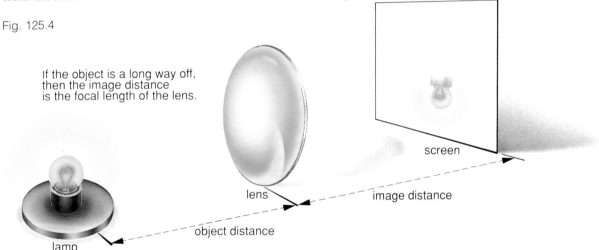

If the object is a long way off, then the image distance is the focal length of the lens.

lamp   object distance   lens   image distance   screen

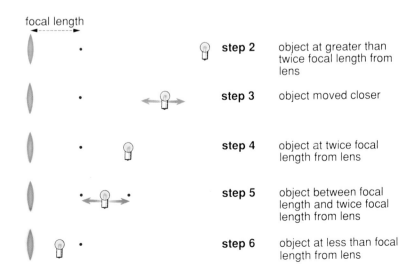

focal length

step 2	object at greater than twice focal length from lens
step 3	object moved closer
step 4	object at twice focal length from lens
step 5	object between focal length and twice focal length from lens
step 6	object at less than focal length from lens

DID YOU KNOW?

Gravity can bend light. The gravity of a star could be used to focus light from stars.

126 MORE ABOUT LENSES

Diverging lenses spread light rays outwards. Ray diagrams can be used to find positions of images formed by converging or diverging lenses.

A concave lens spreads out light rays

A concave lens is thinner in the middle than at the edges. It spreads light rays outwards, or makes them **diverge.** A concave lens is a diverging lens.

If you look at the rays coming out from the lens, your brain sees them as coming from a single point behind the lens. You think that there is an object behind the lens. Really, there is nothing there. A concave lens produces a virtual image.

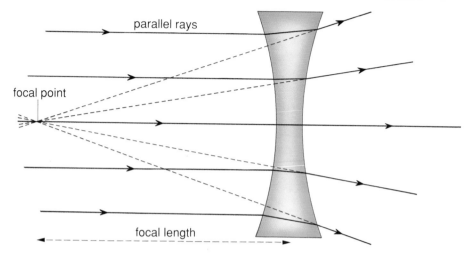

Fig. 126.1 Virtual image formation in a diverging lens. The solid lines show the paths taken by the light rays. An eye assumes the rays have travelled in straight lines, along the paths shown by the dotted lines.

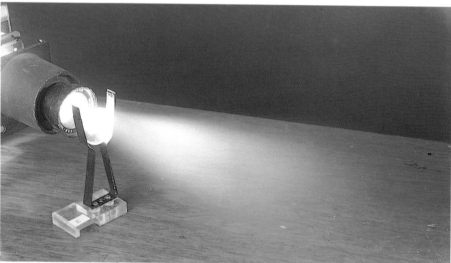

Fig. 126.2 Light diverges after passing through a concave lens.

Questions

1 A lamp is placed 10 cm from a lens of focal length 5 cm. Use graph paper to draw ray diagrams to find the position of the image if the lens is:
   a converging
   b diverging.
2 A microscope slide 22 cm from a 20 cm focal length lens forms a focused image on a wall. Draw a scale diagram (1 mm = 1 cm) to find the distance of the wall from the lens.

— *EXTENSION* —

INVESTIGATION 126.1

Finding the focal length of a diverging lens

The focal length of a diverging lens is not as easy to find as the focal length of a converging lens. When you want to find the focal length of a converging lens, you can use a screen to find where the image is. But with a diverging lens, the image is not real, so you cannot see it on a screen.

Instead, you have to trace the rays back to the focus. This is very hard to do in three dimensions! But if you use a cylindrical lens, the light is only bent in one direction, so you can work in just two dimensions.

1 Set up a ray box to provide three beams of light across a white piece of paper.
2 Place a cylindrical diverging lens in the path of the beams. Mark the position of the lens on the paper.
3 Mark crosses on the paper to show the positions of the rays entering and leaving the lens.
4 Remove the lens and continue the 'exit' lines backwards to find the focus.
5 Measure the focal length of the lens.

6 Find the focal length of a cylindrical converging lens.
7 Identify a diverging and a converging lens with the same focal length. Put the two lenses close together, and place them in the path of the light rays. What effect do they have?

Fig. 126.3   How to draw diagrams. A ray ▶
diagram is a drawing showing how light rays
travel. Ray diagrams can be used to find the
positions of images formed by lenses. Some rays
are much easier to draw than others. In this
figure, the following three rays have been drawn:
1 a ray of light approaching the lens, parallel to
its principal axis. This ray will be bent by the lens,
and will pass through the focal point of the lens;
2 a ray of light passing through the focal point
of the lens, which will emerge running parallel to
the principal axis of the lens;
3 a ray of light passing exactly through the
centre of the lens, which will not be bent at all.

In fact, you need to draw only two of these rays
to find the position of the image, but it is a good
idea to draw three, to make sure you have not
made a mistake.

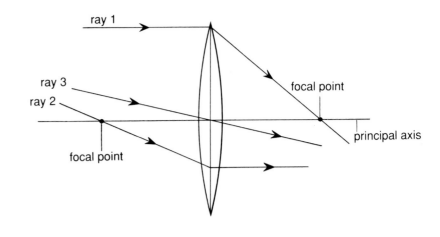

Fig.126.4   Where is the image of a pin, placed ▶
6 cm from a converging lens of focal length 4 cm?
(This ray diagram is shown at half actual size.)
1 Draw the positions of the object, lens, and
focal points on the principal axis. Notice that the
size of the lens is not important, but the line
marking its *position* is. Measure the distances
from the centre of the lens.
2 Draw two rays from the top of the pin to
locate the image. Draw a third ray to check the
accuracy of your drawing. The point at which
these rays cross is the image of the top of the pin.
3 In this example the image is upside-down. It is
magnified and it is 12 cm from the centre of the
lens.

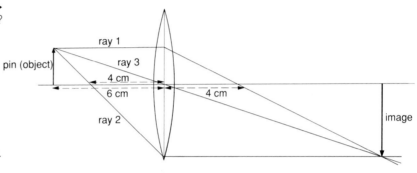

Fig. 126.5   Drawing ray diagrams for a diverging lens. Draw
three rays in the same way as before.
1 A ray of light approaching the lens parallel to its principal
axis. This is bent as though it had come from the focal point of
the lens.
2 A ray travelling towards the focal point on the far side of the
lens. This is bent to run parallel to the principal axis.
3 A ray passing straight through the centre of the lens.
▼

Fig. 126.6   What kind of image is formed by a diverging lens?
Three rays are drawn from the top of the object. To an observer,
these three rays seem to come from the point I. This is where you
see the image. The image is on the same side of the lens as the
object. It is the right way up and smaller than the object. It is a
virtual image, and could not be focused on a screen.
▼

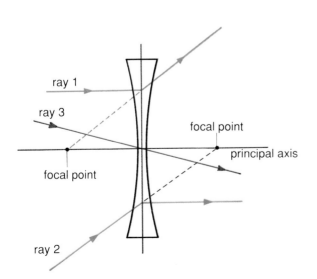

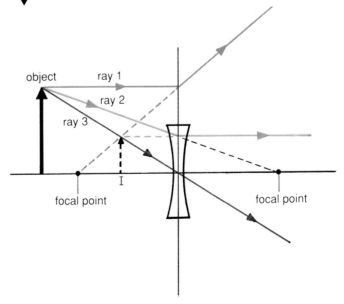

OPTICAL INSTRUMENTS

Optical instruments enable us to see objects which are too distant, too small, or in the wrong position for us to see with the naked eye.

Fig. 127.1   A magnifying glass is held closer to the text than its focal length. A magnified virtual image is formed. Looking through the lens, the writing appears larger.

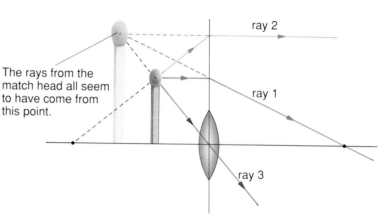

The rays from the match head all seem to have come from this point.

Fig. 127.2   A ray diagram shows how a magnifying glass produces an enlarged image. The object must be placed between the convex lens and its focal point. A magnified virtual image is formed on the object side of the lens. You have to look through the lens, towards the object, to see it. The image is the right way up, and can be very much larger than the object.

Fig. 127.3   An astronomer's telescope. The image formed is upside-down, which does not matter if you are looking at stars or planets.

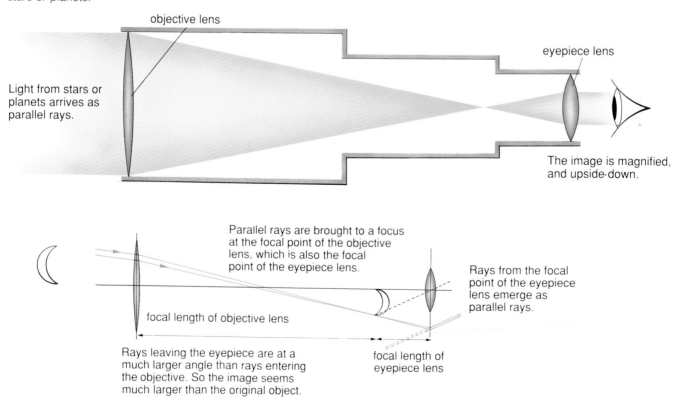

objective lens

eyepiece lens

Light from stars or planets arrives as parallel rays.

The image is magnified, and upside-down.

Parallel rays are brought to a focus at the focal point of the objective lens, which is also the focal point of the eyepiece lens.

Rays from the focal point of the eyepiece lens emerge as parallel rays.

focal length of objective lens

focal length of eyepiece lens

Rays leaving the eyepiece are at a much larger angle than rays entering the objective. So the image seems much larger than the original object.

Fig. 127.4   A ray diagram for an astronomical telescope

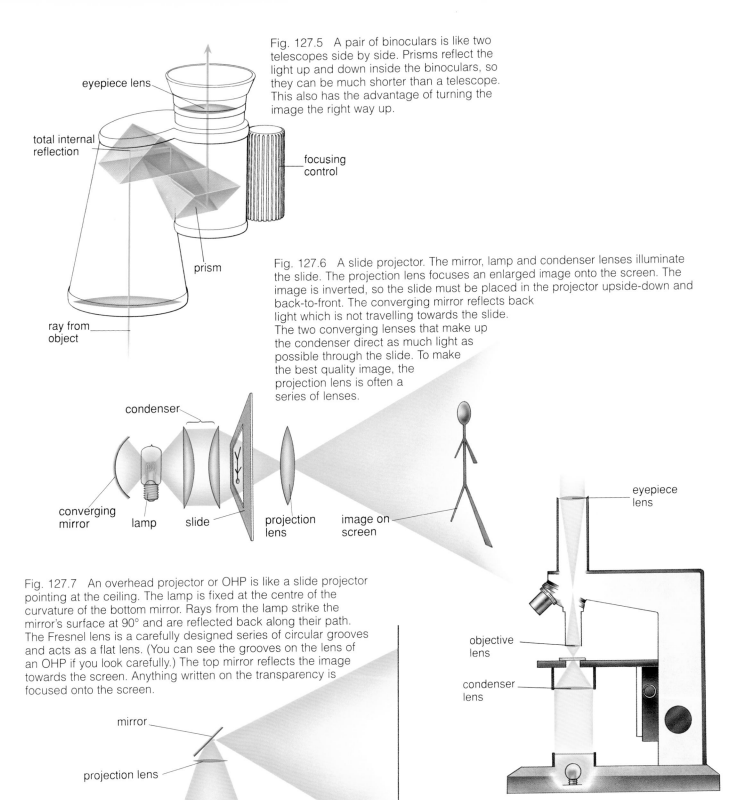

eyepiece lens

total internal reflection

prism

ray from object

focusing control

Fig. 127.5　A pair of binoculars is like two telescopes side by side. Prisms reflect the light up and down inside the binoculars, so they can be much shorter than a telescope. This also has the advantage of turning the image the right way up.

Fig. 127.6　A slide projector. The mirror, lamp and condenser lenses illuminate the slide. The projection lens focuses an enlarged image onto the screen. The image is inverted, so the slide must be placed in the projector upside-down and back-to-front. The converging mirror reflects back light which is not travelling towards the slide. The two converging lenses that make up the condenser direct as much light as possible through the slide. To make the best quality image, the projection lens is often a series of lenses.

condenser

converging mirror

lamp

slide

projection lens

image on screen

Fig. 127.7　An overhead projector or OHP is like a slide projector pointing at the ceiling. The lamp is fixed at the centre of the curvature of the bottom mirror. Rays from the lamp strike the mirror's surface at 90° and are reflected back along their path. The Fresnel lens is a carefully designed series of circular grooves and acts as a flat lens. (You can see the grooves on the lens of an OHP if you look carefully.) The top mirror reflects the image towards the screen. Anything written on the transparency is focused onto the screen.

mirror

projection lens

transparency

lamp

concave mirror

image on screen

Fresnel lens

eyepiece lens

objective lens

condenser lens

Fig. 127.8　A microscope. The condenser and objective lenses behave like a projector, forming an image near the eyepiece. The eyepiece acts as a magnifying glass on this image.

DID YOU KNOW?

The first microscopes used a drop of water for the objective lens.

Questions

1 A periscope has two mirrors. It is arranged to allow you to see over high walls, or over people's heads in a crowd.

a Copy the diagram, and add an object and an eye.

b Draw in a ray of light from the top of the object, showing how this ray is reflected by the mirrors and reaches the eye. Draw a second ray from the bottom of the object.

c Is the object the right way up, or upside-down?

d Is the object the right way round, or laterally inverted?

2 The diagram shows two light rays leaving an object O and striking a mirror. Copy the diagram, draw in the reflected rays, and show where the image would be.

3 The diagram shows a side view of an electric fire.

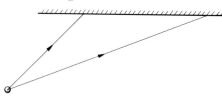

element ———   ——— reflector

a What is the shape of the reflector?

b What would be the best shape for this reflector?

c Suggest a material from which the reflector could be made. Give reasons for your choice.

d What essential safety feature is missing from the fire?

e What types of waves will be emitted from the element of the fire?

f Draw in two rays to show how this radiation is reflected into the room.

4 The diagram shows part of the rear surface of a bicycle reflector. Copy and complete the diagram, to show what happens to each of the three rays from the different cars. (Each ray will be reflected twice.)

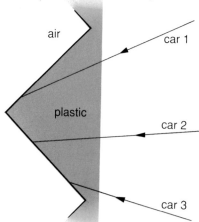

air

plastic

car 1

car 2

car 3

5 A man with 1 m tall waders steps into a stream which looks 90 cm deep. He gets very wet. How was he tricked by refraction?

6 The diagram shows two spotlights shining onto a stage. What colours will be seen in the regions marked A to G?

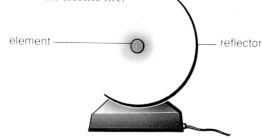

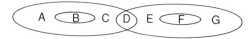

A B C D E F G

7 As sunlight travels down through deep water, different colours are gradually absorbed. The chart shows the depths to which different colours can penetrate in clear water.

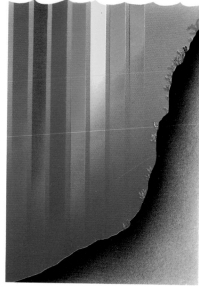

a Which type of light can penetrate deepest?

If you investigate a rocky shore at low tide, you will probably find several kinds of seaweed. Green seaweeds tend to grow high up on the shore. Even when the tide comes in, they will not be covered with deep water. Red seaweeds tend to grow low down on the shore. When the tide comes in, they will be deep under water. The red and green colours of the seaweeds are caused by their light-absorbing pigments.

b Why do seaweeds need light?

c What colours of light are likely to be most available to red seaweeds when the tide is in?

d Why are the seaweeds growing low down on the shore red and not green?

USEFUL FORMULAE + CIRCUIT SYMBOLS

Useful formulae

moment = force × perpendicular distance from pivot

mechanical advantage = $\dfrac{\text{load force}}{\text{effort force}}$

efficiency = $\dfrac{\text{useful output energy}}{\text{input energy}}$

average speed = $\dfrac{\text{distance moved}}{\text{time taken}}$

acceleration (m/s²) = $\dfrac{\text{change in velocity (m/s)}}{\text{time taken (s)}}$

displacement = velocity × time

acceleration = $\dfrac{\text{force}}{\text{mass}}$

momentum = mass × velocity

impulse = change in momentum

kinetic energy = $\dfrac{1}{2}$ mass × velocity²

Electricity

current = $\dfrac{\text{charge passing}}{\text{time}}$

energy = velocity × charge

potential difference (voltage) = current × resistance

resistance = $\dfrac{\text{potential difference (voltage)}}{\text{current}}$

total resistance in series = $R_1 + R_2 + R_3 + \ldots\ldots$

$\dfrac{1}{\text{total resistance in parallel}} = \dfrac{1}{R_1} + \dfrac{1}{R_2} + \dfrac{1}{R_3} + \ldots\ldots\ldots$

power = potential difference × current

energy transferred = potential difference × current × time

power = $\dfrac{\text{voltage}^2}{\text{resistance}}$

power = resistance × current²

$\dfrac{\text{input voltage}}{\text{output voltage}} = \dfrac{\text{turns on input coil}}{\text{turns on output coil}}$

number of electricity units used = power × time
(in kW)   (in hours)

Circuit symbols

connection

switch

push switch

relay

battery

cell

power supply (connections)

a.c. supply

lamp

lamp

resistor

variable resistor (rheostat)

variable resistor (potentiometer)

thermistor

capacitor

capacitor (electrolytic)

ammeter

voltmeter

motor

galvanometer

aerial

amplifier

loudspeaker

bell

diode

transistor, npn

transistor, pnp

photo diode

LED

LDR

photo voltaic cell

inductor

OR gate

NOR gate

AND gate

NAND gate

NOT gate

The Elements

Element	Symbol	Element	Symbol
Actinium	Ac	Mercury	Hg
Aluminium	Al	Molybdenum	Mo
Americium	Am	Neodymium	Nd
Antimony	Sb	Neon	Ne
Argon	Ar	Neptunium	Np
Arsenic	As	Nickel	Ni
Astatine	At	Niobium	Nb
Barium	Ba	Nitrogen	N
Berkelium	Bk	Nobelium	No
Beryllium	Be	Osmium	Os
Bismuth	Bi	Oxygen	O
Boron	B	Palladium	Pd
Bromine	Br	Phosphorus	P
Cadmium	Cd	Platinum	Pt
Caesium	Cs	Plutonium	Pu
Calcium	Ca	Polonium	Po
Californium	Cf	Potassium	K
Carbon	C	Praseodymium	Pr
Cerium	Ce	Promethium	Pm
Chlorine	Cl	Protactinium	Pa
Chromium	Cr	Radium	Ra
Cobalt	Co	Radon	Rn
Copper	Cu	Rhenium	Re
Curium	Cm	Rhodium	Rh
Dysprosium	Dy	Rubidium	Rb
Einsteinium	Es	Ruthenium	Ru
Erbium	Er	Samarium	Sm
Europium	Eu	Scandium	Sc
Fermium	Fm	Selenium	Se
Fluorine	F	Silicon	Si
Francium	Fr	Silver	Ag
Gadolinium	Gd	Sodium	Na
Gallium	Ga	Strontium	Sr
Germanium	Ge	Sulphur	S
Gold	Au	Tantalum	Ta
Hafnium	Hf	Technetium	Tc
Helium	He	Tellurium	Te
Holmium	Ho	Terbium	Tb
Hydrogen	H	Thallium	Tl
Indium	In	Thorium	Th
Iodine	I	Thulium	Tm
Iridium	Ir	Tin	Sn
Iron	Fe	Titanium	Ti
Krypton	Kr	Tungsten	W
Lanthanum	La	Uranium	U
Lawrencium	Lr	Vanadium	V
Lead	Pb	Xenon	Xe
Lithium	Li	Ytterbium	Yb
Lutetium	Lu	Yttrium	Y
Magnesium	Mg	Zinc	Zn
Manganese	Mn	Zirconium	Zr
Mendelevium	Md		

Quantities given are those needed per group if the experiment is performed by students, or the quantity needed to perform a demonstration.

It is expected that teachers will provide safety goggles whenever students are handling potentially dangerous materials, or heating substances.

Teachers should also carry out their own risk assessments, bearing in mind the nature of the equipment, chemicals, environment and people involved.

1.1 Using a smoke cell to see Brownian motion
smoke cell (including light source, lens, cell and cover slip)
power pack to suit bulb
leads
microscope; check that there is sufficient clearance between stage and objectives to accommodate smoke cell
waxed paper straw or string to make smoke

3.1 How quickly do scent molecules move?
Students will ask for their own apparatus, but are likely to need:
perfume - any will do
stopclock
metre ruler or tape

3.2 Diffusion of gases
The apparatus is shown in the diagram on page 13. Concentrated hydrochloric acid and ammonia solution give the best results, but care must be taken not to expose students to fumes at close quarters, or for very long. This is especially important for students with asthma. A small amount of each liquid can be poured into a watch glass in a **fume cupboard**, and a piece of cotton wool (held with forceps) dipped into each. If the soaked cotton wool is quickly pushed into the tube ends, and then immediately held in position with a bung, only a relatively mild smell will be noticed by students. (Consider carrying out the whole demonstration in a fume cupboard where practical.)

3.3 How small is a potassium permanganate particle?
10 test tubes in rack
a few crystals of potassium permanganate
syringe or pipette to measure 1 cm^3
access to water

4.1 Cooling wax
Bunsen burner, mat tripod, gauze
Pyrex beaker
pieces of wax (paraffin)
test tubes and holders
retort stand, boss and clamp
access to water

4.2 Measuring the melting point of a solid
As for 4.1, but not wax
stearic acid
thermometer
This experiment could also be used to check the purity of different substances. Suitable examples might include $CaCl_2$ $6H_2O$ (m.p. 30 °C); hard paraffin wax (52 to 56 °C); butter (28 to 33 °C); lard (36 to 40 °C).

13.1 Comparing the strength of different metals
Samples of wire to include some or all of: copper, nichrome, iron, constantan or Eureka, tinned copper. The wires should be of at least 4 different thicknesses per metals.
masses and hangers (a hook of 24 swg copper can support about 5.5kg)
retort stands, clamps and bosses
something for masses to fall onto, e.g. a carpet tile

13.2 Breaking a wire
Samples of wire as 13.1
masses and hangers (a hook of 24 swg copper can support about 5.5kg)
retort stand, clamp and boss
hand lens
something for masses to fall onto, e.g. a carpet tile

19.1 Comparing ionic and covalent compounds
safety spectacles, spatula
A sodium chloride, ceramic paper
B copper sulphate, tongs
C zinc oxide, 1,1,1-trichloroethane
D sugar, test tubes in rack
E paraffin wax, 100 cm^3 beaker
F cetyl alcohol, circuit for testing for electrical conductivity, Bunsen burner

22.1 Investigating the radiation levels from a gamma source
Students must not handle radioactive sources.
scalar/counter or ratemeter, and Geiger-Müller tube
tongs for handling gamma source
gamma source (mounted in such a way that students do not handle it)
ruler
clamp or similar so that GM tube can be

'held' remotely or moved by the teacher according to student instruction.
The gamma source should not be handled by students, or pointed directly at anyone.
Students could also plan investigations into the penetration of aluminium/lead by gamma rays; or compare the penetration of gamma rays with that of alpha and beta radiation.
Datalogging equipment could be used for this investigation.

22.2 The effect of radiation on living organisms
Students will ask for their own apparatus, but are likely to need:
barley seeds, normal and irradiated, about 10 of each
containers for growing them, e.g. margarine tubs with drainage holes
compost
labels
The seeds will germinate and grow faster if kept in warm conditions.

25.1 Using cubes to simulate radioactive decay
At least 100 cubes, each with one face marked differently from the others.

30.1 Using a microscope
any student microscope (with lamp if needed)
a prepared slide

30.2 Making a microscope slide
clean microscope slide and cover slip
microscope
object to mount on slide, e.g. filamentous alga
dropper pipette
access to water
mounted needle or similar for lowering cover slip
filter paper

30.3 Looking at plant cells
clean microscope slide and cover slip
microscope
scalpel, fine forceps and mounted needle
dropper pipette
access to water
filter paper
small piece of onion bulb
small piece of pond weed
Iodine solution can be used instead of water for mounting the plant tissues.

31.1 The effect of heat on plant cell membranes

fresh raw beetroot
Bunsen burner, mat, tripod and gauze
sand tray with sand
access to water
cork borer
Pyrex beaker
thermometer
The sand tray is not essential, but does make sure that the water heats gently. This makes it easier for students to catch the exact temperature at which colour begins to leak out of the beetroot.

33.1 Osmosis and visking tubing

a piece of visking tubing, roughly 15 cm long
cotton
dropper pipette
a fairly strong sucrose solution
capillary tubing – fairly long if possible
retort stand, clamp and boss
beaker
access to water
stopclock, or the sight of a clock with a second hand

34.1 Osmosis and raisins

a few raisins
Petri dish
access to water
This experiment can be made quantitative. Each student should be given ten raisins, and asked to measure them. An average length is then calculated. The raisins are then soaked for a set length of time, and remeasured. Different methods of measuring could be tried – e.g. measuring the lengths of each individual raisin, measuring the length of a line of raisins, or massing the raisins.

36.1 Looking at a mammal's heart

Any mammal's heart. Sheep, pig and cattle hearts are readily available from most butchers. If asked for in advance, you may be able to obtain one with all attached vessels intact.
dissecting board
instruments

37.1 How does exercise affect the rate at which your heart beats?

stopwatch

41.1 Using a potometer to compare rates of water uptake

Any style of potometer will do – the simplest version is a capillary tube to which wide plastic tubing is attached in an air tight manner. The tube is filled with water and the plant stem inserted in the plastic tubing. More complex versions may have graduated capillary tubing and a reservoir.
secateurs for cutting plant stems
leafy plant; young, strong shoots from shrubs or trees often work well
stopclock
vaseline

42.1 Measuring the extension of a spring

spring, masses and hanger, masses up to 40 g should be sufficient
retort stand, boss and clamp
A similar experiment could be performed with elastic bands, and a comparison made between them and a spring. Careful measurements should show hysteresis (Balanced Science 2, page 49) with the rubber band.
Elastic limit can also be investigated. This should be attempted with care, and safety goggles must be worn. If the bottom spring loop gives way, the spring can fly up.
Students could also try making their own springs from copper wire and investigating how the size of wire and the size of the spring affects its strength.

44.1 Finding centres of gravity

card shapes
cotton
pins
retort stand, boss and clamp
weight or piece of plasticine

45.1 Equilibrium in animals

plasticine
cocktail sticks

47.1 Finding the density of an object

object
thread
balance to give mass or weight, newton balances
Either:
beaker
measuring cylinder
container to catch water
Or:
displacement can
beaker
measuring cylinder

47.2 Finding the density of sand

measuring cylinder
beaker
balance
dry 'silver' sand
access to water

49.1 Plotting magnetic fields

bar and horseshoe magnets
plotting compass
sheet of paper
pencil
Students could also try plotting fields in three dimensions or use a flux meter to investigate the strength of the field.

49.2 Comparing the strength and permanence of iron and steel magnets

iron wire
steel wire, e.g. paper-clips
bar magnet
iron filings
Students could also investigate methods for demagnetising magnets, using a magnetised paper-clip – for example, heat, shock, another magnet etc.

54.1 Calculating power output

stairs or exercise bicycle – it makes a good comparison if students try both and think about the energy wasted in each case.
scales for weighing students
ruler
newton meter
stopwatch

59.1 Penguins

Students will ask for their own apparatus, but are likely to require:
test tubes
beakers
access to hot water – almost boiling water from a kettle is best
thermometers
stopwatches
fan
paper or card
Datalogging equipment can be used for this investigation.

63.1 Measuring the percentage of oxygen in air

apparatus as in diagram
small pieces of copper wire, or copper turnings
two retort stands, clamps and bosses to support barrels of gas syringes

67.1 The effect of evaporation on temperature

5 boiling tubes and single hole bungs to take thermometers
cotton wool
clamps and stands
small quantities of ethanol
stopwatches
N.B. If datalogging equipment is available, this is an ideal experiment as the cooling curve can be acquired directly.

70.1 Finding the solubility of different substances

Students will ask for their own apparatus. They are likely to need:
test tubes and a rack
measuring cylinder, pipette or syringe for measuring volumes of water
stirring rod
distilled water
spatula
access to top pan balance
suitable substances to investigate include:
various sugars, such as glucose, sucrose, fructose etc; sodium chloride (0.3 g per gram of water), potassium sulphate (0.111 g per gram of water), copper carbonate (virtually insoluble)

71.1 How does detergent affect surface tension?

small piece of fabric such as cotton, cotton/polyester mix etc.
dropper pipette
access to water
small amount of liquid detergent such

as washing-up liquid
This experiment can also be performed using a bird's feather. The effect of thoroughly washing the feather with a detergent could be investigated, and the implications for the treatment of oiled seabirds can then be considered.

74.1 Strong and weak acids
12 clean test tubes
2 short strips of magnesium ribbon
2 small marble chips
Universal indicator solution and colour chart (full range or pH 1–7 is best)
0.1 mol/dm³ hydrochloric acid
0.1 mol/dm³ ethanoic acid
10 cm³ measuring cylinder

74.2 Measuring the pH of soil
At least two different samples of soil. The local soil should be used, as well as something which will contrast with it. Chalk soil, loam, peat and a commercially produced compost could be tried.
4 or 5 test tubes with corks to fit
small filter funnel and papers
rack or beaker to support tubes
syringe or measuring cylinder to measure 10 cm³
access to water of pH approximately 7
spatula
Universal indicator paper or solution with colour chart

75.1 Neutralising a weak acid with a weak alkali
2 clean boiling tubes
1 clean teat pipette
full range universal indicator solution and colour chart
0.1 mol/dm³ ethanoic acid
0.1 mol/dm³ ammonia solution
glass rod

75.2 Neutralising a strong acid with a strong alkali
a clean burette
clean conical flask
thermometer
2 mol/dm³ hydrochloric acid
2 mol/dm³ sodium hydroxide solution
clean 20 cm³ measuring cylinder or pipette
Bunsen burner

77.1 Metal hydroxide + acid → metal salt + water
clean burette
clean 25 cm³ pipette and filler
clean conical flask
0.1 mol/dm³ sodium hydroxide solution
0.1 mol/dm³ hydrochloric acid
phenolphthalein indicator

77.2 Ammonia solution + acid → ammonium salt + water
As for 77.1, but substitute ammonia solution for sodium hydroxide solution, and screened methyl orange for phenolphthalein.

77.3 Metal oxide + acid → metal salt + water
beaker
2 mol/dm³ sulphuric acid
copper oxide
spatula
glass rod
Bunsen burner etc. for warming acid
thermometer
universal indicator paper
filter paper and funnel

77.4 Metal carbonate + acid → metal salt + water + carbon dioxide
beaker
2 mol/dm³ sulphuric acid
magnesium carbonate powder
spatula
glass rod
filter paper and funnel

77.5 Metal + acid → metal salt + hydrogen
beaker
2 mol/dm³ hydrochloric acid
magnesium ribbon or powder
spatula if powder used
glass rod

78.1 The effect of sulphur dioxide on plants
4 margarine tubs or other containers, with drainage holes
general purpose compost, e.g. John Innes Number 2
viable barley and maize seeds
means of labelling containers
4 watch glasses
cotton wool
sodium metabisulphite solution
access to water
4 large, transparent polythene bags
ties or rubber bands for bags

80.1 Exothermic and endothermic changes
beaker
25 cm³ measuring cylinder, or pipette and filler
spatula
thermometer
stirring rod
copper sulphate solution
2 mol/dm³ hydrochloric acid
iron filings
potassium sulphate
ammonium nitrate
magnesium ribbon
anhydrous copper sulphate
access to water

83.1 The thiosulphate cross
0.1 mol/dm³ sodium thiosulphate
1 mol/dm³ hydrochloric acid
distilled water
100 cm³ beaker
50 cm³ measuring cylinder
10 cm³ measuring cylinder
paper with dark pencil cross drawn on it
stopclock

83.2 Factors affecting the rate of reaction
7 marble chips of similar size
measuring cylinder
apparatus as in diagram
thermometer
100 cm³ beaker
pestle and mortar
500 cm³ of 2 mol/dm³ hydrochloric acid
stopclock
Bunsen burner, mat, tripod and gauze
crystallising dish

84.1 Catalysing the decomposition of hydrogen peroxide with an enzyme
20 volume hydrogen peroxide solution
syringe, measuring cylinder or pipette and filler to measure 20 cm³
4 boiling tubes
glass marking pen
fresh liver
kitchen knife
tile
pestle and mortar
Bunsen burner, mat, tripod and gauze
glass beaker
stopclock
forceps
stirring rod
splint
Fresh liver reacts very fast, and it may be found that the froth rises too quickly to be measured. Most living tissues contain enough catalase to make this experiment work successfully, so pieces of apple or potato, or a yeast suspension, could be used as an alternative.

87.1 The reaction between magnesium and sulphuric acid
100 cm³ measuring cylinder
50 cm³ measuring cylinder
gas generator and delivery tube (see diagram)
crystallising dish
magnesium ribbon
access to balance
0.1 mol/dm³ sulphuric acid
distilled water
stopclock

87.2 The digestion of starch by amylase
3 boiling tubes
glass marking pen
1% starch solution
0.1% amylase solution
10 cm³ syringe
water bath at 35 °C
test tube rack
access to refrigerator
spotting tile
iodine in potassium iodide solution, with dropper
glass rod
stopclock
thermometer
Amylase from different sources behaves differently, and it is worth checking the best concentration beforehand.
If no refrigerator is available, students could use a beaker of iced water as a water bath for tube C.

90.1 Is light needed for photosynthesis?

a healthy potted plant; pelargoniums are
ideal
glass beaker
Bunsen burner, mat, tripod and gauze
boiling tube
forceps
glass rod
white tile
ethanol
iodine in potassium iodide solution with
dropper
black paper
scissors
paper-clips
access to a dark cupboard and a sunny
window sill

90.2 Is carbon dioxide needed for photosynthesis?

two healthy pot plants, e.g. pelargoniums
starch testing apparatus as for 90.1
two conical flasks fitted with split bungs
two retort stands, clamps and bosses
two small containers to fit inside flasks
sodium hydroxide solution
access to a dark cupboard and a sunny
window sill

91.1 Does light intensity affect the rate of photosynthesis?

apparatus as in diagram
sharp scalpel
stopclock
lamp, preferably with 100 W bulb
material for shading the lamp if necessary

92.1 Getting energy out of a peanut

raw or dry-roasted peanut
mounted needle
Bunsen burner and mat
retort stand, clamp and boss
boiling tube
thermometer
access to water

93.1 The structure of sheep lungs

A set of lungs – these are obtainable from
most butchers, especially if you give
them a few days' notice.
dissecting board and instruments
length of Bunsen burner tubing

94.1 Pressure and volume changes in a model thorax

See diagram for apparatus.

94.2 Does breathing rate increase with exercise?

Students will ask for the apparatus they
require. They should need little more
than a stopclock.
Check that no-one risks illness if they
exercise strenuously. Asthmatics will
probably be able to do this experiment
but with care.

95.1 Comparing the carbon dioxide content of inspired and expired air

See diagram for apparatus.
limewater

hydrogen carbonate indicator solution
Make sure that the breathing tube is
thoroughly disinfected after use.

95.2 Comparing the temperature and moisture content of inspired and expired air

thermometer
blue cobalt chloride paper
forceps
Warn students not to put the thermometer
into their mouth. Homemade cobalt
chloride paper often gives a much
more convincing colour change than
commercially prepared paper.

97.1 Anaerobic respiration in yeast

See diagram
boiled, cooled water
dried or fresh yeast
glucose or sucrose
limewater
paraffin oil
If the yeast is not very active, the experiment may need to be left for more than
30 min. It can be speeded up by using a
water bath at around 38 °C.

99.1 Testing foods for carbohydrates

small samples of foods, some which do
and some which do not contain
carbohydrates
knife or scalpel
tile
Petri dish to hold food samples
several boiling tubes and rack
Bunsen burner and mat
Benedict's solution
iodine in potassium iodide solution, plus
dropper
Heating is safer if a water bath is used (a
beaker of boiling water on a tripod and
gauze). However, the ability to boil a tube
of liquid safely in a Bunsen burner
flame is useful, and this gives a good
opportunity to learn or assess this skill.

99.2 Testing foods for fats

clean, dry filter paper
a range of foods, some which do and
some which do not contain fat
Different syllabuses may specify different
fat tests. The other one commonly used at
this level is the emulsion test. The food is
'dissolved' in absolute ethanol, and the
ethanol then poured into distilled water.
A milky emulsion indicates the presence
of fat.

100.1 Testing foods for proteins

a range of foods, some which do and
some which do not contain protein
knife or scalpel
tile
Petri dish for holding food samples
several test tubes with corks or bungs to
fit
20% potassium hydroxide solution
1% copper sulphate solution
The difference between the purple
'positive' colour and the blue 'negative'
one is not always easy to see. Students
could begin by testing water to give the

blue colour, and cheese to give the purple
one. These two tubes can be left in a rack
for comparison against other results.

100.2 Testing fruit juices for vitamin C

Students will ask for the apparatus they
require. It is likely to include:
test tubes or beakers
syringe or measuring cylinder
dropper pipette
They will also need:
a standard ascorbic acid (vitamin C)
solution – this can be made up using
vitamin C tablets, obtainable from
chemists, and normally stating the
amount of vitamin C in each tablet
1% DCPIP solution, freshly made up, and
covered to prevent excess contact with
air
a range of fruit juices, e.g. different types
of lemon squash, lemon juice, freshly
squeezed fruit juices etc.
Different types of juice produce different
colours with DCPIP; some of them
produce a completely clear solution
while others cause a pink colour. The
disappearance of the blue colour should
be considered to be the end point.

103.1 The absorption of carbohydrates from a model gut

two lengths of visking tubing, each about
15 cm long
starch solution
glucose solution
cotton
access to warm water
dropper pipette
beakers
iodine in potassium iodide solution
Benedict's solution
Bunsen burner and mat
boiling tube

104.1 Human teeth

mirror

106.1 Investigating a pendulum

bob, string and support
ruler
stopwatch
Some types of position sensor can be
connected to a pendulum and used
with a datalogger to record the motion
of the pendulum.

107.1 Standing waves

apparatus as shown in Figure 107.7 using
a signal generator and vibrator. The
masses and pulleys should be arranged
to ensure that there is no sideways force
on the vibrator shaft.
ruler to measure string
various types of string – thick, thin, and of
different materials
assorted masses – enough to maintain the
balance on each side as they are
changed

108.1 Measuring the speed of sound in air

Long tape for measuring length of school field – the longer the distance which can be used, the more accurate the results.

A distance of 100 m will give a delay of 0.3 s.

starting pistol (banging blocks and/or blown whistles co-ordinated with waved arms are generally less satisfactory, but can yield surprisingly accurate readings if averaged over ten or more 'timers' and multiple timings)

stop watch

109.1 How is the speed of a wave affected by the depth of water through which it travels?

tray – the type which is used for storage in many schools is ideal (a long length of square guttering is a useful alternative if the experiment can be performed outside)

ruler to measure mm

stopwatch

110.1 Investigating oscilloscope traces of different types of sound

oscilloscope and microphone

various sources of pure sounds – tuning forks, radio, musical instruments, electronic keyboard etc.

A fast response datalogger and sound meter can be used to capture sound traces.

110.2 The range of human hearing

oscillator and loudspeaker (a good quality loudspeaker is needed so that the test is of hearing and not the frequency response of the loudspeaker)

112.1 Building a loudspeaker

thin cardboard sheet, at least A4 size (cereal boxes are ideal)

cardboard coil-former

sticky tape

scissors

thin insulated wire

commercial loudspeaker to study

signal generator or other 'sound' producing system (for example a radio and amplifier) (an a.c. power pack can produce a convincing hum)

4 Ω resistor to protect amplifier from short circuit

bar magnet

leads and crocodile clips

hole punch and rubber bands to mount the cone between lab stands

116.1 Reflection of light rays by a plane mirror

ray box set up as in diagram

plane mirror and supports

partial blackout

white paper

ruler

set square, protractor

117.1 Images in a plane mirror

white paper

plane mirror and supports

pin with a large head

ruler

119.1 Refraction of light rays in a semicircular Perspex block

semicircular Perspex block

white paper

ray box producing a single ray (as figure 116.4a with single slit)

partial blackout

ruler, protractor

121.1 Bending light

optical pin (the end of which is best pushed in to piece of cork)

card to make screen

light bulb and power source

125.1 Investigating a convex lens

partial blackout, but access to a window

convex lens with a focal length of between 10 and 20 cm

piece of white paper and support to act as a screen

light source to use as an object

metre rule and shorter rule

If graph paper is used as the screen, the image size can be measured and magnification can be investigated.

126.1 Finding the focal length of a diverging lens

partial blackout

ray box providing three parallel rays (as figure 116.4a with triple slit)

white paper

cylindrical diverging and converging lenses of similar focal lengths

ruler

ANSWERS TO QUESTIONS

Topic	Question	
5	**4**	273 K
6	**1**	10
15		
page 42		1 million chloride ions
page 43		2 million chloride ions
25	**1**	4 510 000 000 or 713 000 000 years

3a $^{226}_{88}Ra \longrightarrow ^{222}_{86}Ra + ^{4}_{2}He$

$^{222}_{86}Rn \longrightarrow ^{218}_{84}Po + ^{4}_{2}He$

$^{234}_{90}Th \longrightarrow ^{234}_{91}Pa + ^{0}_{-1}e$

	4a	15 m³
	4b	300 Bq
	5	7.65 days
27	**4f**	5600 years

End of topic (page 68)

	5b	110.5
	5c	40.5 if background remains constant
	7b	12.5 min
	7c	12.5 min
	7d	12.5 min
	7e	49 min from the start or an extra 9 min

8 $^{235}_{92}U + ^{1}_{0}n \longrightarrow ^{144}_{56}Ba + ^{90}_{36}Kr + 2^{1}_{0}n$

only two neutrons are released

9 $^{60}_{27}Co \longrightarrow ^{60}_{28}Ni + ^{0}_{-1}e$

$^{60}_{28}Ni \longrightarrow ^{60}_{28}Ni + \gamma$

Topic	Question	
40	**a**	1s
	b	60
	c	3/10
	d	136 mm mercury
	e	14.5 mm mercury
42	**2a**	6 cm
	2b	No – the spring would probably have passed its elastic limit.
	3a	1500 N
	3b	3000 N
	4	8 cm
	5a	6.67 mm
	5b	The scales would no longer show an accurate reading.
43	**1**	20 N, 46 N, 8.5 kg
	2a	97 000 kg
	2b	970 000 N
	2c	970 000 N
	3	Earth: weight = 10 N Jupiter: mass = 1 kg, g = 24.9 N/kg Earth: mass = 25 kg, g = 10 N/kg Sun: weight = 274 N Moon: 3340 N
45	**3**	35°

Topic	Question	
46	**1a**	10 000 N
	1b	3000 N
	2a	hydrogen
	2b	salt water
	2c	the same each time
47	**1**	540 g
	2	888.9 cm³
	3a	7 g/cm³
	3b	450 g
	4	17.04 g
	5a	2000 N
	5b	200 kg
	5c	0.2 m³
	6a	100 g
	6b	125 cm³
	6c	125 cm³
	6d	500 g

End of topic (page 128)

	2a	650 N
	4b	1.67 N/kg
	4c	The 1.8 kg mass on Earth
	4d	The 10 kg mass
	8a	4 000 000N
	8b	4 000 000N
	8c	400 kg/m³
	8e	more
	8f	less
53	**1**	125 N
	3a	600 N
	3b	1500 J
	3c	About 6000 J but allow a wide range
	4	294 000 J
	5a	300 J
	5c	No
54	**1a**	person A 105 W, person B 300 W
	1b	person A
	2a	1800 J
	2c	280 W
	2d	1.8 s
	2e	The conveyor, as the builder does not also have to be lifted
	3a	44.9 kW or 44 900 W
	3b	1750 N
55	**1a**	5.1 MJ or 5 100 000 J
	1b	About half as much
	2a	0.6°C
	2b	25 200 J
	2c	4.2 J/g°C
	3a	10 g
	3b	52.5 g
	3c	210 cm³
56	**2a**	2000 J
	2c	25 MJ or 25 000 000 J every second
	2d	25 MW or 25 000 000 W
57	**1**	(light bulbs) 20 %
	2	80 %
	1a	(kettles) 240 kJ
	1b	95 %
	2a	(gas fire) 60.6 %
	3a	74 074 J

Topic	Question	
	3b	13 888.9 J
	4a	5 revs per second
	4b	314 cm
	4c	300 W
	4d	300 J
	4e	95.5 N
	5	Engine 40 % efficient, generator 75 % efficient, motor 80 % efficient, so total 24 % efficient
60	**2a**	0.2 J
	2b	0.4 W
61	**2a**	30 000 N/m² or Pa
	2b	20 000 000 N/m² or Pa
	3a	6 000 000 Pa
	3b	6 000 000 Pa
	3c	2400 N
62	**1b**	136.5 kPa
	1c	0 °C or 273 K
	1d	melting ice
	2	15 MPa Force of 1500 N so probably no
	4a	40 000 N
	4b	4000 kg
	6	250 ml
	7	10 mm³
	8	Volume is 60.1 litres PV = 6 010 000 Pa 60.1MPa 45.1 MN
	9	237 kPa
80	**1a**	2640
	1b	3338
	1c	698
	2a	2328
	2b	2252
	2c	76
106	**1**	1 Hz
	2	approx 25 cm
page 255	**1a**	4 Hz
	1b	0.25 s
	2a	50 times
	2b	1/300 or 0.0033 s
108	**1a**	3 s
	2a	3 m
	3	375 000 km
	4a	Fish at 3.3 m stationary. Killer whale at 330 m and moving away
	4b	sea bed
109	**1a**	275.5 m; 909.1 kHz; 300 105 000; 299 970 000; 300 000 000; 3.25 m
	2a	17 m and 17 mm
	2b	0.2 and 200
112	**3a**	4 ms
	3b	250 Hz
	3c	20 ms
	3e	50 Hz

Topic	Question	
114	**2a**	4950 m
	2b	0.0008 %
	2a	3 W
	2b	20 W
	2c	20 %
	3c	falls to $^1/_{32}$
	4b	£525
	4c	£165 if 1 tube for every 3 lights
121	**1**	11 cm and 5.6 m
	2	0.0000006 m or 0.0006 mm
126	**1a**	10 cm from other side of lens
	1b	3.3 cm from same side of lens
	2	2.2 m from lens

INDEX

ACKNOWLEDGEMENTS

The authors and publishers would like to thank the following for their permission to reproduce their photographs:

Part-title photograph for 'Atoms' Mike McNamee/Science Photo Library; 1.1 Dr Jeremy Burgess/Science Photo Library; 2.3 NASA/Science Photo Library; 3.2 Andrew McClenaghan/ Science Photo Library; 4.4 Telstar Transport Engineering Ltd; page 27 courtesy Anglo-Australian Observatory; 9.3 (left) Andrew McClenaghan/ Science Photo Library; 9.3 (right) Philips Lighting; 9.4 photograph supplied courtesy of Robert Giovanelli; 9.5 courtesy Anglo-Australian Observatory; 12.5 F. Walther/Zefa; 13.1 David Lean/Science Photo Library; page 39 Dr Jeremy Burgess/Science Photo Library; 20.4 Zefa/Eigen/Zefa; 20.5 Philippe Plailly/Science Photo Library; 24.1 AEA Technology; 24.3 James Stevenson/Science Photo Library; 24.4 AEA Technology; 24.5 Marie Curie Cancer Care; 26.1 courtesy British Nuclear Fuels plc; 26.2 courtesy United Kingdom Atomic Energy Authority; 26.3 and 26.4 NOVOSTI/ Science Photo Library; 27.1 Philip Hayson/Colorific!; 27.4 photo G. Enrie, from the Audiovisual «La Sindone», Editrice Elle Di Ci, Leumann (Torino) Italy; part-title photograph for 'Cells and Transport' Chuck Brown/Science Photo Library; 28.1 Oli Tennent/ Allsport; 29.4b, c & e John Adds; 29.4d Gene Cox; 30.3 Sinclair Stammers/ Science Photo Library; 31.1 and page 77 John Adds; page 79 Q4 Mark Colyer; page 79 Q6 Biophoto Associates; 34.4 and 34.7 M.R. Crow; 37.3 Leicestershire Health Authority, Health Education Video Unit; 38.6 Biophoto Associates; 39.1 Dr Tony Brain/Science Photo Library; 39.3 John Adds; 39.5 Biophoto Associates; 39.6 and 40.3 St Bartholomew's and The Royal London School of Medicine and Dentistry; page 100 Q1 Gene Cox; page 100 Q5 Biophoto Associates; part-title photograph for 'Forces' Japan

National Tourist Organization; 42.1 (right) Goodfellow Metals Ltd; 42.1 (left) BSP International Foundations Ltd; 44.3a Australian Tourist Commission; 44.3b J. Heydecker/Zefa; 45.4 E.M. Bordis/Zefa; 45.5 Zefa; 46.2 Zefa/Kalt; 47.5 Port of Tilbury; 48.1 (top) Alex Bartel/Science Photo Library, (bottom) Transport Research Laboratory; 48.5 Takeshi Takahara/ Science Photo Library; 50.4 J. Highfield; 51.2 Trevor Scotcher/ Stenton Associates, courtesy PC Flying Doctors; 51.3 Austin Brown/Aviation Picture Library; 51.4 Pascal Rondeau/ Allsport; 51.7 Rover Group; 51.8 National Power plc, Research & Technology; part-title photograph for 'Energy' Soames Summerhays/Science Photo Library; 52.1 SEUL/Science Photo Library; 54.2 National Power; 56.2 JET Joint Undertaking; 56.3b BSP International Foundations; 58.5 electric fire supplied by Alan Power; 58.7 Zefa; 59.2 Planet Earth Pictures/ Rod Salm; 60.3 Ford Motor Company Ltd; 61.5 JC Bamford Excavators Ltd; part-title photograph for 'Air and Water' Mark Boulton/ICCE; 64.1 NASA GSFC/Science Photo Library; 66.1 NASA; 67.1 Planet Earth Pictures/ Robert Jureit; 68.2 Planet Earth Pictures/P.J. Palmer; 68.3 Claude Nuridsany and Maria Perennou/ Science Photo Library; 70.2 CNRI/ Science Photo Library; 70.3 courtesy of Sodastream Ltd; 70.4 James Stevenson/ Science Photo Library; 71.3 Justin Munro; 71.5 Greenpeace/Eriksen; 72.2 M. Wendler/Okapia/Oxford Scientific Films; 72.3 Colorific!/Penny Tweedie; 72.4 Landform Slides; part-title photograph for 'Chemical Reactions' Lawrence Migdale/Science Photo Library; 77.2 Andrew Lambert; 77.3 Farmers Weekly Picture Library; 78.1 Frank Lane Picture Agency; 78.2 reproduced by kind permission of the Dean and Chapter of Lincoln; page 193 Q10 National Power PLC, Research & Technology; 79.2 R. Bond/Zefa; 79.3 Manfred Kage/Science Photo Library; 79.4 Chuck Fell/Colorific!; 80.2 Adam

Hart-Davis/Science Photo Library; 82.1 Valor Heating; 82.2 Bob Thomas Sports Photography; 82.3 Pentaprism; 82.4 James Holmes, Hays Chemicals/Science Photo Library; 85.3 (bottom) WWF/UK/Mike Corley; 85.3 (top) WWF/UK/J. Plant; 85.4 Malcolm Fielding, Johnson Matthey PLC/ Science Photo Library; 86.2 ICI Chemicals & Polymers Ltd; part-title photograph for 'Chemical Reactions in Living Organisms' Stephen Dallon/ NHPA; 88.1 Planet Earth Pictures/ Andrew Westcott; 88.2 Australian Tourist Commission; 89.1 Planet Earth Pictures/Mark Mattock; 90.2 Dr Jeremy Burgess/Science Photo Library; 92.1 Bob Thomas Sports Photography; 96.2 CNRI/Science Photo Library; 96.3 Dr Arnold Brody/Science Photo Library; 96.4 James Stevenson/Science Photo Library; 96.5 BOC Gases; 96.6 St Bartholomew's and The Royal London School of Medicine and Dentistry/ Science Photo Library; 97.1 Dr Jeremy Burgess/Science Photo Library; 97.3 Brewers and Licensed Retailers Association; 97.4 R.Bond/Zefa; 97.5 Bob Thomas Sports Photography; 98.1 Mushroom Growers' Association; 98.2 Vaughan Fleming/Science Photo Library; 98.3 Biophoto Associates/ Science Photo Library; 99.1 Leicestershire Health Authority, Health Education Video Unit; 100.3 St Mary's Hospital Medical School/Science Photo Library; 100.4 Biophoto Associates/ Science Photo Library; 101.1 Martin Dohrn/Science Photo Library; 101.2 Zefa; 101.3 J. Hartley/Panos Pictures; 102.3 St Bartholomew's and The Royal London School of Medicine and Dentistry; 104.3 Science Photo Library; 105.2 CAFOD/Liba Taylor; part-title photograph for 'Waves' Martin Dohrn/Science Photo Library; 106.3 Black & Decker Ltd; 106.4 British Cement Association; 106.5 Chris Priest/Science Photo Library; 108.3 Planet Earth Pictures/Hugh Aldridge; 108.4 P & P.F. James (Photography) Ltd/Heathrow Visions 1989; 108.5 Chris Priest/Science Photo Library;

111.1 Marlow & Company Ltd; 111.4 courtesy Ford Motor Company Ltd; 112.1 Schlapfer/Zefa; 113.3 AGEMA Infrared Systems; 113.4 Kim Taylor/ Bruce Coleman Ltd; 113.5 Heather Angel; 113.6 John Moss/Colorific!; 113.7 P & P.F. James (Photography) Ltd/Heathrow Visions 1989; 113.8 St Bartholomew's Hospital, Department of Medical Illustration; 114.3, 114.4 and 114.5 BT Pictures; 115.2 Alexander Tsiaras/Science Photo Library; 116.3 Brian & Cherry Alexander; 116.5 Richard Bryant/Arcaid; 117.5 Ford Motor Company Ltd; 118.3 Zefa; 118.4 Ellis Optical Company; 118.5 Tony Craddock/Science Photo Library; 122.2 Phil Jude/Science Photo Library; 123.2 Carol Farneti/Partridge Films Ltd/ Oxford Scientific Films;

All other photographs were taken and supplied by Graham Portlock, Pentaprism. All artwork is by Geoff Jones.